ANCIENT MATHEMATICS

The theorem of Pythagoras, Euclid's *Elements*, and Archimedes' method to find the volume of a sphere are all parts of the invaluable legacy of ancient mathematics. Its discoveries and insights continue to amaze and fascinate the modern reader. But ancient mathematics was also about counting and measuring, surveying land and attributing mystical significance to the number six.

This volume offers the first accessible survey of the discipline in all its variety and diversity of practices. The period covered ranges from the fifth century BC to the sixth century AD, with the focus on the Mediterranean region. Topics include:

- mathematics and politics in classical Greece
- the formation of mathematical traditions
- the self-image of mathematicians in the Graeco-Roman period
- mathematics and Christianity
- the use of the mathematical past in late antiquity

There are also segments on historiographical issues – such as the nature of the evidence on early Greek mathematics, or the problem of the authentic text of the *Elements* – as well as individual sections on ancient mathematicians from Plato and Aristotle to Pappus and Eutocius. Fully illustrated with plates, drawings and diagrams, and with an extensive bibliography, *Ancient Mathematics* will be a valuable reference tool for non-specialists, as well as essential reading for those studying the history of science.

S. Cuomo is a lecturer at the Centre for the History of Science, Technology and Medicine, Imperial College, London. She is the author of a book on Pappus, and of articles on Hero and Frontinus.

SCIENCES OF ANTIQUITY
Series Editor: Roger French
Director, Wellcome Unit for the History of Medicine,
University of Cambridge

Sciences of Antiquity is a series designed to cover the subject matter of what we call science. The volumes discuss how the ancients saw, interpreted and handled the natural world, from the elements to the most complex of living things. Their discussions on these matters formed a resource for those who later worked on the same topics, including scientists. The intention of this series is to show what it was in the aims, expectations, problems and circumstances of the ancient writers that formed the nature of what they wrote. A consequent purpose is to provide historians with an understanding of the materials out of which later writers, rather than passively receiving and transmitting ancient 'ideas', constructed their own world view.

ANCIENT ASTROLOGY
Tamsyn Barton

ANCIENT NATURAL HISTORY
Histories of nature
Roger French

COSMOLOGY IN ANTIQUITY
M.R. Wright

ANCIENT MATHEMATICS
S. Cuomo

ANCIENT MATHEMATICS

S. Cuomo

London and New York

First published in 2001
by Routledge
11 New Fetter Lane, London EC4P 4EE

Simultaneously published in the USA and Canada
by Routledge
29 West 35th Street, New York, NY 10001

Routledge is an imprint of the Taylor & Francis Group

Typeset in Garamond by
HWA Text and Data Management, Tunbridge Wells
Printed and bound in Great Britain by
TJ International Ltd, Padstow, Cornwall

British Library Cataloguing in Publication Data
A catalogue record for this book is available from the British Library

Library of Congress Cataloging in Publication Data
Cuomo, S. (Serafina)
Ancient mathematics / S. Cuomo.
p. cm. – (Sciences of antiquity)
Includes bibliographical references and index.
1 Mathematics, Ancient. I. Title. II Series.
QA22 .C85 2001
510′.93–dc21 2001019499

ISBN 0–415–16494–X (hbk)
ISBN 0–415–16495–8 (pbk)

CONTENTS

FIGURES AND TABLES

Figures

Tables

ABBREVIATIONS

CAG	*Commentaria Aristotelis Graeca*
CMG	*Corpus Medicorum Graecorum*
CIL	*Corpus Inscriptionum Latinarum*
DK	H. Diels and W. Kranz (eds), *Die Fragmente der Vorsokratiker*, 6th edn, Zurich 1952
DSB	*Dictionary of Scientific Biography*, New York 1969–
IG	*Inscriptiones Graecae*
ILS	H. Dessau, *Inscriptiones Latinae Selectae*, Berlin 1892–1916
LS	H.G. Liddell and R. Scott, *A Greek-English Lexicon*, 9th edn, rev. H.S. Jones, Oxford 1968
O. Bodl.	J.G. Tait, *Greek Ostraka in the Bodleian Library and Other Collections*, 1, London 1930
OCD	*The Oxford Classical Dictionary*, Oxford 1996
P. Amh.	*Amherst Papyri*, B.P. Grenfell and A.S. Hunt (eds), 2 vols, London 1900–1
P. Berol.	*Berlin Papyri*
P. British Museum	*Greek Papyri in the British Museum*, F.G. Kenyon and H.I. Bell (eds), London 1893–1917
P. Cairo	B.P. Grenfell and A.S. Hunt (eds), *Greek Papyri, Catalogue général des Antiquités égyptiennes du Musée du Caire*, vol. 10, Oxford 1903
P. Cairo Isidor.	*The Archive of Aurelius Isidorus*, A.E.R. Boak and H.C. Youtie (eds), Ann Arbor 1960
P. Cairo Zen.	C.C. Edgar, *Zenon Papyri, Catalogue général des Antiquités égyptiennes du Musée du Caire*, 79, 4 vols, Cairo 1925–31
P. Col. Zen.	W.L. Westermann and L. Sayre Hasenoehrl (eds), *Zenon Papyri. Business Papers of the Third Century B.C. Dealing with Palestine and Egypt*, 2 vols, New York 1934–40

P. Ent.	*Publications de la Société royale égyptienne de Papyrologie, Textes et Documents*, ed. O. Guéraud, 1, Cairo 1931–32
P. Fayum	B.P. Grenfell, A.S. Hunt, D.G. Hogarth, *Fayum Towns and their Papyri*, London 1900
P. Flind. Petr.	*The Flinders Petrie Papyri*, J.P. Mahaffy and J.G. Smily (eds), Dublin 1891–1905
P. Freib.	*Mitteilungen aus der Freiburger Papyrussammlung*, in *Sitzungsberichte der Heidelberger Akademie der Wissenschaften, Phil.-hist. Klasse*, 1914–16
P. Herc.	*Papyri Herculanenses*, see M. Capasso, *Manuale di papirologia ercolanese* 1991
P. Lips.	L. Mitteis, *Griechische Urkunden der Papyrussammlung zu Leipzig*, 1, 1906
P. Lond.	*Greek Papyri in the British Museum*, F.G. Kenyon and H.I. Bell (eds), London 1893–1917
P. Mich.	*Michigan Papyri*
P. Oxy.	*Oxyrhynchus Papyri*, B.P. Grenfell and A.S. Hunt (eds), London 1898–
P. Rev. Laws	B.P. Grenfell, *Revenue Laws of Ptolemy Philadelphus*, Oxford 1896
P. Tebt.	*Tebtunis Papyri*, B.P. Grenfell, A.S. Hunt, J.G. Smyly, E.J. Goodspeed and C.C. Edgar (eds), 3 vols, London/New York 1902–38
P. Vindob.	*Papyrus Vindobonensis*
SelPap	*Select Papyri*, Engl. tr. A.S. Hunt and C.C. Edgar, 5 vols, Cambridge, MA/London 1932–34

LIST OF PLACES MENTIONED

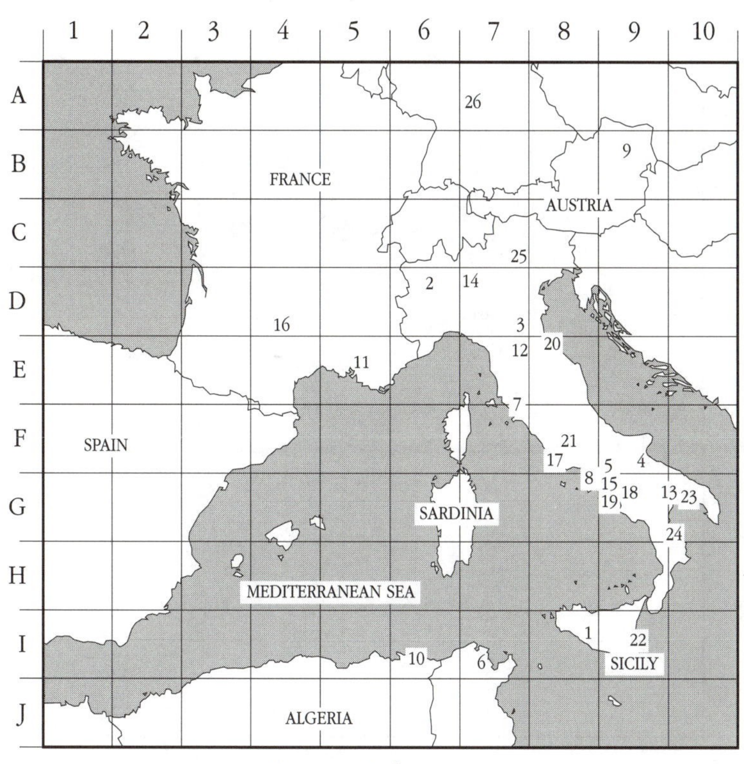

Western Mediterranean

1 Agrigentum I8
2 Aosta (Augusta Taurinorum) D6
3 Bologna D7
4 Canosa (Canusium) F9
5 Capua F9
6 Carthage I7
7 Cosa F7
8 Cumae G9
9 Enns B9
10 Hippo I6
Màlaga off the map towards I1
11 Marseille (Massilia) E5
12 Marzabotto E7
13 Metapontum G10
14 Milan D7
15 Naples G9
16 Orange (Arausius) D4
17 Ostia F8
Osuna (Urso) off the map towards I1
18 Paestum G9
19 Pompeii G9
20 Rimini E8
21 Rome F8
22 Syracuse I9
23 Tarentum G10
24 Thurii G10
25 Verona C7
26 Worms A7

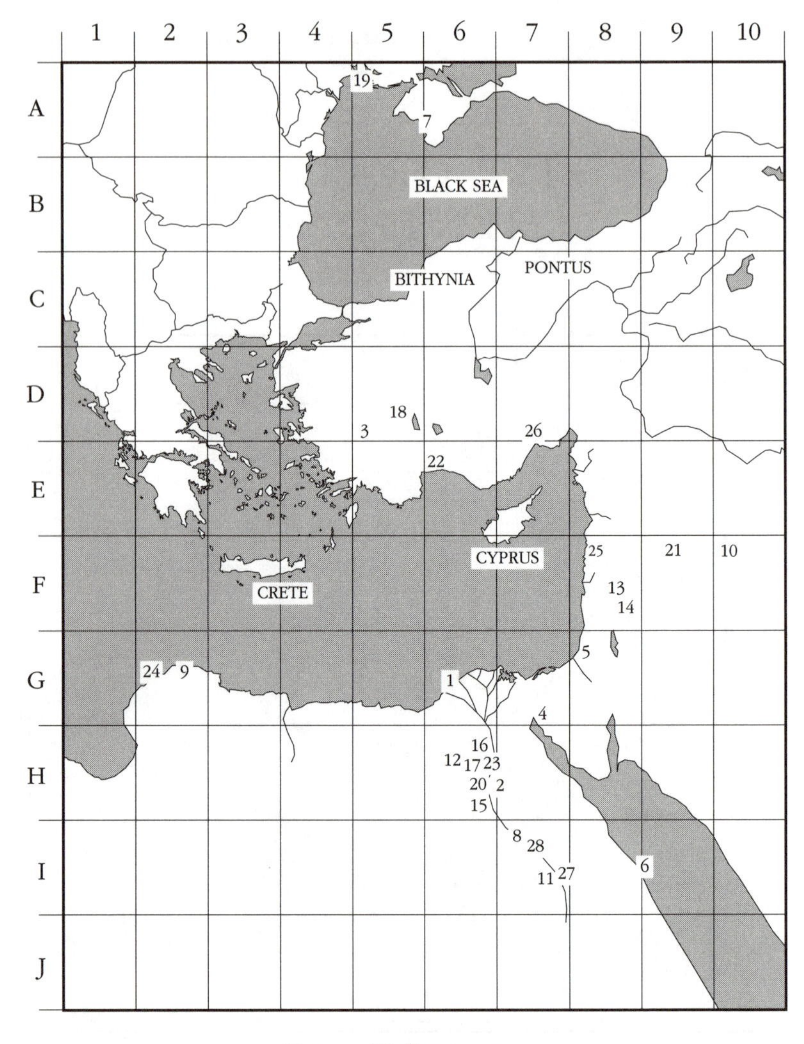

Eastern Mediterranean

1 Alexandria G6
2 Antinoopolis H6
3 Aphrodisias D5
4 Arsinoe G7
5 Ascalona G8
6 Berenice I9
7 Chersonesos on the Black Sea A5
8 Coptus I7
9 Cyrene G2
10 Doura Europus F10
11 Elephantine Island I7
12 Fayum, region of H6
13 Gadara F8
14 Gerasa F8
15 Hermopolis H6
16 Karanis H6
17 Krokodilopolis H6
18 Laodicea D5
19 Olbia on the Black Sea A5
20 Oxyrhyncus H6
21 Palmyra F9
22 Perga E6
23 Philadelphia H6
24 Ptolemaïs G2
25 Sidon F8
26 Soli E7
27 Syene I7
28 Thebes in Egypt I7

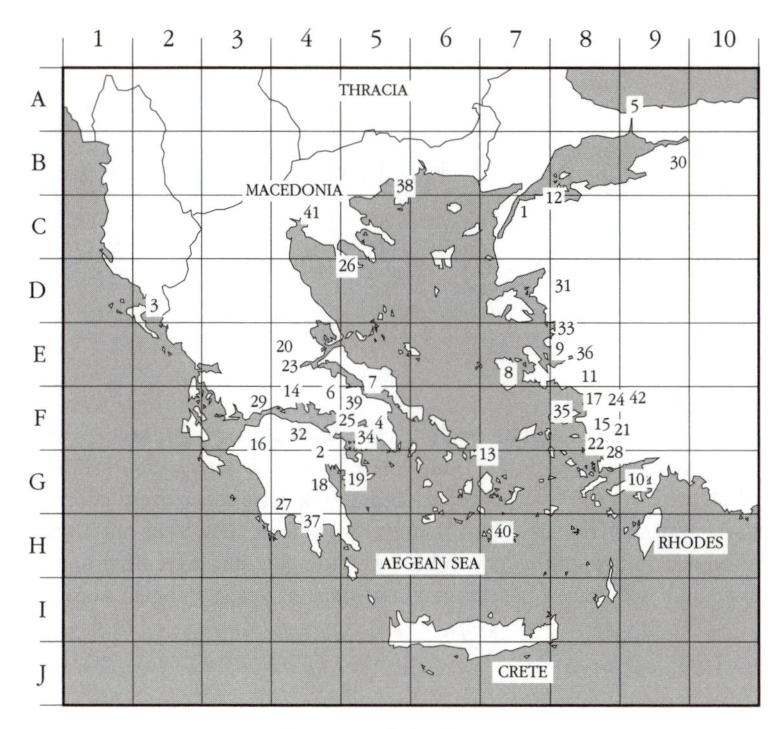

Greece and the Aegean

1 Abydus C7
2 Argos G4
3 Arta, Gulf of D2
4 Athens F5
5 Byzantium A9
6 Chaeronea F4
7 Chalcis F5
8 Chios E7
9 Clazomene E8
10 Cnidus G9
11 Colophon E8
Constantinople, see Byzantium
12 Cyzicus B7
13 Delos G6
14 Delphi F4
15 Didymus F8
16 Elis F3
17 Ephesus F8
18 Epidaurus G4
19 Halieis G5
20 Heraclea E4
21 Heraclea on Latmus F8
22 Iasus F8
23 Lamia E4
24 Magnesia F8
25 Megara F5
26 Mende D5
27 Messene G4
28 Miletus G8
29 Naupactus F3
30 Nicaea B9
31 Pergamum D8
32 Phleious F4
33 Pitane E8
34 Salamis F5
35 Samos F8
36 Smyrna E8
37 Sparta H4
38 Thasos C5
39 Thebes F5
40 Thera H7
41 Thessalonika C4
42 Tralles F9

ACKNOWLEDGEMENTS

The publishers and author would like to thank all copyright holders who have given permission to reproduce material in their possession in *Ancient Mathematics*. While the publishers and author have made every effort to contact all copyright holders of material printed, they have not always succeeded, and would be grateful to hear from any they have been unable to contact with a view to rectifying omissions in forthcoming editions.

INTRODUCTION

Years ago, I was in a taxi somewhere in the British Isles. Being asked, by way of small talk, what I did for a living, I said I was a historian of ancient mathematics. When my driver stopped the car, he got a small notebook out of the glove compartment. As it turned out, he liked to keep a record of the weirdest jobs his customers did, and I (he assured me) could qualify for the top ten.

That was an instructive taxi ride. If they ask me now, I say I am a teacher or, better still, a tourist. Besides, I had to admit that the driver had a point. The history of ancient mathematics is not the most mundane of subjects. Traditionally, it has concentrated on advanced or high-level Greek mathematics – something that combines the wide appeal of a complex subject-matter with the charm of a pretty abstruse language. Again traditionally, the history of ancient mathematics has tended to focus on internal issues (textual analysis, links between various texts and authors, heuristics) rather than trying to relate mathematical practices to their historical contexts. Which, apart from being questionable from a historiographical point of view, is not very likely to make converts from neighbouring fields. In sum, to put it in brutal terms, ancient mathematics can be perceived as mostly incomprehensible and largely irrelevant; a double challenge which this volume will strive to address.

So, I will cast the net wider than usual, and give an account not just of the advanced, high-brow practices, but also of 'lower' and more basic levels of mathematics, such as counting or measuring. I think such a choice will pay off in at least two senses: we will achieve a better-balanced picture, with a full spectrum of activities rather than an isolated upper end; and, since counting and measuring affect a greater section of the population than squaring the parabola or trisecting the angle, looking at them will give us more insight into everyday, everyperson, 'popular'[1] views of mathematics. I will avoid unnecessary technicalities. The reader will find sections (denoted by rules above and below) with samples of mathematical texts: they are meant to offer a taste of what the more advanced end of the spectrum was all about. Non-mathematically-inclined readers can just ignore them, while mathematically-inclined readers who, their appetites whetted, are hungry for more, can refer to the bibliography. Finally, I will endeavour to relate mathematical practices to their times and places, and to other contemporary cultural activities. This will occasionally lead me outside strictly scientific contexts, so we are going to look for mathematics in some strange places, including history and theology.

My choices have come at a cost: given the restrictions of space, I had to leave something out, and the axe has often fallen on 'advanced' mathematics. Thus, some readers may be scandalized to find that I devoted nearly as much space to Iamblichus as I did to Apollonius of Perga. Then again, they can learn more about Apollonius' mathematics from several other books, which is more than can be said in the case of Iamblichus. This volume aims to fill a few gaps – it is to be seen as complementary, not as substitutive, of previous treatments. I also had to omit pre-Hellenistic Egyptian and Mesopotamian mathematics, both for reasons of space, and of personal ignorance (I cannot read the primary sources in the original languages).[2] Moreover, there is a risk that much of what I discuss in the book will not be recognized as mathematics by some of the readers. I trust that I have used actors' categories throughout: if a land-surveyor says that mathematics is part of his job, that is enough for me to take him into account, whether he did use mathematics at all or not. The *claims* are of as much historical value to me as the actual practice, especially since actual practice is almost impossible to reconstruct, and claims tend to be all we have. Finally, I am very aware that a lot of what I say is tentative and vague. The usual excuse ancient historians give for the fuzziness of their work is that the evidence does not allow us to be certain about anything. It is a very good excuse, and, together with the space restrictions and the impossibility of being a specialist for the whole millennium covered by this book, I adopt it as my official excuse wholeheartedly. My aim has been to open paths and stimulate interest, rather than provide fully thought-out arguments for each topic I discuss or question I raise.

The volume is organized chronologically – an odd-numbered chapter dealing with the 'facts' is followed by a companion, even-numbered chapter which problematizes them. Facts comes with quotes because, even though my aim in the odd-numbered chapters has been to provide as exhaustive a survey of the material as possible, selectiveness has crept in. Ideally, odd-numbered chapters should give the reader a full view of what we know about the mathematics of that period, what sources are available, a sketch of what is in the sources. The sections are divided according to the type of evidence, with particularly significant authors (Plato, Euclid, Ptolemy) getting individual sections. Each first section of an odd-numbered chapter is about what I have improperly called 'material' evidence: archaeological data, but also epigraphical, papyrological and, in chapter 7, legal sources. They are 'material' only in the sense that I do not have a better term to distinguish them from literary ones. The even-numbered chapters deal each with two questions arising from the evidence collected in the previous companion chapter. Thus, chapter 4 will explore the issue of the authentic text of Euclid, who has a whole section devoted to him in chapter 3. Of course, the questions I ask are only a fraction of the questions that could be asked of the evidence in each case, and they reflect my interests in, for instance, the historiography of mathematics, the relation between mathematics and politics, the self-image of mathematicians. The reader will notice that I tend not to answer my own questions in any final way, but then that is probably a good thing. Hopefully, and this especially if the book is used for teaching, people will ask their own questions, and give their own answers.

A final disclaimer. My own research deals especially with the mathematics of later antiquity (the AD years), with so-called applied mathematics, and with mathematics in non-mathematical sources. Consequently, I feel even more responsible for what I wrote about those topics than I do for the rest of the book. I have of course relied extensively on the publications, comments, encouragement and criticisms of others, who unfortunately cannot be blamed for any of my mistakes. In fact, I would like to thank a few of those brave and generous people: Richard Ashcroft, Domenico Bertoloni Meli, Paul Cartledge, Sarah Clackson, Silvia De Renzi, John Fauvel, Marina Frasca Spada, Roger French, Peter Garnsey, Campbell Grey, Victoria Jennings, Geoffrey Lloyd, Reviel Netz, Robin Osborne, Richard Stoneman, Bernard Vitrac, Andrew Warwick. And, of course, God bless that taxi driver, wherever he may be now.

Notes

1 I put the term in quotes because, given the state of the evidence, many times it means views of mathematics held by wealthy, educated, socially respected people who would not have identified themselves as mathematicians.

2 For Egyptian and Mesopotamian mathematics, see Ritter (1995a) and (1995b). For other accounts of Greek and Roman mathematics, see Heath (1921), Smith (1951), van der Waerden (1954), Knorr (1986), Dilke (1987), Fowler (1999), Netz (1999a).

1

EARLY GREEK MATHEMATICS: THE EVIDENCE

> There is nothing in all the state that is exempt from audit, investigation, and examination.[1]

Our story officially begins in the Greek-speaking world around the late sixth–fifth century BC, during a period also known as the Classical Age of Greece. Most of what we know about mathematics from this period comes from Athens. The most powerful and among the richest Greek states at the time, Athens was also *the* cultural centre for art, rhetoric, philosophy. Consequently, it attracted people, including mathematicians, from all over the Mediterranean. Moreover, Athens was a democracy, in that its citizen body (comprising only free native adult males), over a period of time and through various upheavals, including periods of tyranny, had gained right to political representation. There was a general public assembly; many public offices were in principle open to any citizen; the law was administered to a significant extent by jury courts, filled by lot from the assembly. The citizen body also manned the army and navy.

Athenian public life, conducted in spaces such as the theatre, the marketplace, the courts, was particularly lively. The Athenians, both rich and poor, landowners and cobblers, come across in our literary sources as particularly fond of debate and discussion about the most various topics: politics, justice, beauty. It has been argued that such a context affected intellectual life in general, and mathematics in particular:

> in the law-courts and assemblies many Greek citizens gained extensive first-hand experience in the actual practice of argument and persuasion, in the evaluation of evidence, and in the application of the notions of justification and accountability.[2]

Whereas the mathematics we find earlier in Egypt or Mesopotamia consisted of specific exercises with verification of the result but no justification of the method employed, Greek mathematics introduced the quest for general propositions which could be proved in such a way as to be objectively persuasive. In other words, where the Egyptians had been able to calculate the volume of a certain cylinder and verify that the result was correct, or at least suitable for their, usually practical, purposes, the Greeks found the general formula for the volume of any cylinder and proved why that formula was right. The public life and political circumstances of Athens and other contemporary Greek states have been seen as a fundamental factor in creating this difference.

Although the context for the emergence of early Greek mathematics has been well described, its details remain fairly obscure, because no strictly mathematical text has survived from the fifth or fourth century BC. I say 'strictly' mathematical because we do have a good number of texts which have a lot to say about mathematics, written by people like Plato or Aristotle, who can be more properly called philosophers. They were less interested in providing an accurate depiction of contemporary mathematicians and mathematics than they were in making philosophical points – what they say cannot be taken as 'neutral' information. Also, both Plato and Aristotle had very strong views about mathematics and its value as a form of knowledge, and we do not know to what extent those views were shared by other people, even from their same social and economic background. Other evidence about our topic is found in historians such as Herodotus, playwrights such as Aristophanes, orators such as Lysias, who were not particularly interested or versed in mathematics, but used it as an example, or for simple operations such as counting ships in a naval battle. Herodotus, Aristophanes, Lysias can be taken as representative of the educated common person, who was probably numerate but not an expert in mathematics, and, although like Plato and Aristotle they had their own agenda, major mathematical or philosophical tenets do not seem to have figured on it.

Hundreds of public documents which have survived from this period are also evidence for the use of mathematics in official contexts such as compiling inventories, recording tribute or accounting for the expense of building a temple. Finally, we have some archaeological material, for instance traces of geometrical town planning or objects which can be identified as counting-boards or abaci (from the Greek *abax*, 'tablet').

I have already remarked on the problems linked to the nature of the evidence in the introduction, but let me repeat myself here. The sources I have collected in this chapter are fragmentary, scattered over time and place, or so concentrated in one place (Athens) as to make any generalization dangerous. They are biased, unclear and silent on many points (mathematics in education, for instance) on which one would love to know more. In fact, chapter 2 will be partly devoted to an analysis of later testimonies about early Greek mathematics, and we will see that most of what we know (or we think we know, or we thought we knew) comes in fact from authors who in some cases write many centuries after the event. Readers will see for themselves how the work of the historian approaches the fluky clues-gathering and educated-guessing of the detective.

Material evidence

To begin with, a little exception to our division of evidence by type. I quote the earliest Greek story on the origins of geometry. According to Herodotus, the Egyptian king Sesostris

> divided the country among all the Egyptians by giving each an equal square parcel of land, and made this his source of revenue, appointing the payment of a yearly tax. And any man who was robbed by the river of a part of his land would come to Sesostris and declare what had befallen him; then the king would send men to look into it and measure the space by which the land was diminished, so that thereafter it should pay in proportion to the tax originally imposed. From this, to my thinking, the Greeks learnt the art of measuring land (*geometria*).[3]

In the use Sesostris made of it, geometry was both part of the procedure of exacting taxes and a guarantee to his people that the assessment of the amount of tax to be exacted was fair. This story encapsulates the strict connections between mathematics and practical tasks, between mathematics and the political sphere, and between mathematics and persuasion; connections which we find again in early Greek instances of geometry in the literal sense.

From as early as the eleventh century BC, a combination of land hunger, social unrest and sense of opportunity prompted several Greek communities to branch out in the Mediterranean, founding new cities or colonies, among them Miletus in Turkey, Cyrene in North Africa, Massilia (Marseille) in France. The colonists aspired not just to a change in location, but to a new life, ideally with improved status and economic circumstances. The land newly 'acquired' – the presence of indigenous people was variously negotiated – was often to be equally divided between the settlers, ignoring any distinctions there may have been among them in the mother community. For instance, we learn from one of the foundation accounts of Cyrene (a colony of Thera) that

> [the Therans] are to sail on equal and similar terms, by family, one son to be chosen † those in their prime and free men from the rest

> of the Therans † to sail. If the settlers establish a settlement, any of their relatives who later sails to Libya is to receive citizenship and rights and a share of undistributed land.[4]

Archaeological surveys of the sites of several Greek colonies (Metapontum, Chersonenos in the Black Sea, Massilia, perhaps Pharos in Croatia) have revealed land-division grids which follow regular geometrical patterns (see Figure 1.1).

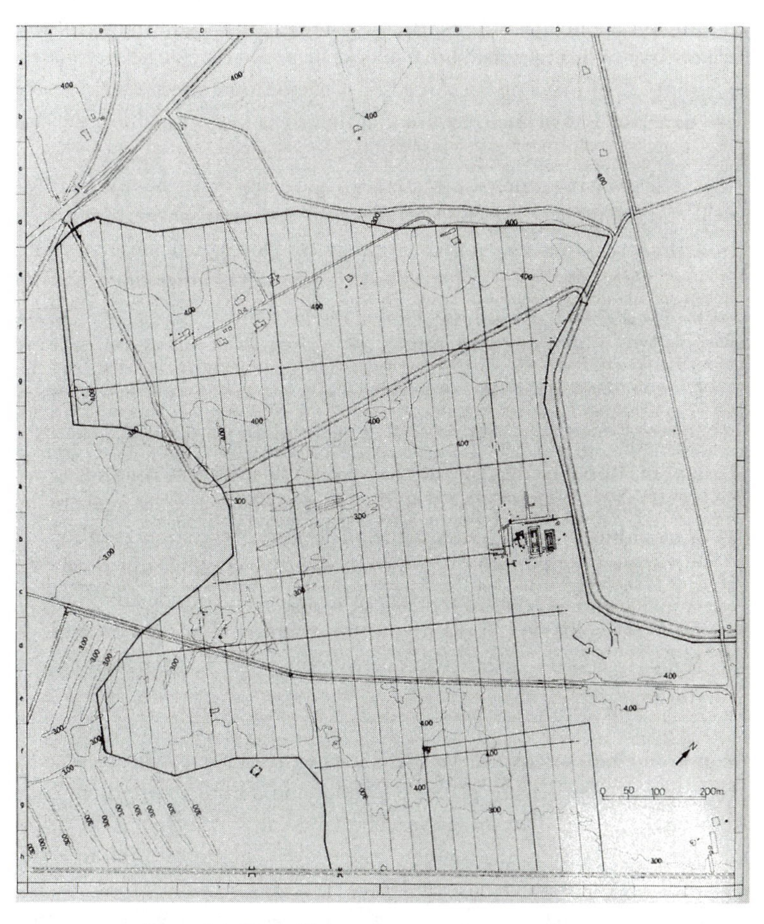

Figure 1.1 Plan of Metapontum and surrounding area (reproduced with permission from *Atti dell'Accademia Nazionale dei Lincei, Notizie degli scavi di antichitá,* Supplement to 1975, 8th series, 29 (1980), fig. 10 p. 32)

It is very difficult to date this kind of evidence, but at least in some cases the grids seem to be contemporary with the foundation of the city. This suggests an orderly land distribution, where equality of share was guaranteed by the equality of the geometrical figures (usually rectangles), in its turn dependent on the accuracy of the relevant measurements.[5]

Better attested still is urban geometrical land division. Many ancient cities, mostly from the fifth and fourth century BC, although there are earlier examples, exhibit an orthogonal plan – Greek foundations, such as Rhodes, Agrigentum, Paestum; Etruscan ones such as Marzabotto and Capua, and, in the fourth and especially third century BC, Roman colonies such as Alba Fucentia (near Rome) and Cosa. All these cities have straight streets cutting each other at right angles, and urban blocks shaped like rectangles or squares, sometimes all of the same size. Sometimes two or more streets seem to have been designated as 'main' streets, in that they meet in the middle of the city area or are recognizably larger than the 'side' streets.

As well as having concrete advantages (e.g. facility of orientation), orthogonal city planning was imbued with social and political meanings: in a geometrically laid-out city, unlike a casually-conglomerated one, everything had a proper place. Spatial order suggested orderings of other kinds. Mathematics thus served both as a tool for solving practical tasks, and as a carrier of meanings, a combination confirmed by Aristotle's discussion on how best to organise and arrange a city. He mentioned Hippodamus of Miletus,

> the same who invented the art of planning cities, and who also laid out the Piraeus [...] The city of Hippodamus was composed of 10,000 citizens divided into three parts – one of artisans, one of husbandmen, and a third of armed defenders of the state. He also divided the land into three parts, one sacred, one public, the third private [...] The arrangement of private houses is considered to be more agreeable and generally more convenient if the streets are regularly laid out after the modern fashion which Hippodamus introduced.[6]

Apart from the port of Athens (the Piraeus), later ancient sources credit Hippodamus with having planned Rhodes and Thurii in Southern Italy (founded in 443 BC). His provisions for an ideal city went from geometrical land division to social and economic tripartitions, all probably informed by a general notion of order and regularity.

Hippodamus may be taken as an early specimen of 'applied' Greek mathematician, together with others whose memory has come down to us in connection with their buildings or other achievements. The names of architects like Kallikrates or Philo have been preserved in inscriptions which link them to the Parthenon and to the Arsenal in Athens, respectively.[7] We can get an idea of the possible mathematical content of the activities of those men by looking at some features of ancient Greek architecture. It has been argued that Greek temples, from as early as 540 BC (date of the temple of Apollo at Corinth) were built so that their width was the mean proportional between their length and height. Many fifth-century works, such as the Parthenon in Athens (*c.* 447–420 BC), incorporated refinements 'to correct optical illusions which would make a truly regular temple look irregular'.[8] For instance, the axes of the columns incline inwards; the platform on which the temple stands is not perfectly horizontal but curved; the corner columns are slightly thicker than the others. It is thought that these refinements were obtained through a combination of mathematical rules of proportion and rule of thumb. Also, the dimensions of each element of a building could be determined proportionally on the basis of a pre-established standard or 'building block', through arithmetical manipulations.[9] Plato testifies that architecture enjoyed a 'superior level of craftsmanship over other disciplines' because of 'its frequent use of measures and instruments which give it high accuracy'.[10]

Another stunning example of the accuracy that could be achieved by Greek builders is linked to the name of Eupalinus of Megara.[11] The so-called Eupalinus tunnel, built *c.* 550–530 BC, is part of a water-supply system for the city of Samos, and runs for some 1036 m under a mountain. Excavations have revealed that the tunnel was made by two teams of workers, who started to dig at the two opposite sides of the mountain and managed to meet in the middle (see Figure 1.2).

When trying to understand how this could be achieved, archaeologists have focussed on three main problems: where to start the tunnel, how to have the two ends start on the same level, so as to meet on the same plane, and how to have the two ends meet in the middle. Most scholars agree that mathematics was used to solve the second and third problem. In order to start the two ends on the same level, the builders may have used 'a lot of very basic arithmetic (i.e. additions) to measure the height up one side of the mountain and down the other', together with sighting instruments and stakes. Many different solutions to the third problem have been proposed, some simpler, some more complex, but all basically involving a system of triangulations, effected with the help of instruments such as the *gnomon* or the *chorobates*, and drawing on geometrical notions about the properties of similar triangles. We can imagine that Eupalinus knew enough mathematics to plan the tunnel and instruct his workforce adequately.

But there is more; series of numbers have been found on the walls of the tunnel, underground. The most complete sequence appears on both stretches of the conduit; starting from near the entrances, the numbers, situated at

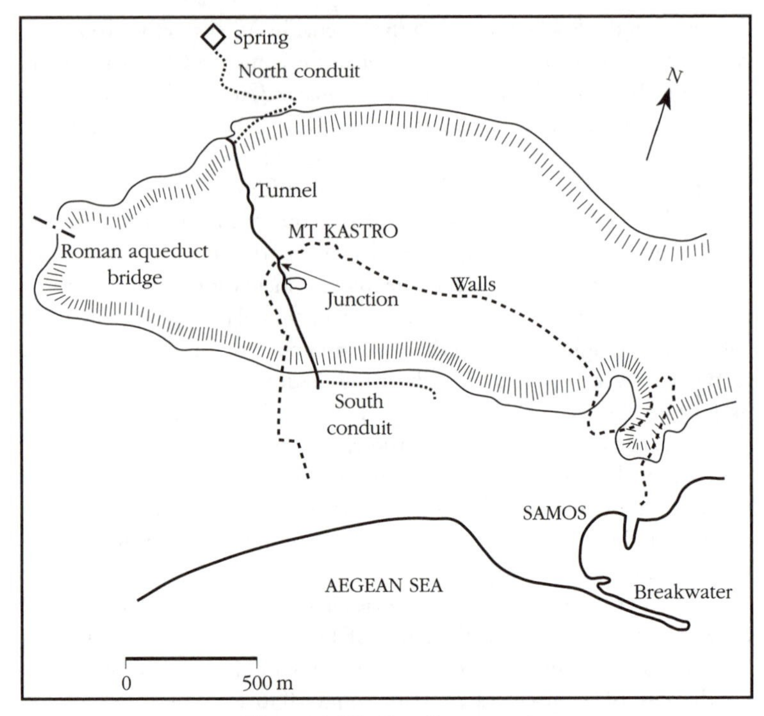

Figure 1.2 The Eupalinus tunnel
(reproduced with permission from Rihll and Tucker (1995), p. 406, fig. 18.2)

regular intervals, go like this: 10 20 30 40 50 60 70 80 90 100 10 20 30 ... 200 (the south-end sequence stops at 200) 10 20 ... 300 (here the north-end sequence stops). It looks as if the builders were assessing the distance and keeping tabs on their work by means of simple counting.[12]

The numbers on the walls of the Eupalinus tunnel are one of the earliest example of the so-called Milesian or Ionian notation, where numbers are represented by letters of the alphabet as shown in Table 1.1.

Commonly used in mathematical texts, including papyri, the Milesian notation gained ground from about the middle of the third century BC and eventually replaced an earlier system, the so-called acrophonic or Attic notation, which is attested from the seventh century and seems to have disappeared by the end of the first century BC.[13] As can be seen in Table 1.2 (and this is what acrophonic means), the signs usually derive from the initial of each number's name in Greek: e.g. 10=Δ for *deka*, 1,000= X for *chilioi*.

The acrophonic notation included signs not only for numbers, but also for amounts of money, notably for the obol, one of the main currency units in Attica, and some of its subdivisions, for the drachma (worth six obols)

Table 1.1 Milesian or Ionian notation

α = 1	ια = 11	κα = 21	τ = 300
		κβ = 22	
β = 2	ιβ = 12	λ = 30	υ = 400
γ = 3	ιγ = 13	μ = 40	φ = 500
δ = 4	ιδ = 14	ν = 50	χ = 600
ε = 5	ιε = 15	ξ = 60	ψ = 700
ς = 6	ις = 16	ο = 70	ω = 800
ζ = 7	ιζ = 17	π = 80	ϡ = 900
η = 8	ιη = 18	ϙ = 90	͵α = 1,000
θ = 9	ιθ = 19	ϱ = 100	͵β = 2,000
ι = 10	κ = 20	σ = 200	͵ι or M = 10,000

Table 1.2 Acrophonic or Attic notation

I = 1	ΔΔΔ = 30	𐅄 = 50	⊦ = 1 drachma	𐅈 = 5 talents
II = 2	H = 100	𐅅 = 500	I = 1 obol	𐅉 = 10 talents
Γ = 5	X = 1,000	𐅆 = 5,000	C = half obol	𐅋 = 100 talents
Δ = 10	M = 10,000	𐅇 = 50,000	T = 1 talent	𐅍 = 1,000 talents

and for the talent (worth 6,000 drachmae), and occasionally for areas of land and grain measures. This fact is a strong reminder of the contexts where numbers were used: trade and commerce, transactions involving land and produce. The acrophonic system, unlike the Milesian, is not attested in literary texts: we find it above all in inscriptions, as we have mentioned, and on counting-boards.

The so-called Salamis table (Figure 1.3) is an example of what a Greek abacus may have looked like.[14]

Similar slabs of stone inscribed with signs for numbers and/or for money have been found in Thyrium and in the Aegean islands of Naxos and Amorgos.[15] We have no direct information as to how calculations were carried out on these counting-boards, but reconstructions are possible on the basis of medieval European or modern Chinese equivalents. Pebbles, beans or purpose-made tokens would have been used as counters, the lines would have served as markers for tens or hundreds, and the signs would have indicated the value of counters laid on them.[16]

We have an illustration of a somewhat different counting-board from a fourth-century BC vase (Figure 1.4).

The scene depicted has been interpreted as the payment of tribute to the Persian king Darius, and the person with a beard is probably a tax collector. It has been suggested that he is sitting at a normal table with coins scattered on it, but the presence of letters would rather indicate that this was a a sort of desktop abacus, with a frame around it and legs to make it into a small

Figure 1.3 The so-called Salamis abacus
(marble, 1.5 m × 0.75 m, now in the Epigraphical Museum at Athens, reproduced with permission from Pritchett (1965))

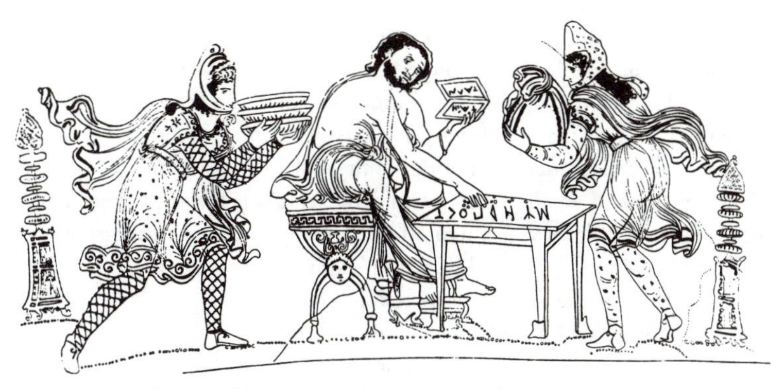

Figure 1.4 The so-called Darius vase, found at Canosa (in Apulia) in Southern Italy, 1.3 m high, now in the Archaeological Museum at Naples

table. The man is holding a wax tablet with the abbreviated words 'A hundred talents', probably the record of a transaction.[17]

The second main source for the acrophonic notation, i.e. inscriptions containing numbers, is much more abundant, with examples from all over the Greek-speaking world, although the great majority comes from Attica. Many of these documents are inventories of temple property or building records; we also have contribution lists; judicial sentences that imply payment of a fine; records of sale of property; acknowledgements of loans or of the repayments of loans, where one of the parties involved is often the state or a temple. Given their sheer quantity, I will focus on three examples: a tribute list, a public account from Athens and a building record from Epidaurus.

The tribute list (425 BC) concerns a re-assessment of the sums that the Greek cities politically subject to Athens had to pay each year.[18] The decree established the procedure for the periodical assessment of tribute – one thousand jurors (drawn by lot) would join with the council at a certain time of the year, and work intensively until the quotas to be paid had been expressed. It also assigned suitable punishment to the magistrates who failed to carry out their duty and contemplated the possibility that the subject cities may complain or be unable to pay. The inscription ends with a list of all the cities which were to pay tribute, with the amount due by each. An excerpt (imagine this arranged in a tall column):

> Tribute from the islands/Paros 30 talents/Naxos 15 talents/Andros 15 talents/[...] Thera 5 talents/Ceos 10 talents/[...] Pholegandros 2,000 drachmae/Belbina 300 drachmae/[...] Myrina on Lemnos 4 talents/Imbros 1 talent/Total of the tribute from the islands 163 talents 413 drachmae/Tribute from the Ionian region/[...]

Even though the text of the tribute quotas is fragmentary, it is possible to observe that the cities are grouped by region, and that the total amount of tribute for each regional group is summed up at the end of each subsection of the list. Notice that the inscription has no mention of a census of the resources of each tributary community: the Athenian jurors and councillors 'shall not assess a smaller amount of any city than it was paying before, unless because of impoverishment of the country there is a manifest lack of ability to pay more' (21–2). There is an explicit indication that the assessment has to be made 'in due proportion' (18).[19]

The second document from Athens (426/5 to 423/2 BC) records the loans given out by the temple treasuries of Athena and other gods:[20]

> The accountants calculated the following dues in the four years from Panathenaea to Panathenaea; the treasurers handed over the following, Androcles of Phlyeus in charge together with the financial secretaries ... and in charge together with the commanders Hippocrates of Cholargus and in charge together with Cecropis being the prytany for the second time, the council had been in function for four days, Megakleides first was secretary, under the archon Euthunos, 20 talents; the interest on these produced 5695 talents and one drachma. The second donation under the second prytany of the Cecropis, the remainder were, seven days into the prytany, 50 talents, interest to this 2 talents and 1970 drachmae. [...]

The list is punctuated by sum totals of moneys due, usually at the end of each year and at the end of each four-year period (in correspondence with the festival of the Great Panathenaea). The sums are sometimes written out both in full and in numerical signs, and include the calculation of the interest that accrues to them. The text also mentions state accountants (*logistai*), who seem to have had two main tasks: to supervise the accounts of public financial bodies, as exemplified by this inscription, and to supervise the accounts that each public official had to render at the end of his term of service. In the latter case, they were drawn by lot from the assembly of all the citizens and formed boards of thirty; in the former, they were drawn from the more restricted council of five hundred and formed boards of ten.[21] It is possible that in either case the actual task of producing the accounts was left to secretaries, while the accountants went through the final result to make sure that the records were correct. The *logistai* were able to impose fines or bring charges of peculation, and were very much in the public eye: on at least one occasion, a person who had held several public posts had his time as accountant singled out, because of the opportunities the job provided for bribery and illegal profits.[22]

The third inscription comes from Epidaurus, where extensive building activities were carried out throughout the fourth and third centuries BC. Since at least the end of the sixth century, the city had been one of the main sites for the cult of Asclepius, a healer god, to whose sanctuary people would travel for a cure. The oldest set of inscriptions (*c.* 370 BC) is a comprehensive account of building expenses, who did what bit of the work, and how much they were paid for it.[23] For instance,

> To the plasterer Antiphilos, for models, 60 drachmae. Journey money to Ariston, 10 dr. To Isodamos, for nails for the fencing, 7 dr. 3 ob. To Aristonos for bricks, 1 dr. To the herald for Thebes, 2 ob. To Aristaios for pitching the workshop doors, 2 dr. 5 ob. [...] For a lock and key, to Isodamos, 15 dr.[24]

A later Epidaurus account contains, like the Athenian documents, recurrent recapitulations, with the totals of the sums spent up to a certain point.

Inscriptions involving numbers are important evidence of the role of mathematics in the public life of several Greek states. The use of inscriptions as historical sources, however, is not unproblematic. For instance, their abundance in fifth- and fourth-century Athens was seen as linked to the presence of a democratic government, to the open character and accountability of its administration, to the high level of literacy of its population, and to the relation between all these factors. More recently, it has been shown that no simple equation can be made between public written documents and form of government – we have inscriptions from non-democratic states, including those in other parts of Greece – and that Athens remained a largely oral culture, where only very few people could read and write. Thus, revised interpretations of the significance of inscriptions in the classical age try to recapture the complexity of their meaning, their 'utilitarian' (recording and publishing data, providing information, allowing public scrutiny) as well as their symbolic functions: commemorative, celebratory, religious, intimidatory, even magical. Indeed, 'there were certainly cases in fifth-century Athens when the symbolism of the record was of as much importance as the ritual recorded'.[25]

With particular reference to the inscriptions we have discussed, then, we can speculate that, if indeed at least some accountants were chosen by lot, your average Athenian citizen must have been sufficiently numerate to supervise, if not produce, accounts. Moreover, and perhaps more significantly, these documents testify that mathematics had a significant public presence. To quote Robin Osborne again:

> The accounts [...] preserved, and made visible to all, rituals of the sort which the Council so frequently oversaw, rituals of counting and handing over in which what was seen to be done was far more important than the intrinsic significance of the act.[26]

Although it would have often been impossible actually to check all the figures on an inscription, they were out there in the public domain. I have emphasized that in several cases lists of figures were punctuated by subtotals – a way for the maker of the document to keep count, but also to make reading easier for a potential public. If anyone at all went through these lists to check that everything added up, expressing subtotals would have been an indication that the administrators of the state shared the process of accounting with the rest of the citizen body. We can also infer that people, even those who were not fluently numerate, were familiar with both the utilitarian and symbolic functions of mathematics, and that they associated it with a number of public areas: the administration of the state, its financial wheelings and dealings, foreign policy, war, religion, owing, distributing and receiving.

Historians, playwrights and lawyers

Herodotus is the earliest literary source we will consider. Of course, mathematics existed in Greece, and in Greek literature, before him. For instance, Homer compared a thick battle engagement with two men who, measuring-rod in hand, fight about the boundary of their common field, and 'in a narrow space contend each for his equal share', while Theognis associated truthfulness and justice with measuring instruments – a man sent as envoy must 'take care to be more true than scale or rule or lathe'.[27] But Herodotus, as we have seen with his account of the Egyptian origins of geometry, is the first to report on the history of mathematics. His work also contains a good deal of counting: he counted the days in a man's life, the generations of Egyptians, the tribute due to the Persian king Darius, the length of Darius' journey to the West (which he undertook with an aim to conquering Greece), the troops and various armaments of Darius' successor Xerxes, and the soldiers in the Greek army.[28] All these figures have at least two purposes: one, they reinforce some impressions Herodotus wants to convey (that the Egyptians are a really ancient people, that the Persians are much more numerous, richer and more powerful than the Greeks, who nevertheless manage to win against the odds); two, they add to Herodotus' construction of himself as an accurate reporter of facts. From the point of view of the present volume, they are again evidence that mathematics was seen as a means of rendering an account of things that were of common interest, and affected the body politic. Accurate figures, where the accuracy was guaranteed by carrying out the operation in full detail, appeared more trustworthy than just qualifying the tribute of 'huge' or the troops of 'countless'.

A rather different type of story about the origins of mathematics comes from the first playwright we will look at, Aeschylus or whoever is the author of *Prometheus Bound.* The Titan, chained to a rock at the edge of the world, his liver perpetually eaten by an eagle, explains the reason for his punishment thus:

PROMETHEUS: For the suffering race of humankind [Zeus] cared nothing, he planned to wipe out the whole species [...] I saved humanity from going down smashed to bits into the cave of death. For this I'm wrenched by torture: painful to suffer, pitiable to see. [...] Humans used to foresee their own deaths. I ended that. [...] What's more, I gave them fire.

CHORUS: Flare-eyed fire!? Now! In the hands of these things that live and die!?

PROMETHEUS: Yes, and from it they'll learn many arts (*technai*) [...] hear what wretched lives people used to lead, how babyish they were – until I gave them intelligence, I made them masters of their own thought. [...] they knew nothing of making brick-knitted houses the sun warms, nor how to work in wood. They swarmed like bitty ants in dugouts in sunless caves. They hadn't any sure signs of winter, nor spring flowering, nor late summer when the crops come in. All their work was work without thought, until I taught them to see what had been hard to see: where and when the stars rise and set. What's more, I gave them numbering, chief of all the stratagems. And the painstaking, putting together of letters: to be their memory of everything, to be their Muses' mother, their handmaid! [...] In a word: listen! All the arts are from Prometheus.[29]

Mathematics, like the other arts an offspring of Prometheus, is, like him, associated with craft, resourcefulness and even trickery, the ability to prevail in a conflict against a stronger adversary. At the same time, it epitomises the passage of humankind from a feral state to civilization, it is both a sign and a cause of its newly-acquired intelligence and self-awareness. The 'chief of all the stratagems' is depicted in action in this passage from a comedy staged in 422 BC:

> And not with pebbles precisely ranged, but roughly thus on your fingers count the tribute paid by the subject states, and just consider its whole amount; and then, in addition to this, compute the many taxes and one-per-cents, the fees and the fines, and the silver mines, the markets and harbours and sales and rents. If you take the total result of the lot, 'twill reach two thousand talents or near. And

> next put down the Justices' pay, and reckon the sums they receive a year: six thousand Justices, count them through, there dwell no more in the land as yet, one hundred and fifty talents a year I think you will find is all they get.[30]

The author, Aristophanes, was one of the most successful playwrights of his day. The passage above is like a comically distorted version of the inscriptions we have seen in the previous section: the character in the play even refers to a tribute list. It sends up a situation where your average citizen read through the inscription, rehearsed the calculations in his head or on his fingers, and had the enlightening and very democratic experience of finding out where the money was going. Here, to the people's jury courts, the target of satire in this particular play. In general, Aristophanes' works reflect contemporary concerns, ranging from bitter commentaries on the long war between Athens and Sparta, to irreverent portraits of magistrates, philosophers, poets and mathematicians. In the *Birds*, performed in 414 BC, a new city has to be founded from scratch. The main character, Peisthetaerus, is visited by various people who offer their services.

Enter

METON: I come amongst you –

PEISTHETAERUS: Some new misery this! Come to do what? What's your scheme's form and outline? What's your design? What buskin's on your foot?

METON: I come to land-survey this Air of yours, and mete it out by acres.

PEISTHETAERUS: Heaven and earth! Whoever are you?

METON: Whoever am I? I'm Meton, known throughout Hellas and Colonus.

PEISTHETAERUS: Aye, and what are these?

METON: They're rods for Air-surveying. I'll just explain. The Air's, in outline, like one vast extinguisher; so then, observe, applying here my flexible rod, and fixing my compass there, – you understand?

PEISTHETAERUS: I don't.

METON: With the straight rod I measure out, that so the circle may be squared; and in the centre a market-place; and streets be leading to it straight to the very centre; just as from a star, though circular, straight rays flash out in all directions.

PEISTHETAERUS: Why, the man's a Thales![31]

Meton combined what are the now separate skills of an engineer, an astronomer and (judging from his role in the play) a land-surveyor, town-planner or architect. We know from later sources that he was responsible for a reform of the Athenian calendar, that he set up an instrument for the observation of solstices on the Pnyx (the hill in Athens where the general assembly met), and that he built a fountain and/or a time-keeping device and/or a water-operated time-keeping device in the Agora, on a hill called Colonus Agoraios (hence the reference in the passage). Meton must have been famous enough to be recognized by the general public attending the theatre; the same goes for Thales, a sixth-century BC mathematician and philosopher from Miletus with whom he is ironically equated. The audience must have also been able to understand the comic allusion to the problem of squaring the circle, which was considered impossible to solve, or absurd.

Stretching our interpretation, we could suggest that Meton was made into a figure of fun because he engaged in a type of mathematics which not everybody recognized as relevant or important, or which was simply too sophisticated for everybody to understand. In another play where Aristophanes lampoons the new intellectual education made popular by the sophists, the seminal question is asked:

STREPSIADES: And what's this?
STUDENT: Geometry.
STREPSIADES: So what's that useful for?
STUDENT: For measuring land.
STREPSIADES: You mean land for cleruchs?
STUDENT: No, land generally.
STREPSIADES: A charming notion! It's a useful and democratic device.[32]

A cleruch was a settler-soldier who received land as a reward for his service. A pun is lost in translation: the expression 'land generally' also means 'land for everybody', as if a land distribution bonanza was intended, but all the student is saying is that the geometry they do at his school concerns itself with general and abstract questions: squaring the circle perhaps, solving the economic problems of the multitude definitely not.

A contemporary of Aristophanes and together with Herodotus the greatest historian of the classical period, Thucydides also makes interesting use of numbers. He describes how, in a situation of extreme emergency during the Peloponnesian War (*c.* 431 BC), Pericles, the Athenian leader, tried to persuade his fellow citizens of a certain course of action. Part of his argument consisted of a long list of revenues, which once again reads like one of the financial inscriptions:

> And he bade them be of good courage, as on an average six hundred talents of tribute were coming in yearly from the allies to the city, not counting the other sources of revenue, and there were at this time still on hand in the Acropolis six thousand

> talents of coined silver (the maximum amount had been nine thousand seven hundred talents, from which expenditures had been made [...]).

Next, Pericles quantified defence structures ('the length of the Phalerian wall was thirty-five stadia to the circuit-wall of the city, and the portion of the circuit-wall itself which was guarded was forty-three stadia') and human resources ('the cavalry [...] numbered twelve hundred, including mounted archers, the bow-men sixteen hundred, and the triremes that were sea-worthy three hundred.').[33] Those items would not have normally been on an account, but including them as quantifiable resources adds to the sense that collective counting, reckoning one's strengths and taking decisions consequently amounts to acquiring self-awareness as a community. Another episode from the same war involved the people of Plataea, who found themselves allied to Athens but closely threatened by Spartan troops, whose envoy,

> endeavouring to reassure them [...], said: 'You need only consign the city and your houses to us, [...] pointing out to us the boundaries of your land and telling us the number of your trees and whatever else can be numbered; [...] as soon as the war is over we will give back to you whatever we have received; until then we will hold it all in trust'.[34]

Trust whose only support seems to have been a list of whatever could be numbered – once again, something that would have looked like an inventory inscription.

Mathematics appeared in yet another public context: the legal courts. In Athens a great number of cases were tried in front of a jury drawn by lot from the entire citizen body or those of them who volunteered. The number of people in the jury varied from case to case, with the final verdict reached by majority of votes. Tokens – solid ones if the juror found for the defendant, perforated tokens otherwise – were collected by attendants into vessels, emptied out onto an abacus and counted. The abacus seems to have been of a special type, with holes on it to host the tokens; in particular it had as many holes as there were jurors, to make sure that everybody's vote was accounted for. One of our main sources for this procedure, Aristotle, specifies that this was so that the tokens would be 'set out visibly and be easy to count, and that the perforated and the whole ones may be clearly seen by the litigants'.[35] Transparency of procedure was signified by a shared calculation, by doing mathematics together.

The legal courts were of course an arena for the performance of speeches, many of which are extant. No professional lawyers were supposed to exist (being paid to give a defence or accusation speech was illegal), but speeches were often written by third parties with outstanding rhetorical skills. Since the audience was, like that for Aristophanes' plays, a general one, their language and allusions needed to be sophisticated but comprehensible, persuasive but also somewhat pandering to common tastes.

A great many court cases were about financial matters, so that the relevant speeches contain accounts, inventories, and other evidence involving numbers.[36] An example is Lysias' *On the Property of Aristophanes*, delivered around 388–87 BC. Some relatives of the defendant had been condemned to death and their property confiscated. The amount confiscated was found to be much less than expected, and the defendant's father was accused of having hidden property. When he died, the charge was inherited, as it were, by the defendant, his only son. He had to make the case that the property in question was actually much less than the state had anticipated – a case, basically, about quantifiable value. He (or rather Lysias, who wrote the speech for him) started by establishing the good character of his father, who, far from witholding property from the state, was a generous benefactor: 'Yet, gentlemen, my father in all his life spent more on the state than on himself and his family, – twice the amount that we have now, as he often reckoned in my presence'.[37] A 'public' account whose only witness was the defendant – but, if that sounds unconvincing, an itemized account is also produced:

> Now, Aristophanes had acquired a house with land for more than five talents, had produced dramas on his own account and on his father's at a cost of five thousand drachmae, and had spent eighty minae on equipping warships; on account of the two, no less than forty minae have been contributed to special levies; for the Sicilian expedition he spent a hundred minae, and for commissioning the warships [...] he supplied thirty thousand drachmae to pay the light infantry and purchase their arms. The total of all these sums amounts to little short of fifteen talents.[38]

Many of the expenses are in fact for the benefit of the community, from subsiding the theatre to helping out in the Peloponnesian War. The jurors are taken through the calculations, so that they can not only appraise the actual amount of Aristophanes' father's generosity, but also appreciate his civic commitment.

In some cases, the audience could be faced with competing calculations. In the speech *Against Diogeiton*, also written by Lysias around 400 BC, a guardian was accused by his wards of having robbed them of their fortune. According to the plaintiff, Diogeiton had been given

> five talents of silver in deposit; [...] seven talents and forty minae † and two thousand drachmae invested in the Chersonese. [The father of the wards] charged him, in case anything should happen to himself, to dower his wife and his daughter with a talent each [...]; he also bequeathed to his wife twenty minae and thirty staters of Cyzicus. [...] He was killed at Ephesus [...] [Diogeiton] gave [the dead man's wife] in marriage with a dowry of five thousand drachmae, – a thousand less than her husband had given her. Seven years later the elder of the boys was certified to be of age; [...] Diogeiton summoned them, and said that their father had left them twenty minae of silver and thirty staters.[39]

Diogeiton's deceit was first exposed by the widow, who claimed that he had received five talents in deposit, and

> convicted him further of having recovered seven talents and four thousand drachmae of bottomry loans, and she produced the record of these [...]. She also proved that he had recovered a hundred minae which had been lent at interest on land mortgages, besides two thousand drachmae and some furniture of great value.[40]

More figures follow:

> Gentlemen of the jury, I ask that due attention be given to this calculation [...] [Diogeiton] has had the face to [...] make out a sum of seven talents of silver and seven thousand drachmae as receipts and expenses on account of two boys and their sister during eight years. So gross is his impudence that, not knowing under what headings to enter the sums spent, he reckoned for the viands of the two young boys and their sister five obols a day; for shoes, laundry and hairdressing he kept no monthly or yearly account, but he shows it inclusively, for the whole period, as more than a talent of silver. For the father's tomb, though he did not spend twenty-five minae of the five thousand drachmae shown, he charges half this sum to himself, and has entered half against them. Then for the Dionysia [...] he showed sixteen drachmae as the price of a lamb [...] for the other festivals ad sacrifices he charged to their account an expenditure of more than four thousand drachmae; and he added a multitude of things which he counted in to make up his total, as though he had been named in the will as guardian of the children merely in order that he might show them figures

> instead of the money [...] I will now base my reckoning against him on the sum which he did eventually confess to holding.[41]

Diogeiton did produce an account – only, such a blatantly inflated one that it was possible to unmask him by going through the numbers verifying at each step their implausibility. Accurate accounting is identified by the speaker with good behaviour – bad accounting, in all its details concerning crucial moments of Athenian civic life (the festivals, burying one's father, coming of age, dowries), is associated not only with greed and lack of family feeling, but also with transgression of religious and social rules and, in other cases, with corruption and/or undemocratic political leanings. Aeschines, in *Against Ctesiphon* (330 BC), stressed the importance of the Athenian law according to which every magistrate had to give account: 'In this city, so ancient and so great, no man is free from the audit who had held any public trust'.[42] Being prepared and willing to give an account showed not just financial honesty, but also political transparency, the recognition that one's actions could be checked, scrutinized, and discussed by the public.

The identification of jurors and accountants was again invoked by Demosthenes, in a speech delivered against Aeschines as part of the same case against Ctesiphon.

> I shall prove without difficulty that [Aeschines] has no right to ask you to reverse that opinion – not by using counters, for political measures are not to be added up in that fashion [he refers to a *logismos*, a calculation], but by reminding you briefly of the several transactions, and appealing to you who hear me as both the witnesses and the auditors (*logistai*) of my account.[43]

Peppered with puns, the plea Demosthenes was making in his defence was for the jurors to go through the steps of his argument the way they would go through the steps of a calculation, except that his speech did not involve any counters or abacus. Compare Aeschines' response:

> But if such a statement as I have just made, falling suddenly on your ears, is too incredible to some of you, permit me to suggest how you ought to listen to the rest of my argument: When we take our seats to audit the accounts of expenditures which extend back a long time, it doubtless sometimes happens that we come from home with a false impression; nevertheless, when the accounts have been balanced, no man is so stubborn as to refuse, before he leaves the room, to assent to that conclusion, whatever it may be,

> which the figures themselves establish. I ask you to give a similar hearing now.[44]

The jurors were called upon to be not just witnesses, but also calculators – in principle, a *logistes* could tell with absolute certainty whether the account presented to him worked out or not, whether everything added up. It was this kind of persuasiveness, mathematical persuasiveness, that both speakers wished to claim for their arguments. In chapter 2, I will further discuss some of the implications of the evidence collected in this section. For now, let me observe that accounts – collective, public counting – were a pervasive practice in classical Athens. We find them not only in inscriptions, but also in various genres of literature. Along with its practical functions, public counting was associated with political accountability, and in fourth-century legal speeches it seems increasingly to symbolize the role itself of the Athenian citizen.

Plato

It is very difficult to put Plato's philosophy in a nutshell, both because it is extremely complex, and because in different works he sometimes says rather different things on the same question. He wrote in the form of dialogues, with his teacher Socrates usually cast as the main speaker and mouthpiece of his views. Each and every part of Plato's thought has been the object of much debate ever since antiquity, although perhaps none as much as the so-called theory of forms. Generalizing and simplifying, this is the belief that there are two levels of reality, corresponding to two levels of knowledge of that reality: the sensible world which we live in, and of which we have knowledge primarily through the senses, and a world of changeless entities, called forms, of which we have true and certain knowledge through our intellect. Things in the world are imperfect reproductions of the forms, which constitute 'thing in themselves', i.e. not this or that horse, but *the* horse, not this or that act of justice, but justice *itself*.

Most of Plato's work is devoted to the discussion of particular forms (Justice, Love, Pleasure, the Good), to the relation between forms and things in the world, and to how true knowledge can be attained. Mathematics was of special interest to him, because, while it dealt with sensible things and was employed in fields such as architecture or the military, it also concerned itself with the general, the abstract, the unchanging. In a calculation, for instance, one operates not just with three oxen or five fingers, but with 'three' and 'five', which have their own characteristics (e.g. they are both odd numbers) quite independently of the objects they are assigned to. Again, one measures a specific triangular field, but determines how the area of a *general* triangle is to be found. Plato claimed that, by practising mathematics, the mind got used to turning from sensible particulars towards abstract concepts, and was thus in a better position to gain knowledge of the forms. Moreover, the results of mathematics were convincing – everybody was prepared to believe that three and five made eight, whereas opinions about, say, the good state were bound to differ. Mathematics thus provided an example of persuasive discourse. There is also a third aspect: in one of his dialogues, the *Timaeus*, Plato described how the universe and all the things in it, including humans, came to be, and how the physical elements were shaped following a geometrical pattern (more details below). He thus posited mathematics at the very foundations of the world around us.

Now, there are two basic ways in which Plato can be used as a source for the history of early Greek mathematics: as a philosopher and as a historian. In the first capacity, he engaged with questions such as what kind of objects numbers and geometrical figures are, what is their relation to forms on the one hand and things in the world on the other, what is the value of mathematical knowledge. Plato addressed all these issues, but did not answer them unequivocally. According to some interpretations, he considered mathematical entities to be forms; then again, he may have seen them as *intermediate* between the realm of forms and that of sensibles; some evidence can also be adduced that he thought of *all* forms as mathematical entities, or that mathematical entities were the only true forms there were. His view of mathematical knowledge, on the other hand, seems more straightforward: he gave it a crucial role in the training of a good philosopher, because of the above-mentioned habit it imparts, of turning the mind to the general and the abstract. At the same time, mathematics for him was not perfect, because it still relied on undemonstrable principles (e.g. geometers *assume* 'the odd and the even, the various figures, the three kind of angles [...] as if they were known'), and it still concerned itself with objects (e.g. triangles drawn on a whiteboard). If one visualized knowledge as a line divided into segments (an analogy Plato adopts in the *Republic*) with total ignorance at one end and perfect knowledge at the other, disciplines such as mathematics would fall immediately short of perfect knowledge, in a segment labelled *dianoia*, argumentative reasoning.[45]

In his capacity as a historian, Plato provided information about mathematicians contemporary to, or earlier than him, and referred to issues that were being discussed, had been solved or constituted common knowledge. Sometimes he reported whole mathematical passages to make a philosophical point, thus giving us insight into what his readers would have been expected to know or understand. In this book, I will only deal with this second aspect of Plato as a source – I will not discuss his philosophy of mathematics at any greater length than I have done above. That said, I am very aware that it is not possible sharply to separate Plato the philosopher from Plato the historian of mathematics. We just have to be alert to the fact that any historical information we may glean is embedded within a much wider and more complicated context.

The picture Plato presents of the mathematics of his day is both diverse and vast. First of all, he talks about the mathematicians' method, *how* they did what they did. He characterizes their way of proceeding as hypothetical: they used claims that were not proved but only assumed to be valid, and on which other claims were then based. Thus, proving a mathematical statement consisted of working back from that statement to other known statements of which the first was a consequence, until one found an accepted hypothesis which needed no further proof and could then serve as the starting-point (reversing the logical process that led to it). By 'hypothesis' (literally something that is put under) Plato seems to have meant an assumption or starting-point in quite a general sense, including both definitions and fully-fledged propositions:

> For example, if [geometers] are asked whether a specific area can be inscribed in the form of a triangle within a given circle, one of them might say: 'I do not yet know whether that area has that property, but I think I have, as it were, a hypothesis that is of use for the problem, namely this: If that area is such that when one has applied it as a rectangle to the given straight line in the circle it is deficient by a figure similar to the very figure which is applied, then I think one alternative results, whereas another results if it impossible for this to happen. So, by using this hypothesis, I am willing to tell you what results with regard to inscribing it in the circle – that is, whether it is impossible or not. [...]'[46]

Plato distinguishes several branches of mathematics according to their main object of enquiry: arithmetic and logistics (the science of calculation), which both studied numbers, albeit in different ways; geometry and stereometry, which dealt with geometrical objects in two and three dimensions, respectively; harmonics and astronomy. As for the content of the various mathematical disciplines, we are told that arithmetic and logistics study the 'odd and even', a phrase Plato seems to use as a synonym for 'numbers', and he also mentions square and solid numbers and the harmonic and arithmetical mean, plus intervals of various kinds.[47] As for geometry, Plato often deals with the issue of incommensurable or irrational magnitudes, which he also calls 'unaccountable' (*alogos*) or 'inexpressible' (*arhetos*). The classical example of an incommensurable line is the diagonal of a square, which cannot be measured by the same unit as its side. Plato observed that incommensurables may be such when taken individually but not such when taken together,[48] and seemed to put great store by a grasp of their complexities. In the *Laws* we are told that if

> all we Greeks believe [lines, surfaces and volumes] to be commensurable when fundamentally they are incommensurable, one had better address these people as follows (blushing the while on their behalf): 'Now then, most esteemed among the Greeks, isn't this one of those subjects [...] it was disgraceful not to understand [...]?'.[49]

It could be that at the time Plato wrote this, his last dialogue, a knowledge of incommensurables had indeed yet to become common, but it could also be that he overemphasized the ignorance of his fellow Greeks as a way of contrasting the superficial opinions held by the many and the truth, which is fully understood only by a few. The same issues underlie a famous passage in the *Meno*, where a slave, asked the right questions by the right person (Socrates), successfully tackles a geometrical problem involving, again, incommensurables. Anybody can gain, or, as Plato argues, retrieve correct mathematical knowledge, but it is important not to trivialize even apparently simple operations such as doubling a square, and to recognize the need for appropriate guidance.

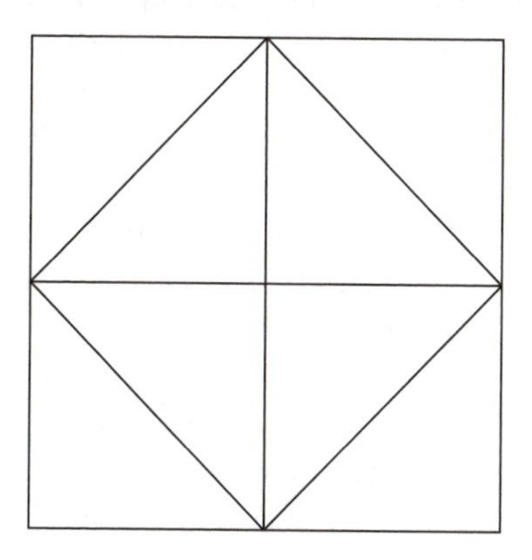

Diagram 1.1

> [Socrates wants to illustrate to Meno his idea that knowledge is recollection, by taking an uneducated slave and showing that, with appropriate questioning, even the slave can exhibit mathematical knowledge (see Diagram 1.1)] *Socrates*. Tell me now, boy, do you know that a square is a figure like this? *Boy*. Yes. *S*. A square then

is a figure that has all these sides equal, and they are four? *B*. Yes indeed. *S*. Does it not also have these lines through the middle equal? *B*. Yes. *S*. [...] If [...] this line was two feet, and this two feet, how many feet would the whole be? Look at it this way: if there were two feet here, but only one foot here, would the figure not be once two feet? *B*. Yes. S. Since then there are two feet here, would it not become twice two feet? *B*. It would. [...] *S*. How many feet then is twice two feet? Calculate and tell me. *B*. Four, Socrates. *S*. Now there could be another figure double this one, which has all the sides equal like this one. *B*. Yes. *S*. How many feet will it be? *B*. Eight. *S*. Come on then, try to tell me how long each of its sides will be. The side of this is two feet; what about each side of the double one? *B*. It is clear, Socrates, that it is the double. *S*. [...] Tell me, are you saying that the double figure is based on the double line? [...] *B*. I do. *S*. Now the line becomes double if we add another the same from here? *B*. Yes indeed. *S*. Say, the eight-foot figure will be based on this, if four sides the same are generated? *B*. Yes. *S*. Let us draw from this four equal sides then. Would this not be what you say is the eight-foot figure? *B*. Yes. *S*. Now, within this there are four, each of which is equal to the four-foot one? *B*. Yes. *S*. How big is it then? Is it not the quadruple of this? *B*. And how not? *S*. Is this, the double, then the one which is quadruple? *B*. No, by Zeus. *S*. But how many times larger is it? *B*. Four times. *S*. Therefore, boy, the figure based on the double line will be not double but quadruple? *B*. You speak the truth. *S*. Now, four times four is sixteen, is it not? *B*. Yes. *S*. The eight-foot figure, on how long a line should it be based? [...] The four-foot figure is on this half line here, is it not? *B*. Yes. *S*. Very well. Is the eight-foot figure not double this and half that? Will it not be on a line greater than this and lesser than that? Is that not so? *B*. I think that it is so. *S*. Good, you answer what you think. And tell me, was one not two feet, and the other four feet? *B*. Yes. *S*. Therefore the side of the eight-foot figure must be greater than this one of the two-feet figure, and lesser than that of the four-feet figure? *B*. It must be. *S*. Try to express then how long you say this is. *B*. Three feet. *S*. Now if it is three feet, let us add the half of this, and it will be three feet; for these are two, and this is one. And from here in the same way these are two and this is one, and so the figure that you say is generated? *B*. Yes. *S*. Now if there are three feet here and three feet here, the whole figure be three times three feet? *B*. It is evident. *S*. How much is three times three feet? *B*. Nine feet. *S*. And the double had to be how many feet? *B*. Eight. *S*. So the eight-foot figure cannot be based on the three-foot line? *B*. Certainly not. *S*. But on how long a line? Try to tell us with precision, and if you do not want to calculate with numbers, then show me on how long a line. *B*. By Zeus, Socrates, I do not understand. *S*. [...] Tell me, is this not for us a four-foot figure? You understand? *B*. Yes. *S*. We add to it this other figure which is equal to it? *B*. Yes. *S*. And we add this third figure equal to either of them? *B*. Yes. *S*. Now could we fill in the corner like this? *B*. Certainly. *S*. Would these four equal figures not originate? *B*. Yes. *S*. Well then, how many times is the whole larger than this? *B*. Four times. *S*. But it should have come out double – do you not remember? *B*. Certainly. *S*. Now, does not this line from corner to corner cut each of these figures in two? *B*. Yes. *S*. Now these are four equal sides surrounding this figure. *B*. They are. *S*. Look now: how large is this figure? *B*. I don't get you. *S*. Does not each internal line cut half of each of these four figures? *B*. Yes. *S*. How many areas this big are there in this? *B*. Four. *S*. How many in this? *B*. Two. *S*. What is the four to the two? *B*. Double. *S*. How many feet in this? *B*. Eight feet. *S*. On what line? *B*. This one. *S*. On the line stretching from corner to corner of the four-foot figure? *B*. Yes. *S*. The sophists call this diagonal; so that if diagonal is its name, do you say, o Meno's boy, that the double figure would be based on the diagonal? *B*. Most certainly, Socrates.[50]

Plato's *Theaetetus*, set in 399 BC, introduces us to two experts on the topic of incommensurables: Theodorus of Cyrene[51] and his pupil Theaetetus, who died in 369 BC. We are told that Theodorus of Cyrene had studied with the sophist Protagoras, but had 'very soon inclined away from abstract discussion to geometry',[52] and that he had explored some particular cases of squares with area 3, with area 5, and so on, observing that their sides were not commensurable with the unit. Theaetetus, at the fictional time of the dialogue still a whiz kid, is credited with broader enquiries. He found a general principle that allowed him to identify which lines were incommensurable in length but not when squared. For instance, as in the case Theodorus himself was studying, two squares whose sides are (in modern symbolism) $\sqrt{3}$ and $\sqrt{5}$ will be commensurable in area (3 and 5) but not in length, i.e. as far as their sides are concerned. This much seems clear, but debate has raged over the precise details of the procedures used by Theodorus and Theaetetus, their respective contributions, and the actual state of the issue at the time (had incommensurables just been discovered, or had they only just been brought to Plato's notice?).[53]

Another topic about which Plato provides precious information is that of the five regular bodies, also called Platonic bodies. In the *Timaeus*, which is a story of the origin of the universe, he introduces a sort of creator figure, the demiurge (in Greek, that means 'craftsman' or 'artisan'). The demiurge moulds the *cosmos* using the forms as models, and gives air, earth, fire and water (the four traditional elements of early Greek natural philosophy) a geometrical structure. The basic constituents of matter are two types of right-angled triangle: isosceles and scalene; everything else can be put together using those two. Thus, fire is made from minuscule pyramids, made from equilateral triangles, in their turn made from six scalene right-angled triangles; air is made from octahedra, water from icosahedra, earth from cubes. Dodecahedra are used to 'decorate' the universe.[54] Why those five solids and not others? Well, the universe in the *Timaeus* is spherical, and pyramid, cube, octahedron, icosahedron and dodecahedron are the only five regular solids which can be inscribed in a sphere. A proof of this statement is not found until Euclid. On the other hand, the geometry contained in the *Timaeus* is put in very vague terms, and possibly deliberately so, since the dialogue claims to be a 'likely story', not an accurate report. Plato may have *chosen* not to include a more rigorous mathematical justification of his grouping of the five bodies. In any case, it seems that geometrical research by the time he was writing the *Timaeus* had identified the solids involved, and probably noticed their common properties.[55]

Apart from having both caught Plato's attention, incommensurables and regular bodies have another thing in common: both have been linked by later sources to the Pythagorean school. Some sources even accused Plato of having plagiarized Pythagorean writings to produce the *Timaeus*. Plato indeed does mention the Pythagoreans, but primarily about their theories on the soul and perhaps (in a passage where no names are named) for their research into music. Indeed, everything we know about their *mathematical* discoveries and interests comes from later, often much later, centuries and is generally thought to be unreliable – which is why the reader will not find much about Pythagoras and the Pythagoreans in this chapter. Most scholars will agree that there was a Pythagorean school of philosophy from the sixth until probably the fourth century BC, that they were involved in politics and that they had certain beliefs about life and the universe, including perhaps the tenet that 'everything is number', or that number holds the key to understanding reality. But most scholars today also think, for instance, that

Pythagoras never discovered the theorem that bears his name.[56] I will return to this topic in chapter 2, and more briefly in the next section.

In sum, what information we can get from Plato is fascinating and rich, and fraught with problems. Beyond issues of detailed reconstructions and precise attributions, which may never be solved with any certainty, he definitely is testimony to the vitality of mathematics at the time, and to the interest that mathematical issues aroused in the educated public at large. Plato will return in the next chapter – let us now turn to another illustrious member of that educated public at large.

Aristotle

Aristotle was Plato's most famous pupil, and one of his most stringent critics: among other things, he disagreed with his views about knowledge, the way the universe worked and whether forms really did exist. For him, mathematical knowledge was not of things that existed in the real world – it provided certainty, surely, but about entities which were *abstractions* from physical objects rather than objects with a separate existence (a belief this which he attributed to the Platonists). Since, grossly put, knowledge for Aristotle was ultimately knowledge of physical reality, the validity of mathematics as a form of knowledge, while granted for its strict domain of application, i.e. mathematical objects, was jeopardized by the fact that its relation with physical reality was at best problematic. At the same time, however, Aristotle considered mathematics an important model for scientific discourse, and devoted large sections of his work to a discussion of the nature of mathematical objects. He explored the logical structure of mathematics, the way it started from undemonstrated principles, on which consensus had been gained, and then proceeded rigorously to conclusions that held generally rather than particularly, thus becoming unassailable by objections or criticisms.

As we have done with Plato, we can artificially distinguish the information we can obtain from Aristotle as a philosopher, and Aristotle as a historian. Again, I will focus on the latter rather than the former, and again, I am aware that no such distinction is really possible. There seems to be a consensus, however, that Plato was more of a mathematician than Aristotle, that he got more involved and participated more directly in the subject. Consequently, what Aristotle says about mathematics, even in his capacity as a philosopher, would seem to reflect historical circumstances to a greater degree than Plato. But we will be cautious anyway.

Like Plato, Aristotle was very interested in the mathematicians' method, and it is on mathematical procedures that some of his philosophical discussions concentrate. He provides the earliest extant account of what criteria should be followed to obtain demonstrative validity, and several of his works are devoted to an analysis of demonstration, argumentation, forms of discourse – not just how knowledge is to be gained, but how it is to be organized, expressed and defended against objections. Although nowhere did Aristotle explicitly state that the mathematics of his day had reached the ideal described in his works, or that it was his discipline of choice, many of the examples deployed in the *Posterior Analytics*, which dealt with scientific demonstration particularly, were taken from mathematics. For instance, he stated that all scientific argumentations must have indemonstrable starting-points or principles: subject-matters such as unit or magnitude, which are assumed to exist; statements such that 'if equals are taken from equals, the remainders are equal', which are assumed to be true; definitions of 'odd' or 'even' or 'irrational' or 'verging', which are assumed to be understood in the same way by all the participants.[57] An element of necessity is also required: one has to be aware that what is in a such-and-such way could not be otherwise, for instance the diagonal cannot be commensurable. In fact, the proof that the diagonal cannot be commensurable (of which more below) is used by Aristotle to exemplify the so-called privative demonstration, also known as *reductio ad absurdum* or proof *per impossibile*: if the consequences of a certain statement are absurd, the statement itself is false. Finally, universality, another characteristic a scientific demonstration should possess, is discussed in these terms:

> Even if you prove of each triangle either by one or by different demonstrations that each has two right angles – separately of the equilateral and the scalene and the isosceles – you do not yet know of the triangle that <it has> two right angles [...] nor <do you know it> of triangle universally, not even if there is no other triangle apart from these. [...] So when do you not know universally, and when do you know *simpliciter*? [...] Clearly whenever after abstraction it belongs primarily – e.g. two right angles will belong to bronze isosceles triangle, but also when being bronze and being isosceles have been abstracted. But not when figure or limit have been.[58]

Aristotle's analysis of demonstration stemmed from his, and his contemporaries', interest and enquiries into rhetoric and persuasive discourse. He realized that the method of mathematics was distinctive with respect to other forms of argumentation: 'it is evidently equally foolish to accept probable reasoning from a mathematician and to demand from a rhetorician demonstrative proofs'.[59] Mathematics was thus rather generally associated not only with a certain subject-matter (numbers, geometrical figures), but also with a certain style:

> Some people do not listen to a speaker unless he speaks mathematically, others unless he gives instances, while others expect him to cite a poet as witness. And some want to have everything done accurately, while others are annoyed by accuracy [...] Therefore one must be already trained to know how to take each sort of argument, since it is absurd to seek at the same time knowledge and the way of attaining knowledge, and neither is easy to get. The minute accuracy of mathematics is not to be demanded in all cases, but only in the case of things which have no matter.[60]

The last sentence is a reminder of mathematics' limitations: it may be a powerful tool for persuasion, but its range of applicability, or the suitability of its method, are restricted.

Like Plato, Aristotle often refers to incommensurables, and, like Plato, he uses them to say things about knowledge in general. In a famous passage from the *Metaphysics*, he has philosophy originate from curiosity:

> Everybody begins [...] by wondering that things are as they are, as one does about self-moving puppets, or about the solstices or the incommensurability of the diagonal; for it seems wonderful to all who have not yet seen the cause, that there is something which cannot be measured even by the smallest thing. But we must end in the contrary and, according to the proverb, the better state, as is the case in these instances too when one learns the cause; for there is nothing which would surprise a geometer so much as if the diagonal turned out to be measurable.[61]

It would then seem that, by Aristotle's time, mathematicians had learnt the cause of incommensurability in the square – indeed, elsewhere Aristotle says that they could prove 'that the diagonal of a square is incommensurable with its side by showing that, if it assumed to be commensurable, odd numbers will be equal to even'. In other words, a *reductio ad absurdum* proof was available.[62] Results which may have already been known to Plato, e.g. that the sum of the angles of a triangle is equal to two right angles, or that the angle in a semicircle is a right angle, are used by Aristotle as 'clear stock examples'.[63] He also hinted that proportion theory was by then a well-established field, and specified that some theorems which had previously been proved separately for numbers, lines, solids, and times, had now been proved universally.[64] Not only was Aristotle able to view mathematical developments over time, he also made the double equation, already underlying the *Theaetetus*, between 'before' and 'after' and 'specific' and 'universal', as if the natural course of mathematics was in the direction of greater and greater generality.

On the other hand, Aristotle is testimony that not all was proceeding smoothly in the mathematical field – at least not in his opinion. One problem area he identifies is the theory of parallel lines: the people investigating them 'unconsciously assume things which it is not possible to demonstrate if parallels do not exist'.[65] In other words, their reasoning is circular. Another dodgy field is the squaring of the circle, already encountered as a subject of ridicule in Aristophanes. Aristotle knew of several attempts to solve the problem, none of them in his view successful. He mentions, rather briefly, Bryson, Antiphon (both sophists – Antiphon was also known as a poet) and Hippocrates of Chios: about Antiphon, he says that it is not the business of the geometer to refute his solution because it was not based on geometrical principles. Bryson's quadrature of the circle relied on universal principles, rather than principles proper only to geometry. As for the quadrature by means of lunules or segments, which interpreters traditionally attribute to Hippocrates, although Aristotle's text is fairly ambiguous on the matter, exactly what was wrong with it has been the object of much debate. It seems clear that, unlike Antiphon's and Bryson's attempts, it had to be taken as a serious geometrical effort.[66] The obscurity of Aristotle's mathematical references has tantalized interpreters since later ancient times and, as we shall see in chapter 2, many people tried to reconstruct early Greek solutions to the quadrature of the circle with varying degrees of implausibility.

We may ask the crucial question again: to what extent does Aristotle reflect the actual mathematical practice of his day? The reader will probably know the answer by now: we simply cannot tell for sure. His direct involvement with mathematics may have been negligible and his admiration for the subject remarkable, yet, most of the times Aristotle names mathematicians it is in order to criticize them, either because of fallacious arguments, or on even more exquisitely philosophical grounds. Thus, he attacks the Platonists for believing that mathematical objects have an existence separate from the things in the world. Eudoxus of Cnidus, associated with Plato's school and celebrated by other sources for his many achievements in geometry, is acknowledged by him 'only' as an astronomer and a philosopher.[67] As for the 'so-called Pythagoreans', Aristotle is prepared to acknowledge that they were the first to advance the study of mathematics

> and having been brought up in it they thought its principles were the principles of all things. [...] in numbers they seemed to see many resemblances to the things that exist and come into being [...] such and such a modification of numbers being justice, another being soul and reason, another being opportunity – and similarly

> almost all other things being numerically expressible; since, again, they saw that the attributes and the ratios of the musical scales were expressible in numbers; since, then, all other things seemed in their whole nature to be modelled after numbers, and numbers seemed to be the elements of all things, and the whole heaven to be a musical scale and a number. And all the properties of numbers and scales which they could show to agree with the attributes and parts and the whole arrangement of the heavens, they collected and fitted into their scheme; and if there was a gap anywhere, they readily made additions so as to make their whole theory coherent. E.g. as the number 10 is thought to be perfect and to comprise the whole nature of numbers, they say that the bodies which move through the heavens are ten.[68]

This is just the prelude, however, to a systematic criticism of a whole host of mistaken beliefs, interspersed with information about the so-called Pythagoreans' actual mathematical research: their investigations into music and the connection between accords and numerical ratios, their interest into the properties and definitions of odd, even, square and rectangular numbers. Both with the Pythagoreans and with the followers of Plato, ideas about mathematics are only a part of what Aristotle criticizes: he is clearly not interested in faithfully reproducing his adversaries' theories, and he obviously does not care strictly to separate what is of interest for us (mathematics) from other topics.

We can certainly use his testimony to conclude that the Pythagoreans, whoever they were, took mathematics to be very significant, and attributed moral, political and cosmological meanings to numbers, and that interest in mathematics was rife in Plato's school in his last days and after his death. Yet, any more detailed reconstruction contains, in my view, an excessive element of speculation. Here I prefer to keep to the vagueness of our early sources, and leave the (relatively) brash clarity of our later ones to the next chapter.

Notes

1 Aeschines, *Against Ctesiphon* 22. Here and henceforth I have used Loeb translations, unless otherwise indicated.

2 Lloyd (1990), 8. See also Vernant (1965), chapter 6; Lloyd (1972), (1979), (1987b); Netz (1999a).

3 Herodotus, *Histories* II 109. Debate has raged over the actual links between Egyptian and Greek mathematics, and on the dependence of the latter on the former, see e.g. Kahn (1991), Bernal (1992).

4 This document is an inscription from the fourth century BC, found in Cyrene, and allegedly reproducing a seventh-century BC original. The translation, discussion and references are in Osborne (1996), 10–15.

5 See e.g. Carter (1990); Stancic and Slapsak (1999). For the political significance of rural and urban land division, see Castagnoli (1956); Asheri (1966) and (1975); Boyd and Jameson (1981).
6 Aristotle, *Politics* 1267b, 1330b, tr. B. Jowett, Princeton University Press 1984.
7 See the evidence collected in Svenson-Evers (1996); Philo's work described in *IG* 22.1668 (347–46 BC).
8 Coulton (1977), 109.
9 Coulton (1977), 64.
10 Plato, *Philebus* 56b, tr. D. Frede, Hackett 1997. The 'instruments' in question are specified a few lines later: straight-edge and compass, mason's rule, plumbline and carpenter's square.
11 Mentioned by Herodotus, *Histories* III 60.
12 Comprehensive evidence on the tunnel in Kienast (1995); extensive discussion of the three problems in Rihll and Tucker (1995), quotation at 410; see also Burns (1971).
13 Tod (1911–12), 128. In fact, it would be more correct to refer to the acrophonic system as a group of notations, because we have several versions of it from all over the Mediterranean basin, with some signs differing rather widely from place to place. Attica is the region around Athens.
14 *IG* II^2 2777 and see Smith (1951), II 162 ff. There has been debate as to whether the Salamis object is a gaming table or a counting-board; both Heath (1921), I 46 ff. and Pritchett (1965) incline for the first interpretation. There is literary and archaeological evidence, however, that abaci *were* extensively used and it seems likely, as suggested even by Heath (1921), I 50, that they may have at least looked like the Salamis table.
15 *IG* IX 488; *IG* XII 99; *IG* XII 282, respectively. The abacus from Thyrium has been interpreted as a fragment of the accounts of the state by its editors, but Tod (1911–12), 112, with whom I agree, thinks it 'far more probable that the stone was a counting-board'. See also Leonardos (1925–26); Lang (1968); *IG* II^2 2778, 2779, 2781. Number 2780, from Eleusis, is catalogued as an abacus, but I think it may be a list of some sort, because the same string of numbers is repeated three times, and this is neither typical of other abaci, nor understandable on the basis of the slab itself being an abacus.
16 On counters and early Greek mathematics see Netz (forthcoming).
17 Several other vases and sherds from the same period are inscribed with numbers to indicate their capacity, price or weight. Sherds may have been used as an alternative to abaci when no complicated counting was needed, cf. Lang (1956).
18 The complete inscription in *IG* I^3 71; tr. B.D. Meritt and A. West, Ann Arbor 1934; commentary in Meritt *et al.* (1939), A 9 and in Meiggs and Lewis (1989).
19 An expression also used by the contemporary historian Thucydides to describe requisitions of bakers in times of war, which had to be, again, proportionate to the size of their mills: *The Peloponnesian War*, VI 22.
20 *IG* I^3 369; commentary in Meritt (1932), 128 ff., my translation.
21 Our (somewhat muddled) information comes from Aristotle, *Constitution of Athens* 48.3; 54.2 and from inscriptions like the one above, see Rhodes (1972) and (1981).
22 Aeschines, *Against Timarchus*, 107.
23 A complete study of the temple and of the inscriptions in Burford (1969), whose translation I quote. The earliest set of inscriptions is: *IG* IV^2 102, 104 and 743; *SEG* XI 417a; *SEG* XV 208–9. The latest inscription, also known as the Tholos accounts, is *IG* IV^2 103.
24 *IG* IV 1484 B II 250–4.
25 Osborne (1994), 13. See also Thomas (1992).
26 Osborne (1994), 15.
27 Homer, *Iliad* 12.421–3; Theognis 805 ff., cf. 543, tr. M.L. West, Oxford 1993.
28 Herodotus, *Histories* I 32; II 142; III 89–95; V 52–4; VII 184–7; IX 28–30, respectively.

29 Aeschylus, *Prometheus Bound* 343–378, tr. J. Scully and C.J. Herington, Oxford 1975, with modifications.
30 Aristophanes, *Wasps* 656–63. The reckoning continues until verse 718.
31 Aristophanes, *Birds* 992–1009. For more information on Meton, see Bowen and Goldstein (1988).
32 Aristophanes, *Clouds* 202–5; tr. A.H. Sommerstein, Aris and Phillips 1982.
33 Thucydides, *The Peloponnesian War* II 13.3–9.
34 Thucydides, *The Peloponnesian War* II 72.3.
35 Aristotle, *Constitution of Athens* 69. The abacus on which trial votes were counted is also mentioned by Aristophanes, *Wasps* 332–3.
36 Several of our extant speeches were written for trials involving bankers or their relatives. This is just one indication that banking activities were widespread at the time, and implied a whole host of arithmetical operations, presumably conducted on the abacus, perhaps by specialized slaves: calculation of interest, exchange between different currencies, division of profits between partners with different shares, see Bogaert (1976); Cohen (1992).
37 Lysias, *On the Property of Aristophanes, Against the Treasury* 9.
38 Lysias, *ibid.* 42–3. Similar accounts (going through the calculation in order to defend or accuse) are given in e.g. Lysias, *Defence Against a Charge of Taking Bribe* 1–5 (*c.* 403–402 BC); Isaeus, *On the Estate of Hagnias* 40–6 (between *c.* 396 and *c.* 378 BC); Demosthenes, *Against Aphobus I* 9–11, 34–9, 47 (364 BC); *Against Leptines* 77, 80 (*c.* 355 BC); *For Phormio* 36–41 (mid-fourth century BC).
39 Lysias, *Against Diogeiton* 4–9.
40 Lysias, *ibid.* 13–15.
41 Lysias, *ibid.* 19–28.
42 Aeschines, *Against Ctesiphon* 9–27; quotation at 17. Cf. also e.g. Lysias, *Against Nicomachus* 5 (*c.* 399 BC, he refused to show his accounts for four years), 19–20 (he entered sacrifices to an inflated excess amounting to six talents); *On the Property of Aristophanes* 50–1 (Diotimus is no longer suspected of embezzlement when ready to show his accounts). See Tolbert Roberts (1982).
43 Demosthenes, *On the Crown* 229.
44 Aeschines, *Against Ctesiphon* 59.
45 Plato, *Republic* 509d–511e. The translation of the passage is pretty controversial, for instance, the last phrase of our quotation (510c6) has also been rendered 'since they were known'. A discussion of the divided-line passage with further references in Mueller (1992).
46 Plato, *Meno* 87a f., see also *Republic* 510c.
47 For references on odd and even see Knorr (1975), 106n101. For square and solid numbers, Plato, *Timaeus* 31c–32b. For means and intervals, *ibid.* 35b–36d.
48 Plato, *Greater Hippias* 303b. Cf. also *Republic* 534d (irrational lines are compared to political rulers).
49 Plato, *Laws* 819e–820c, tr. T.J. Saunders, Hackett 1997, with modifications (note the shift from 'we Greeks' to 'these people'). Cf. also *Parmenides* 140c.
50 Plato, *Meno* 82b–85b, my translation. Note that the word for 'line' and 'side' is the same (*gramme*).
51 Theodorus is also mentioned by Xenophon, *Memorabilia* 4.2.10.
52 Plato, *Theaetetus* 165a, tr. M.J. Levett, rev. M.F. Burnyeat, Hackett 1997.
53 Some of the most recent contributions are: Szabó (1969); Knorr (1975); Fowler (1999).
54 Plato, *Timaeus* 54a–57c.
55 See e.g. Sachs (1917); Waterhouse (1972).
56 See Burkert (1972); Huffmann (1993); Zhmud (1997).
57 See especially Aristotle, *Posterior Analytics* 76a ff.

58 Aristotle, *Posterior Analytics* 74a–b, tr. J. Barnes, Princeton 1984.
59 Aristotle, *Nicomachean Ethics* 1094b, tr. W.D. Ross, rev. by J.O. Urmson, Princeton 1984; see also *Posterior Analytics* 79a.
60 Aristotle, *Metaphysics* 995a.
61 Aristotle, *Metaphysics* 983a, tr. W.D. Ross, Clarendon Press 1928, with modifications. An almost comprehensive collection of mathematical passages in Aristotle is Heath (1949). For the passages where Aristotle cites incommensurables, see the references in Fowler (1999), 290–1.
62 Aristotle, *Prior Analytics* 41a, 50a.
63 Cf. Mendell (1984) for references.
64 Aristotle, *Posterior Analytics* 74a–b.
65 Aristotle, *Prior Analytics* 65a.
66 Evidence collected in Heath (1949). For an extensive discussion of the evidence about Hippocrates of Chios, see Lloyd (1987a).
67 See Napolitano Valditara (1988).
68 Aristotle, *Metaphysics* 985b–986a.

2

EARLY GREEK MATHEMATICS: THE QUESTIONS

Early Greek mathematics was not one but many; there were various levels of practice, from calculations on the abacus to indirect proofs concerning incommensurable lines, and varying attitudes, from laughing off attempts to square the circle to using attempts to square the circle as examples in a second-order discussion about the nature of demonstration. In sum, different forms of mathematics were used for different purposes by different groups of people. Perhaps one common feature is clearly distinguishable: mathematics was a public activity, it was played out in front of an audience, and it fulfilled functions that were significant at a communal level, be they counting revenues, measuring out land or exploring the limits of persuasive speech. The first question addressed in this chapter is what I call the problem of political mathematics. I take 'political' in the literal Greek sense, as something that has to do with the *polis*, the city/community/state. When reading fifth- and fourth-century BC philosophical sources, I have always been struck by the frequency with which mathematical images or examples are used to make points which are not related to mathematics at all – often, points about politics. Moreover, Plato has some very interesting statements on the question of who mathematics should be for, and which mathematics ought to be done by whom: he established parallel hierarchies between forms of mathematics and categories of people. Once again, these were deeply political statements. So, having warned the reader in the introduction that I will ask questions rather than answering them, the first section will expand and muse on the theme, what were the political functions of early Greek mathematics?

The second section will tackle a historiographical issue: how later ancient sources depict early Greek mathematics, and what can be done with them. It will be, I am afraid, an exercise in scepticism.

The problem of political mathematics

In chapter 1, we observed that not only was public counting associated with political accountability, it became a symbol of, or a way of talking about, political participation and the role of the citizen. Moreover, accounts were but one of several mathematical activities that took place in a public context: there were also commercial arithmetic, practised by traders and bankers, the geometry of land division, and in general the mathematics associated with the *technai*, for instance architecture. Land division and commercial arithmetic can be connoted as 'democratic' mathematics: the former was a guarantee of equal distribution, whereas the latter was identified with moneyed economical exchange, as opposed to non-moneyed, non-quantified, status-dependent transactions, which had traditionally been dominated by aristocratic value systems.[1] Aristotle, himself a supporter of oligarchy rather than democracy, even put forth what we could call a mathematizing theory of monetary exchange, where the value of a thing can be, in principle, completely reduced to a number, and transactions to arithmetical operations.

> [Aristotle on money as the measure of all things] All things that are exchanged must be somehow commensurable. It is for this end that money has been introduced, and it becomes in a sense an intermediate; for it measures all things, and therefore the excess and the defect – how many shoes are equal to a house or to a given amount of food. The number of shoes exchanged for a house must therefore correspond to the ratio of builder to shoemaker. For if this be not so, there will be no exchange and no intercourse. And this proportion will not be effected unless the goods are somehow equal. All goods must therefore be measured by some one thing [...] Money, then, acting as a measure, makes goods commensurate and equates them; for neither would there have been association if there were not exchange, nor exchange if there were not equality, nor equality if there were not commensurability. Now in truth it is impossible that things differing so much should become commensurate, but with reference to demand they may become so sufficiently. There must, then, be a unit, and that fixed by agreement (for which reason it is called money); for it is this that makes all things commensurate, since all things are measured by money. Let A be a house, B ten minae, C a bed. A is half of B, if the house is

worth five minae or equal to them; the bed, *C*, is a tenth of *B*; it is evident, then, how many beds are equal to a house, that is, five. That exchange took place thus because there was money is evident.[2]

Democratic mathematics, however, was but one version of the possible political uses of mathematics, which in its various forms was used to articulate *conflicting* positions about the *polis*, man, knowledge, and their interaction.

Take the case of accounts. Ancient authors themselves were quick to point out that no simple equation could be made between accounts and honesty or willingness to have one's actions scrutinized, or between accounts and democracy. Legal rhetoric exposed examples of bad accountancy while at the same time relying on accountancy as an image of clarity and objective good judgement. Both Aristotle and Plato retained audits and auditors in their oligarchic ideal states.[3] Aristotle pointed out that one of the conciliatory methods tyrants may adopt in order to secure their power was rendering accounts of receipts and expenditure, thus providing an illusion but not the substance of democracy,[4] and Plato depicted a real-life mathematical expert as the embodiment of the dangers of knowledge inappropriately used. The sophist Hippias of Elis is presented in the eponymous dialogues as a sort of travelling salesman of general culture, who, as well as acting as envoy for his city, instructed (for a fee) the youths of various parts of Greece in the art of persuasive discourse, in grammar and history, and in astronomy, geometry and arithmetic. It is as a skilled calculator, who hangs out in the *agora*, 'next to the tables of the bankers', that Socrates addresses him here:

SOCRATES [...] If someone were to ask you what three times seven hundred is, could you lie the best, always consistently say falsehoods about these things, if you wished to lie and never tell the truth? [...] So we should also maintain this, Hippias, that there is such a person as a liar about calculation and number. [...] Who would this person be? Mustn't he have the power to lie, as you just agreed, if he is going to be a liar? [...] And were you not just now shown to have the most power to lie about calculations? [...] Do you, therefore, have the most power to tell the truth about calculations? [...] Then the same person has the most power both to say falsehoods and to tell the truth about calculations. And this person is the one who is good with regard to these things, the arithmetician?

HIPPIAS Yes.[5]

Mathematics, the transparent, accountable knowledge *par excellence*, the knowledge which you should be able easily to control by running a check on it, in this short passage is blown apart and revealed as the site of contradictions: truth and falsity are almost indistinguishable – the arithmetician, (by extension, and forgive me for speculating) the accountant, the person who embodies democratic control over the workings of the *polis*, is shown to be the potential master of deceit. Further scepticism had been voiced, according to Plato, Aristotle and later reports, by Protagoras, also a sophist. He observed that geometrical objects are not really as the geometers say they are (for instance, a material circle tangent to a material straight line will touch it in more than one point), and in general that in mathematics 'the facts are not knowable, the words not acceptable'.[6]

Not all critics of mathematics were that philosophically sophisticated. Some people simply could not see the point of speculating on the quadrature of the circle and similar things. Aristophanes and his audience were at home with counting or with geometry as land-division, but found Meton and the Socratic student of the *Clouds* a bit of a joke. According to Xenophon, Socrates himself

> said that the study of geometry should be pursued until the student was competent to measure a parcel of land accurately in case he wanted to take over, convey or divide it, or to compute the yield [...] He was against carrying the study of geometry so far as to include the more complicated figures, on the ground that he could not see the use of them.[7]

Plato's contemporary Diogenes the Cynic apparently would wonder that 'the mathematicians should gaze at the sun and the moon, but overlook matters close at hand', and thought that 'we should neglect music, geometry, astronomy, and the like studies, as useless and unnecessary'.[8] Isocrates, while allowing that astronomy and geometry could be practised as gymnastics for the mind, and to keep young men occupied and out of harm's way, reported nonetheless that 'most men see in such studies nothing but empty talk and hair-splitting, since none of these things is useful either in private or in public life'. He also warned against the dangers of an excessive use of mathematics, which, if pursued too intensely, would have impeded the harmonious mental development of the youth, and in any case was too accurate to be of real use in everyday practical applications – too much accuracy was not always necessary.[9] For more views on the role and dangers of mathematics in education, let us turn back to Plato.

Since in Plato's view knowledge and wisdom were the best entitlements to political power, his ideal state was to be ruled by philosophers. They would be brought up from early childhood following an educational curriculum: gymnastics, reading and writing, military training to begin with; later on, between the ages of around twenty-two and thirty-two years old, mathematics: geometry, stereometry, astronomy, harmonics, but, first of all, arithmetic. Not only did this latter have practical applications in war, for which the rulers had to be prepared, it also helped the mind overcome the pitfalls of sensible knowledge. Both the practical and the more philosophical uses of arithmetic were appreciated by Plato; he drew, however, a crucial distinction:

> it would be appropriate [...] to legislate this subject for those who are going to share in the highest offices in the city and to persuade them to turn to calculation and take it up, not as laymen do, but staying with it until they reach the study of the natures of the numbers by means of understanding itself, not like tradesmen and retailers, for the sake of buying and selling, but for the sake of war and for ease in turning the soul around, away from becoming and towards truth and being.

Analogous claims are made for geometry, whose practitioners use a language which is

> very absurd, if very inevitable. [...] They talk as if they were actually doing something and as if the point of all their theorems was to have some actual effect: they come up with words like squaring and applying and adding and so on, whereas in fact the sole purpose of the subject is knowledge.[10]

Mathematics has then a double character. There are two kinds of arithmetic, that of the 'many', the money-oriented traders and merchants, and that of the people who philosophize. The first kind concerns itself with things which are given a number (two oxen, two armies), the second with numbers considered independently of things. A parallel distinction is introduced between geometry for a concrete purpose on the one hand and philosophical geometry on the other. The distinction is not neutral: 'the arts which are stirred by the impulse of the true philosophers are immeasurably superior in accuracy and truth about measures and numbers'.[11]

Plato thus establishes a boundary between 'good' and 'bad', or, more accurately, 'better' and 'worse' mathematics. The two operate in a continuum, in that they both perform the same operations, or generally speaking talk about the same things. Some people use numbers to count money, some others to reflect about Forms. What in his view irremediably separates them, and gives them different value, is the *use* they make of their subject-matter, the purpose they have.

Plato's reflections on the ethics of knowledge involved not only mathematics, but also rhetoric, medicine, and, more generally, the *technai*, which were object of wide debate between the fifth and the fourth centuries BC.[12] Opinions about their nature and status veered between on the one hand equating *techne* and science (*episteme*), or claiming that anybody who wanted a reputation in philosophy had to learn as many of the *technai* as possible, and, on the other hand, denying that some arts could exist at all, or definitely subordinating *techne* to science, for reasons that included the former's variability or its lack of proof. One of the big issues for Plato was whether moral knowledge and politics were *technai*: was there such a thing as an expert in morals, the way there were experts in medicine, gymnastics, horse-rearing? Could happiness and justice be taught and learnt, the way one did with building a house or making a statue? And could one reach a criterion that would enable him or her always to take the best decision?

> What would seem to be our salvation in life? Would it be the art of measurement or the power of appearance? While the power of appearance often makes us wander all over the place confused and regretting our actions and choices, both great and small, the art of measurement, in contrast, would make the appearances lose their power by showing us the truth, would give us peace of mind firmly rooted in the truth and would save our life. [...] What if our salvation in life depended on our choices of odd and even, when the greater and the lesser had to be counted correctly [...] What then would save our life? Surely nothing other than knowledge, specifically some kind of measurement, since that is the art of the greater and the lesser? In fact, nothing other than arithmetic, since it's a question of the odd and even? Would most people agree with us or not?[13]

Despite this passage, nowhere in his works does Plato even attempt to quantify actions, or goods, or pleasures. The attractiveness of a mathematical model for ethics lies not in its actual feasibility, but in the fact that it evokes accuracy and incontrovertibility. There never is any question that four is greater than three, for instance:

> If you and I were to disagree about number, for instance, which of two numbers were the greater, would the disagreement about these matters make us enemies and make us angry with each other, or should we not quickly settle it by resorting to arithmetic? Of course we should. Then, too, if we were to disagree about the relative size of things, we should quickly put an end to the disagreement by

> measuring? Yes. And we should, I suppose, come to terms about relative weights by weighing? Of course?[14]

In other words, if Plato was after a superior *techne* to use as a model for, or somehow transfer to, the ethical and political field, mathematics was a very strong candidate. In the event, he came to the conclusion in his middle and later works that the reduction of political and moral knowledge to a *techne*, especially as promoted by the sophists, had a number of undesirable consequences. First of all, it was not clear whether *anybody* could be an expert in politics (in fact, the *Republic* puts forth that only the philosopher-rulers could); also, it was necessary to distinguish true moral and political knowledge from pseudo-political *technai* such as, above all, rhetoric. Further, an expert in an art was better at lying about it than someone who was simply ignorant of it, and being a technical expert did not amount to knowing good from evil. Even in politics, one may have learnt how to persuade people of a course of action, but not how to determine the *best* course of action. One may have known how to gain knowledge, but not how to put it to its best use. To employ an analogy:

> no part of actual hunting [...] covers more than the province of chasing and overcoming; and when they have overcome the creature they are chasing, they are unable to use it: the huntsmen and fishermen hand it over to the cooks, and so it is too with the geometers, astronomers, and calculators – for these also are hunters in their way [...] – and so, not knowing how to use their prey, but only how to hunt, I take it they hand over their discoveries to the dialecticians to use properly.[15]

Activities that produced or gained knowledge were to be subordinated to more discerning activities, best able to use that knowledge. Accordingly, mathematics was subordinated to philosophy or dialectic.

In sum, Plato constructed a dichotomy between a mathematics that did not let itself be guided by philosophy, and a mathematics which handed over its direction to philosophers. The latter stayed true to the nature of mathematics itself, at least as Plato saw it, and grew and prospered. The former instead pursued goals which were not necessarily good (such as material enrichment) and thus remained at best fundamentally blind, stunted and misguided, at worst, it led to abuse and wrongdoing. Clearly, his reflections about knowledge and its uses are inseparable from a consideration of the *people* involved. As there are two kinds of arithmetic, there are two kinds of arithmeticians: tradesmen, builders, the common and morally undisciplined layman on one side, true philosophers, philosopher-rulers, wise legislators on the other. If some ways of doing mathematics had indeed come to symbolize democratic activities or values, Plato's reminder that mathematics is only good when supervised by an ethically and philosophically informed elite is sending a clear political message, which is perhaps nowhere as clear as in his last dialogue.

The *Laws* is a description of Plato's second-best ideal state, conceived as a colony. Its political leadership, as in the *Republic*, is both restricted to a few individuals, and strictly associated with knowledge, including mathematical knowledge of a certain kind. The number of citizens for the new state is mathematically regulated: 5,040, not a person more, not a person less. The advantage of the number 5,040 is that it has the largest number of consecutive divisors, making it possible to employ it everywhere:

> this is the mathematical framework which will yield you your phratries and demes and villages, as well as the military companies and platoons, and also the coinage-system, dry and liquid measures, and weights. The law must regulate all these details so that the proper proportions and correspondences are observed. [...] [The legislator] will assume it is a general rule that numerical division and variation can be usefully applied to everything – to arithmetical variations and to the geometrical variations of surfaces and solids, and also to those of sounds, and of motions [...] The legislator should take all this into account and instruct all his citizens to hold fast, so far as they can, to this system. For in relation to household administration, to politics and to all the arts (*technai*), no single branch of educational learning has so great a power as the study of numbers. [...] These subjects [number and calculation] will prove fair and fitting, provided that you can remove pettiness and greed [...] otherwise you will find that you have unwittingly produced a rascal [...] instead of a sage: examples of this we can see today in the effect produced on the Egyptians and Phoenicians and many other nations by the petty character of their approach to wealth and life in general.[16]

It is precisely because numbers are so powerful that their use has to be regulated: although everybody has to get a basic smattering of mathematics, the subject is to be studied in depth by only a chosen few, who would know how to use it in the right way.[17] Indeed, not all mathematical notions are beneficial: for instance, simple division of goods in equal shares, which would amount to equality 'according to measures, weights and numbers', is a rash idea, and not a realistic possibility. Political participation depends on another, much preferable, type of equality: the one according to nature, i.e. according to what everybody deserves. Given that, admittedly, 'natural' equality is very difficult to assess for humans who do not have 'the wisdom and judgement of Zeus', this means that some people are more equal than others.

A similar type of mathematical politics, or political mathematics, is found in Aristotle. Like Plato, he established a correspondence between epistemic and social hierarchy. In the *Metaphysics* he put theoretical knowledge, including mathematics, at the top of a value scale whose next steps down were the *techne* of the master-worker, the experience of the manual worker ('we think the manual workers are like certain lifeless things which act indeed, but act without knowing what they do') and finally sense-perception pure and simple, like animals have. For Aristotle, the pursuit of knowledge presupposed leisure, freedom from daily cares and independence, all of them prerogatives of the privileged classes. He emphasized the analogy between knowledge and power:

> of the sciences [...] that which is desirable on its own account and for the sake of knowing [...] is more of the nature of wisdom than that which is desirable on account of its results, and the superior science is more of the nature of wisdom than the ancillary; for the wise man must not be ordered but must order, and he must not obey another, but the less wise must obey him.[18]

Aristotle's version of the origin of mathematics is also quite revealing:

> as more arts were invented, and some were directed to the necessities of life, others to its recreation, the inventors of the latter were always regarded as wiser than the inventors of the former, because their branches of knowledge did not aim at utility. Hence when all such inventions were already established, the sciences which do not aim at giving pleasure or at the necessities of life were discovered, and first in the places where men first began to have leisure. This is why the mathematical arts were founded in Egypt; for there the priestly caste was allowed to be at leisure.[19]

While the geographical attribution, so to speak, is at odds with Plato's picture of the greedy Orientals, the value distribution is the same: the original and true nature of mathematics is detached from any material, common, concrete uses; the wiser sort of mathematician is *not* someone who has to work for a living. The leisured man is also in a better position to cultivate philosophy, and understand what is better for the state. In fact, two more passages in Aristotle blithely associate the right sort of mathematics and the right sort of politics. The first reprises Plato's double notion of equality:

> Justice involves at least four terms, namely, two persons for whom it is just and two shares which are just. And there will be the same equality between the shares as between the persons, since the ratio between the shares will be equal to the ratio between the persons; for if the persons are not equal, they will not have equal shares [...] Justice is therefore a sort of proportion; for proportion is not a property of numerical quantity only, but of quantity in general, proportion being equality of ratios, and involving four terms at least [...] The principle of distributive justice, therefore, is the conjunction of the first term of a proportion with the third and of the second with the fourth; and the just in this sense is a mean between two extremes that are disproportionate, since the proportionate is a mean, and the just is the proportionate. This kind of proportion is termed by mathematicians geometrical proportion [...] But the just in private transactions [...] is not the equal according to geometrical but according to arithmetical proportion. For it makes no difference whether a good man has defrauded a bad man or a bad man a good one [...] the law looks only at the nature of the damage, treating the parties as equal [...] Hence the unjust being here the unequal, the judge endeavours to equalize it [...] if we represent the matter by a line divided into two unequal parts, he takes away from the greater segment that portion by which it exceeds one-half of the whole line, and adds it to the lesser segment. When the whole has been divided into two halves, people then say that they 'have their own', having got what is equal. This is indeed the origin of the word *dikaion* (just): it means *dicha* (in half).[20]

There is a subtle distinction here between criminal justice ('the just in private transactions') and economic, or distributive, justice. The first can afford to apply full equality, or arithmetical proportion. Even supporters of oligarchic forms of government were prepared to subscribe to the principle that everybody is the same in front of the law: even Plato's ideal state has auditors. When it comes to the big divide between rich and poor, however, equality 'by merit' in the form of geometrical proportion rears its head. Let us look at a second Aristotelian passage:

> party strife is everywhere due to inequality, where classes that are unequal do not receive a share of power in proportion [...] for generally the motive for factious strife is the desire for equality. But equality is of two kinds, numerical equality and equality according to worth [...] the proper course is to employ numerical

> equality in some things and equality according to worth in others [...] what is thought to be the extreme form of democracy and of popular government comes about as a result of the principle of justice that is admitted to be democratic, and this is for all to have equality according to number. [...] But the question follows, how will they have equality? Are the property-assessments of five hundred citizens to be divided among a thousand and the thousand to have equal power to the five hundred?[21]

In a sense, the supporters of extreme democracy are making a mathematical mistake: their claims are disproportionate, as anybody can verify, because it is rather absurd to try and divide five hundred by a thousand, or to equal one thousand to five hundred. It is as if political mistakes, and even revolutions, are caused by the wrong application of mathematical concepts, or by the application of the wrong mathematical concepts.

In conclusion, mathematics in classical Athens had a number of practical uses, and was very visible. The city kept accounts of its financial operations, officers were obliged to give accounts, any citizen could be required to do sums for the benefit of the *polis*. The public counting involved was a ritual that celebrated the transparency of, and involvement of the Athenian citizens in, the running of the state; it was constructed as a *deeply democratic* type of mathematics. As such, we find it in another crucial arena, i.e. the law courts, where verdicts were reached by simple counting of votes. Also, as Plato reminds us, we find a lot of counting in public in yet another crucial arena of democracy: the moneyed economy of the marketplace. Yet, traders proverbially cheat under your own eyes, accounts can be fudged, calculations can trick you, mathematics in the wrong hands goes the wrong way, is misapplied, misunderstood and becomes bad. In some texts, a dark side of mathematics emerges, which corresponds to the pitfalls and shortcomings of democracy.

I think it is evident that mathematics was a 'polyvalent symbol within a complex symbolic system' and that there was a struggle over who controlled its signification.[22] What made the stakes particularly high was the fact that early Greek mathematics was, through accounts, commercial transactions, architecture, in the simple and banal forms of measuring and counting, part of the experience of many people, not just of the highly educated; it was a *techne*, which means it could be learnt and taught by virtually anybody. Moreover, already at the time, mathematics was viewed as an objective, certain, persuasive, form of knowledge. It was, to use anachronistic terms, a science. A lot of things can be used and are used as signifiers in all periods to talk about, say, the social and political order – what makes the use of science a particularly strong signifier is that it claims to be objective knowledge, it is used as a signifier *because* it projects that aura of objectivity onto political and social discourse. This is at least part of the reason why Plato, for instance, upheld an ideal of mathematics as a philosophical, detached, elite, pursuit – in order to reappropriate it. Controlling the signification of science, in other words, was, and continues to be, a particularly powerful way to control other areas of discourse.

The problem of later early Greek mathematics

My introductory disclaimer, that, given the limitations of space, many things had to be left out of this book, applies particularly to this section. In an attempt to introduce the novice to the raw business of squeezing reliable evidence out of unlikely informers, I will focus on seven ancient sources (Archimedes, Philodemus, Plutarch, Diogenes Laertius, Proclus, Simplicius and Eutocius), rather than discussing modern reconstructions of early Greek mathematics, especially since they inevitably use those same ancient sources anyway. Each of the chosen seven had his own reasons for citing pieces of mathematics from the past, but we cannot always reconstruct their agenda in full detail; so, while those reasons obviously affect their reliability, this happens to an extent we can only (educatedly) guess.

Archimedes tells us precious little: that some of the ancient geometers devoted many efforts to squaring the circle, without really succeeding because they supported their proofs with inadmissible lemmas. What they did manage to prove was that circles are to each other like the squares on their diameters; that spheres are to each other like the cubes on their diameters; and that pyramids and cones are one third of prisms and cylinders (respectively) with the same base and the same height. All these results relied on the lemma that two unequal surfaces can be continuously subtracted one from the other, until their difference is smaller than any given area. Now, on another occasion, Archimedes explicitly attributes to Eudoxus the proposition about cones and cylinders, adding that 'one should give no little credit to Democritus, as the first to formulate the statement about that figure, without proof'. We can safely assume, then, that Eudoxus was the discoverer of the lemma and of the other two results, too.[23] This attribution is very important because the lemma in question is extensively employed in Euclid's *Elements* – one can infer that the results contained in the *Elements* which depend on the lemma may have also been discovered by Eudoxus. How far one goes with attributions by inference depends of course on how original one thinks Euclid was, or how advanced one thinks mathematics was in the fourth century BC.

Little precious remarks can go a long way, and produce a sort of chain reaction effect, in that they can corroborate the testimony of other authors and reflect positively on their reliability in general. For instance, if we take

Archimedes to be reliable, then information about Democritus can be used to substantiate a passage in Plutarch. The fact that Archimedes seems to concord with Plutarch on one point makes Plutarch rather more reliable in general, perhaps even when he is reporting things for which we have no external corroboration.

In fact, Archimedes is considered a very trustworthy source. Not only was he relatively close in time to Democritus and Eudoxus: he was a mathematician, so we can assume that he understood the material and had access to a wide range of treatises and results, either in person or through his extensive and well-documented contacts with mathematicians in Alexandria. Not only was Archimedes a mathematician, he was a very good, arguably the best ancient mathematician, so again that adds to his reliability because we tend to make an (anachronistic) equation between scientific expertise and professional ethics. We believe that Archimedes was what we could call intellectually honest, that he apportioned praise and blame where it was due, so that when he said Democritus had done something, that is something Democritus had indeed done. Nobody I know (not even me) doubts Archimedes' testimony on Democritus and Eudoxus. Nevertheless, I would invite reflection on one point: Archimedes himself is writing history of mathematics. We are trusting him as a historian, because we trust him as a mathematician. He identifies a development in the research on a particular set of problems, with Democritus formulating the statement but not providing the proof, other ancient geometers coming up with proofs which were incorrect because they relied on inadmissible principles, and finally Eudoxus discovering the right lemma and thus proving a number of results. Archimedes himself builds on Eudoxus' achievement and takes a version of his lemma as a starting-point. It is a cumulative vision of mathematics, which also corresponds to a certain vision of the heuristic process: the statement is arrived at first, then a proof is sought, using the right auxiliary propositions. We know that this was more or less Archimedes' own heuristic process in his own actual practice, because he tells us as much. The historical development of mathematical research thus mirrors, or is made to mirror, the individual experience of the mathematician. History reflects on a larger scale what happens within a lifetime of work.

We next find Eudoxus as one of the main characters in Philodemus' history of Plato's Academy, in its turn based on Dicearchus:

> At that time there was also a remarkable advancement in the mathematical studies, with Plato as architect and propounder of problems, and the mathematicians then researching them with zeal. Thus metrology and the problems about definitions reached an acme for the first time, while Eudoxus and his circle renewed

> completely the original results of Hippocrates. Geometry also advanced greatly. Then originated analysis and the propositions about *diorismoi*, and overall geometry was taken greatly forwards. Nor were optics or mechanics left behind.[24]

Apart from the reference to *diorismoi*, the testimony is rather vague: Plato's Academy fostered mathematical research, Eudoxus and other people were associated with it and they had wide interests. This much is corroborated by other sources. That said, we also have to remember that Philodemus was writing a celebratory history of his philosophical school of choice, that we have no indication that he was himself a mathematician, and that some three centuries separate him from Eudoxus' time. The image of the Academy as a fertile greenhouse of mathematical talent is a recurrent one in Platonist traditions, as is the figure of Plato presiding over the mathematicians, telling them what to research, and keeping their eyes steadily on the real prize. The parallel here between Plato and an architect, someone who directs the work of very skilled *technitai*, while being himself in possession of a superior kind of knowledge, the only one who can see where it is all going, fits in very well with Plato's views as sketched in the previous section. Continuity is stressed by the reference to Hippocrates of Chios, but the emphasis is primarily on renewal, originality and climaxing 'for the first time'. Philodemus, and/or Dicearchus, seem keen to characterize Platonist mathematics as different from analogous, but not equally philosophically-informed, practices.

Both Eudoxus and Democritus figure in Plutarch, who reports a mathematical puzzle attributed to the latter:

> if a cone should be cut by a plane parallel to its base, what one must suppose the surfaces of the segments prove to be, equal or unequal? – for, if unequal, they will make the cone uneven by giving it many step-like notches and asperities; and, if they are equal, the segments will be equal, and the cone, being composed of circles that are equal and not unequal, will manifestly have got the properties of the cylinder – which is the height of absurdity.[25]

While, for Archimedes, Democritus realized the equivalence of a certain cone and a certain cylinder, Plutarch reports a more detailed and polemical discussion of the difficulties involved in studying the cone. No easy connection can be made between the two – it could be that Democritus explored geometrical paradoxes in an attempt to prove what he had simply formulated, or that he happened upon the formulation in the context of his reflections on geometrical paradoxes. Can we trust Plutarch as to the details? Perhaps yes: from the rest of his work it is clear that he was quite knowledgeable about mathematics, and that he had access to many early sources. The testimony is not out of character as far as Democritus, one of the founding fathers of atomism, is concerned, but it also seems to be contained in an intermediate source, rather than being quoted from Democritus' works themselves – they were probably lost by the time Plutarch was writing. If we move on to Eudoxus, we find ourselves again on rather shaky ground. Take this passage:

> Plato himself reproached Eudoxus and Archytas and Menaechmus for setting out to move the problem of doubling the cube into the realm of instruments and mechanical constructions, as if they were trying to find two mean proportionals not by the use of reason but in whatever way would work.[26]

It was indeed to Plato, in Plutarch's narration, that a delegation of people from Delos turned for a solution to the duplication of the cube – once again, Plato appears in the role of ringmaster of mathematical practice.[27] Plutarch was well acquainted, if not with Philodemus/Dicearchus, with other works in the same tradition which depicted the relation between Plato and the mathematicians around him in pretty much the same light. Now, it seems from other sources that Eudoxus, Archytas and Menaechmus did in fact provide solutions to the duplication of the cube, so for some aspects Plutarch appears to be trustworthy. Nevertheless, a certain image of Plato was crucial to Plutarch's view of mathematics as a whole, so it was important for him to emphasize the significance of Platonic input and overall supervision. The fact that our survey now counts two reports, both agreeing on Plato's 'architectural' function, does not necessarily increase their veracity quotient, because the earlier one, or a version of the earlier one, may have been read by the later author, who shared similar interest and philosophical loyalties. In fact, there is a notable tendency for stories like these to accumulate more and more details as they are passed on: Eudoxus and his circle have now become Eudoxus, Archytas, Menaechmus and Helicon of Cyzicus. The past gets suspiciously clearer the further away we get from it.

Diogenes Laertius' main work is a sort of biographical dictionary of great philosophers, from Thales to Epicurus. He gathered as many pieces of information as he could, sometimes in contrast with one another, always indicating his sources (many of which are otherwise totally unknown to us), occasionally assessing their credibility. Of Democritus, for instance, Diogenes confirms that he was 'versed in every department of philosophy', including mathematics, which he had learnt while travelling in Egypt. Eudoxus, 'an astronomer, a geometer, a physician and a legislator' is included among the famous Pythagoreans; he also discovered the properties of curves and learnt geometry from Archytas. Democritus and Eudoxus were not unique in their choice of topics for investigation: Thales, whom earlier sources tend to depict as a rather unspecifically wise man, interested in astronomy, capable of engineering feats, according to Diogenes also learnt geometry in Egypt, where he measured the height of the pyramids by means of their shadow (probably using the properties of similar triangles). He 'was the first to inscribe a right-angled triangle in a circle, whereupon he sacrificed an ox. Others tell this tale of Pythagoras'. In the entry on Pythagoras, however, the sacrifice is motivated by the discovery that in a right-angled triangle the square on the hypotenuse is equal to the squares on the sides containing the right angle, i.e. what is still called Pythagoras' theorem. Eudoxus' alleged teacher, the Pythagorean Archytas of Tarentum, as well as being general of his city seven times in a row, applied mathematical principles to mechanics and, conversely, 'employed mechanical motion in a geometrical construction, namely, when he tried, by means of a section of a half-cylinder, to find two mean proportionals in order to duplicate the cube'.[28]

The wealth of details should not blind us to the fact that there is a general plan, an ideally complete map of philosophy, from its remote origins, in a continuous stream, all the way to more recent times. Diogenes systematized his information in at least two ways: he classified people on the basis of the school they belonged to, in some cases forcibly enlisting them to a school (how else would Eudoxus become a Pythagorean?), and he created teacher-pupil links between them. The detectable presence of a general plan need not make one entirely suspicious of individual elements – Diogenes after all admits to uncertainty about Pythagoras' discoveries, and his information on Archytas may confirm what we know from Plutarch. Yet, one cannot help being struck by the sheer abundance of information about very early figures like Thales, who was already a sort of semi-mythical character by Aristophanes' time.

Thales figures again in Proclus' potted history of mathematics:

> Thales, who had travelled to Egypt, was the first to introduce [geometry] into Greece. He made many discoveries himself and taught the principles for many others to his successors, attacking some problems in a general way and others more empirically. Next after him Mamercus [...] is remembered as having applied himself to the study of geometry [...] Following upon these men, Pythagoras transformed mathematical philosophy into a scheme of liberal education, surveying its principles from the highest downwards and investigating its theorems in an immaterial and intellectual manner. He it was who discovered the doctrine of proportionals and the structure of the cosmic figures. After him Anaxagoras of

> Clazomenae applied himself to many questions in geometry, and so did Oenopides of Chios [...] Following them Hippocrates of Chios, who invented the method of squaring lunules, and Theodorus of Cyrene became eminent in geometry. For Hippocrates wrote a book of elements, the first of whom we have any record who did so. Plato, who appeared after them, greatly advanced mathematics in general and geometry in particular [...] At this time also lived Leodamas of Thasos, Archytas of Tarentum, and Theaetetus of Athens, by whom the theorems were increased in number and brought into a more scientific arrangement. Younger than Leodamas were Neoclides and his pupil Leon, who [...] was able to compile a book of elements more carefully designed [...] He also discovered *diorismoi* [...] Eudoxus of Cnidus, a little later than Leon and a member of Plato's group, was the first to increase the number of the so-called general theorems; to the three means already known he added three more and multiplied the number of propositions concerning the 'section' which had their origin in Plato, employing the method of analysis for their solution. Amyclas of Heracleia, one of Plato's followers, Menaechmus, a student of Eudoxus [...] and his brother Dinostratus made the whole of geometry still more perfect. Theudius of Magnesia [...] produced an admirable arrangement of the elements and made many partial theorems more general. There was also Athenaeus of Cyzicus, who [...] became eminent in other branches of mathematics and most of all in geometry. These men lived together in the Academy, making their enquiries in common. Hermotimus of Colophon pursued further the investigations already begun by Eudoxus and Theaetetus, discovered many propositions in the *Elements*, and wrote some things about locus-theorems. Philippus of Mende [...] also carried on his investigations according to Plato's instructions [...] Not long after these men came Euclid.[29]

Some of the information in this passage is well-substantiated by contemporary sources: Aristotle may be taken to confirm that Hippocrates investigated the quadrature of lunules, Plato tells us that Theodorus and Theaetetus made contributions to geometry. Some other statements match the sources we have surveyed in this section: Thales made a trip to Egypt as in Diogenes Laertius, Eudoxus employed the method of analysis as in Philodemus, a connection is made between Eudoxus and Menaechmus as in Plutarch. As we have pointed out before, however, this indicates a common source and a partly shared agenda, more than constituting mutual corroboration. Several of the mathematicians above are hardly mentioned by sources significantly earlier than Proclus, and when they are, it is for their enquiries into astronomy and natural philosophy rather than for any mathematical contributions they might have made. Above all, many of the names on Proclus' list are not known outside it: Mamercus, Neoclides, Leon, Amyclas, Theudius, Athenaeus, Hermotimus. There is at present no way of establishing what these people did, who they were or (if one wants to be *really* suspicious) whether they existed at all.

Proclus' picture worryingly looks as if he had fleshed out a by then well-established canon of names and lines of transmission, with the aim of constructing a progressive development towards greater generalization and theoretical sophistication. His mathematicians seem to flow more or less seamlessly into each other's lives, inheriting problems from the previous generation; the role of Plato, the Academy and their particular interests is seen as crucial; the production of perfectly-arranged 'elements' is presented ever since Hippocrates as a major concern, culminating with Euclid. Once again, there is no cogent reason to think that individual details are wrong – taking the picture wholesale as historically accurate, however, is a different matter.

In fact, Proclus' list is an interesting case, because of its possible association with Eudemus. A pupil of Aristotle's, Eudemus wrote a history of geometry, now lost, from the unique vantage-point of a contemporary witness of the activity between the fourth and third century BC. Now, in other parts of the same commentary to Euclid, Proclus alleges Eudemus as his source. Although he is generally seen as too late, too philosophically involved and not enough of a mathematician to be fool-proof, this has had the power to give him some credibility. Not a few modern scholars seem to have clung to his report, one of the most complete and detailed pieces of evidence we have about a crucial period of Greek mathematics, as the only rock in a sea of vague, undetailed evidence. Yet, one should sceptically point out that, first, it is fair to expect Proclus to be more interested in putting his own points across than in quoting Eudemus word for word; second, that we have no guarantee that Eudemus' report was itself historically accurate and immune from philosophical biases; third, that it is unlikely that Eudemus' text would survive in its original form until Proclus' time.

Simplicius, next on our list, records another little miracle of his own, also taken from Eudemus' book. You may remember that Aristotle often used mathematical examples. We are not the first readers to find his allusions rather cryptic – a late ancient audience also needed to be filled in on the details. Simplicius was a sixth-century AD writer whose main extant work consists of commentaries to Aristotle's books, including explanations of passages which had by then become unclear. He explains Aristotle's reference to 'squaring by means of sections' thus:

> The squaring by means of sections is the squaring by means of lunules, discovered by Hippocrates of Chios; for the lunule is a section of circle. Eudemus, however, in his history of geometry says that Hippocrates demonstrated the quadrature of the lunule not on the side of a square but in general [...] For every lunule has an outer circumference equal to a semicircle or greater or less, and if Hippocrates squared the lune having an outer circumference equal to a semicircle and greater and less, it would seem that the quadrature was proved in general. I shall set out what was said by Eudemus word for word, adding a few things from Euclid's *Elements* for the sake of clarity since I am reminded of the summary style of Eudemus, who set out the proofs concisely, according to the ancient custom. He says this in the second book of the history of geometry. And the quadratures of lunules, which seemed to be part of the complex propositions because of its kinship with the circle, were first written about by Hippocrates, and seemed to be properly carried out [...] He made a starting-point, and posited as first thing useful for these, that similar segments of circles have the same ratio to each other as the squares on their bases. And this he proved by showing that the squares on the diameters have the same ratio as the circles. [...][30]

Simplicius' text retains at least partly the existence of various layers of transmission; he shows awareness of the difference between what he himself is saying, what Eudemus was saying and what Hippocrates may have been saying in *his* turn, but that is not his only concern, because he also wants his readers to understand him, if necessary by modifying the text for the sake of clarity. Can we take Simplicius to be a 'faithful' testimonial of Hippocrates' mathematical practice? Then again, how can we be sure that he really had Eudemus' text, and that Eudemus really had Hippocrates' text?

To conclude with another little sixth-century AD miracle, Eutocius. In the second book of his *Sphere and Cylinder*, Archimedes takes the solution to the problem of the two mean proportionals as granted. Some nine centuries later, Eutocius of Ascalona, who was writing a commentary on Archimedes' text, decided to expand on that point by reporting several solutions to the problem. His anthology includes a duplication of the cube attributed to Plato, which is carried out by means of a moving ruler (thus contradicting Plutarch's account), a solution by Menaechmus, one by Archimedes' contemporary Diocles, which is not attested elsewhere but has been corroborated by the recovery of Diocles' book in an Arabic version, and a long passage by (allegedly) Eratosthenes:

> It was researched by the geometers in what way to double the given solid, it retaining the same shape, and this problem was called the duplication of the cube; for, having posited a cube, they sought to double it. Nobody having solved it for a long time, Hippocrates of Chios first came up with the idea that, if two mean proportionals taken in continued proportion were discovered between two straight lines, of which the greater was double the lesser, the cube would be doubled, so that the puzzle was by him turned into another puzzle as great as the first. After a time, they say, certain Delians, trying to double a certain altar in accordance with an oracle, were stuck with the same puzzle, and were sent to ask the geometers who were with Plato in the Academy if they could discover what they sought. Having applied themselves diligently and seeking to find two mean proportionals between two given straight lines, Archytas of Tarentum is said to have discovered them by means of the semicylinders, and Eudoxus by means of the so-called curved lines; but it so happened that while all these were written in a demonstrative fashion, it was not possible to make them handy and put them to use, except to a certain small extent Menaechmus, and even that with difficulty.[31]

Again a micro-history of mathematics, this time focussed on the duplication of the cube, and contradicting some elements of Plutarch's story, because, far from being too compromised with practice, Archytas' and Eudoxus' solutions are not considered practical enough. In fact, at least in the former case, Eutocius' readers could judge for themselves, because he reports Archytas' solution in full, and on the authority of Eudemus (see Diagram 2.1):

> Let the two given straight lines be *AΔ*, *Γ*; it is necessary to find two mean proportionals between *AΔ*, *Γ*. Let the circle *ABΔZ* be described around the greater straight line *AΔ*, and let *AB* be fitted in equal to *Γ* and let it be prolonged until it meets the tangent to the circle from *Δ* in the point *Π*, let *BEZ* be drawn parallel to *ΠΔO*, and let a right semicylinder be imagined on the semicircle *ABΔ*, and on *AΔ* a right semicircle posited in the parallelogram of the semicylinder. This semicircle rotating as if from *Δ* to *B*, the end *A* of the diameter remaining at rest, it will cut the cylindrical surface in its rotation and will describe a certain curve in it. Again, if *AΔ* remains at rest and the triangle *AΠΔ* goes around with a motion opposite to the semicircle, it will produce a conic surface with the straight line *AΠ*, which moving around will come across

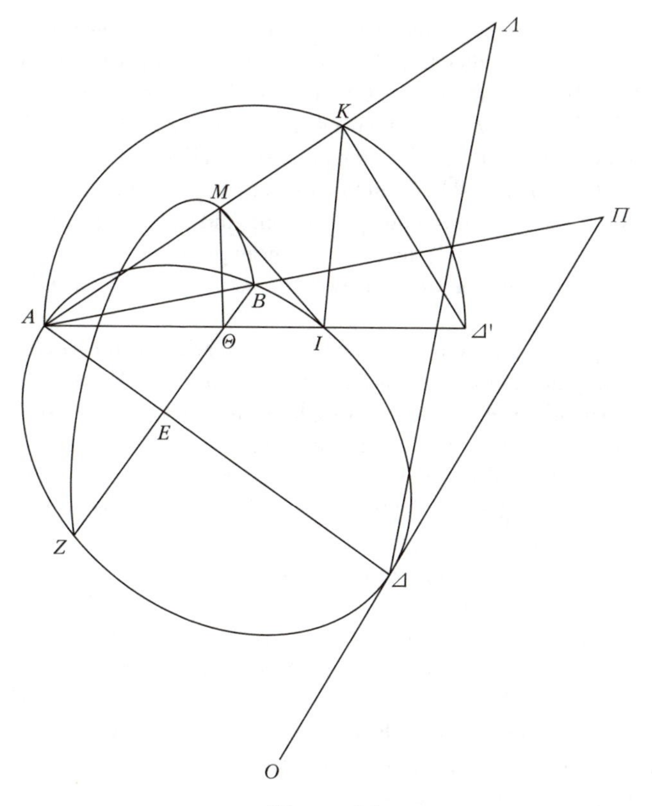

Diagram 2.1

the cylindrical line in a certain point; at the same time B will describe a semicircle on the surface of the cone. Let the moving semicircle have a position according to the point of junction of the curves, such as the position $\Delta' KA$, and the triangle moved in the opposite direction a position $\Delta\Lambda A$; let the point of the said junction be K, and let the semicircle described through B be BMZ, and let BZ be the section common to it and to the circle $B\Delta ZA$, and let there be drawn from K a perpendicular to the plane of the semicircle $B\Delta A$; it will fall on the circumference of the circle because the cylinder is right. Let it fall, and let it be ΛI, and let the line joining I and A come across BZ at the point Θ; let $A\Lambda$ come across the semicircle BMZ at the point M, and let $K\Delta$, MI, $M\Theta$ be joined. Now, since each of the semicircles $\Delta' KA$, BMZ is perpendicular to the underlying plane, therefore their common section $M\Theta$ is also perpendicular to the plane of the circle; so that $M\Theta$ is also perpendicular to BZ. Therefore the rectangle from $B\Theta$, ΘZ, which is the same as the rectangle from $A\Theta$, ΘI, is equal to the square on $M\Theta$; therefore the triangle AMI is similar to each of $MI\Theta$, $MA\Theta$, and the angle IMA is right. The angle $\Delta' KA$ is also right. Therefore $K'\Delta$, MI are parallel, and because of the similarity of the triangles there will be the proportion: $\Delta' A$ is to AK, as KA is AI, as AI is to AM. Therefore the four straight lines ΔA, AK, AI, AM are in continuous proportion. And AM is equal to Γ, since it is equal to AB; therefore two mean proportionals, AK, AI, have been found between the two given straight lines $A\Delta$, Γ.[32]

Strange as it may sound, Hippocrates' squaring of the circle, as in Simplicius, and Archytas' duplication of the cube, as in Eutocius, are the earliest 'extant' Greek mathematics we have, and they are both credited to Eudemus. Are they to be trusted, and the felicity of such transmissional accidents celebrated? Or should we postulate that 'Eudemus' really was a late compilation, perhaps deliberately couched in archaicizing language, perhaps even a forgery?

As I said, I have no ready answers. I will just issue a further warning. All the sources we surveyed, including Archimedes, were writing their own little histories of early Greek mathematics, and all for different reasons. One should bear that in mind, and not view them as mere reservoirs of information. That we should gather more minutely detailed information about fifth-century BC geometry from sixth-century AD authors than we are able to do from their own contemporaries, is only one of the succulent paradoxes of history, and we may never be able completely to unravel it.

Notes

1 von Reden (1995) and (1997); Morris (1996); Kurke (1999).

2 Aristotle, *Nicomachean Ethics* 1133a–b, tr. W.D. Ross revised by J.O. Urmson, Princeton 1984, with modifications.

3 Plato, *Laws* 945b ff.; Aristotle, *Politics* 1318b; 1322b.

4 Aristotle, *Politics* 1314b.

5 Plato, *Lesser Hippias* 367a–c, tr. N.D. Smith, Hackett 1992. Cf. also *Greater Hippias* 281a, 285b–c; *Protagoras* 318d–e and Xenophon, *Memorabilia* 4.4.7.

6 See Untersteiner (1961), chapter 2 and *P. Herc.* 1676 in Gaiser (1988).

7 Xenophon, *Memorabilia* 4.7.2–4.

8 In Diogenes Laertius, *Lives of the Philosophers* 6.28, 73.

9 Isocrates, *Panathenaicus* 26–9 (written *c.*342 BC); *Antidosis* 261–9 (*c.*354 BC); *Helen* 5 (*c.*370 BC); *Against the Sophists* 8 (*c.*390 BC). Cf. Aristotle, *Metaphysics* 995a.

10 Plato, *Republic* 522c–527a. Is Plato here responding to Aristophanes' lampoon of Meton, also a learned man involved in the foundation of an ideal city? For similar themes see also *Philebus* 56d–57a and [Plato], *Epinomis* 990a–991e.

11 Plato, *Philebus* 57d.

12 See Cambiano (1991), to which my discussion is heavily indebted; Lloyd (1963); Roochnik (1996).
13 Plato, *Protagoras* 356d–357a, tr. S. Lombardo and K. Bell, Hackett 1992. See also *Philebus* 19a–b; *Statesman* 283d–285a. A discussion of this in Nussbaum (1986), 89 ff.
14 Plato, *Euthyphro* 7b–c.
15 Plato, *Euthydemus* 290c. Cf. also Sophocles, *Antigone* 332–63 (probably written *c.* 450–40 BC).
16 Plato, *Laws* 746d–747c, Loeb translation with modifications. See Harvey (1965); Cartledge (1996).
17 Plato, *Laws* 817e–818a.
18 Aristotle, *Metaphysics* 981b–982a; *Posterior Analytics* 76a, 87a.
19 Aristotle, *Metaphysics* 981b18–26.
20 Aristotle, *Nicomachean Ethics* 1131a–1132a.
21 Aristotle, *Politics* 1301b–1302a; 1318a.
22 Adapting what Kurke (1999), 23 says about money.
23 Archimedes, *Quadrature of the Parabola* 262.13 ff.; *Sphere and Cylinder I* 4.9 ff.; *Method* 430.1 ff., respectively.
24 Philodemus, *The Academy* col.Y 2–18 in Gaiser (1988), my translation.
25 Plutarch, *Against the Stoics on Common Conceptions* 1079e–f and cf. *Table-Talk* 718e–f.
26 Plutarch, *ibid.* 718e–f, Loeb translation with modifications.
27 Plutarch, *On the E in Delphi* 386e and *On the Genius of Socrates* 579a–d (mentioning Eudoxus and Helicon of Cyzicus).
28 Diogenes Laertius, *Lives of the philosophers* 9.34 ff. for Democritus; 8.86 ff. for Eudoxus; 1.22 ff. for Thales; 8.1 ff. (esp. 11 ff.) for Pythagoras; 8.79 ff. for Archytas.
29 Proclus, *Commentary on the First Book of Euclid's Elements* 65–68, tr. by G.R. Morrow, Princeton University Press 1970, reproduced with permission. A different version of the same list in [Hero], *Definitions* 136.1. On Proclus' list see Vitrac (1996).
30 Simplicius, *Commentary on Aristotle's Physics* 55.26 ff., my translation.
31 Eutocius, *Commentary on Archimedes' Sphere and Cylinder II* 88.4–90.13, my translation.
32 Eutocius, *ibid.* 84.12–88.2, my translation.

3

HELLENISTIC MATHEMATICS: THE EVIDENCE

> Some of the numbers to which I have given a name [...]
> surpass not only the number of grains of sand
> that could fill the Earth [...] but even
> the number of grains of sand that could fill the universe itself.[1]

At the battle of Chaeronea in 338 BC, Philip II of Macedonia defeated a coalition of Greek states, and established control over the peninsula. He was succeeded in 336 BC by his son Alexander, soon to be known as Alexander the Great, who embarked on a military campaign and, in the space of a few years (he died in 323), brought down the already shaky Persian Empire and appropriated its immense former domains, stretching as far as Northern India, and including Asia Minor, Syria and Egypt. What in Alexander's intentions, had he lived long enough, would probably have been a unified mega-empire was eventually divided up among his successors, most of whom had been officers in his army. The members of the new ruling elite (the monarchs, their families, their associates) were prevalently Greek or Greek-speaking and promoted the settlement of Greeks on their newly-acquired territories – hence the term Hellenistic kingdoms for their states, and of Hellenistic age for the period in which they flourished, roughly third to second century BC, when Rome gradually established predominance in the Mediterranean basin.

A few words on the nature of the evidence. What we have seen happening with Athens in the fifth and fourth centuries BC more or less happens in this period with Egypt. Many mathematicians or philosophers from other parts of the world went to, or worked in, or corresponded with, people working in Alexandria, the capital of the new Ptolemaic state. A lot of our material evidence consists of texts written on papyrus, whose very survival is due to dry and hot climatic conditions, such as are rarely found outside Egypt. The most remarkable thing of all about the evidence for Hellenistic mathematics, however, is that (at last!) we have whole real mathematical treatises instead of scrappy, indirect information. So, the sections being once again organized by type, we will look at the material evidence first; then at texts by non-mathematicians, arbitrarily divided into rest of the world and philosophers. To follow, the mathematicians themselves, divided (again, arbitrarily) into little people and Big Guys (Euclid, Archimedes, and Apollonius). The sections on the Big Guys are in their turn sub-divided into a description of contents – the topics they dealt with – and of procedures.

Material evidence

Until the fourth century BC, the main forms of attack against a fortified city seem to have been to starve its inhabitants, bribe someone in order to have them open the gates, or build a giant wooden horse and hope the enemies had not heard the story before. Siege devices such as beams with chains to ram down the gates are mentioned by Thucydides and were probably extensively employed.[2] Around 399 BC, the story changed with the invention of a new type of weapon: the catapult.

Early catapults probably looked like oversized bows, but their initially limited range and power quickly improved. There are no archaeological remains of war engines earlier than the first century AD, but we do have indirect evidence of their presence from changes in fortification construction. Following not upon the discovery of the catapult immediately, but rather with its established and widespread use in the early to mid-third century BC, thicker walls were built, different wall designs adopted, the better to resist the impact of projectiles, and larger and differently-shaped towers erected. These elements are in evidence at sites such as Heraclea on Latmus (*c.* 300 BC, with a second phase of construction around the mid-third century BC), Ephesus (*c.* 290 BC), Fort Euryalus at Syracuse, whose latest phases date from the end of the third century BC, Iasus (late third century BC), and Doura Europus in Syria (*c.* 300 BC). See Figure 3.1.

Some Hellenistic fortifications, for instance Doura Europus, Ephesus or Fort Euryalus, exhibit geometrical patterns or structural details strikingly similar to those described in the earliest extant treatise on military architecture, the second-century BC *Poliorketika* (siege-craft) by Philo of Byzantium.[3] He described a zig-zag-shaped wall whose outline was a regular combination of triangles, and towers of various types, from square and quadrangular to hexagonal to semicircular. The building blocks for those latter had to be shaped with particular accuracy, by measuring the external circumference and preparing models in wood. The text was complemented by figures, now lost. Philo also dealt with the storing of provisions in the eventuality of a long siege: he specified that granaries had to be built symmetrically, with the height in proportion to the size, 'determining the height of the ceiling arcs on the basis of the foundations'.

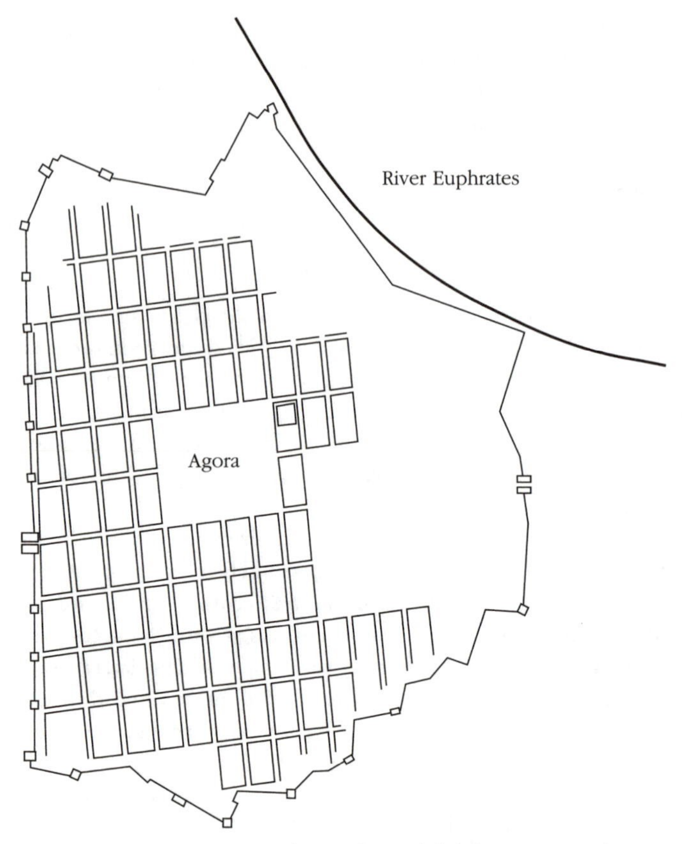

Figure 3.1 Doura Europus with city plan and fortifications in evidence (adapted from Garlam (1974))

Philo also wrote a *Belopoiika* (construction of propulsion instruments), addressed to the same person as the *Poliorketika*. Evidently, the same person could wear the hat both of the military architect and of the machine-maker.[4] A further hat in the wardrobe of people like Philo, perhaps of his addressee Ariston, or of the Polyeidus of Thessaly, Dionysius of Alexandria, Charias of Magnesia mentioned by other third-century sources, would have been that of town-planner and surveyor.[5] It was often the same people who were responsible for devising the city layout and the fortifications around it. While depending on the geographical situation, regular geometric patterns, square or rectangular, have been observed in many Hellenistic town plans such as that of, again, Doura Europus, Kassopeia near the Gulf of Arta (third century BC), Goritzas and Demetrias in Thessaly (also third century BC), and the Roman colonies of Cosa (273 BC) and Rimini (268 BC).[6]

Roman colonists were often retired soldiers, so it is perhaps not surprising that the plan of their town often resembled that of their military camps, which, in the opinion of the historian Polybius, exemplified what could be called a geometrical mind-set:

> the manner in which [the Romans] form their camp is as follows. [...] Fixing an ensign on the spot where they are about to pitch [the general's tent], they measure off round this ensign a square plot of ground each side of which is one hundred feet distant, so that the total area measures four *plethra*. [...] They now measure a hundred feet from the front of all these tents, and starting from the line drawn at this distance parallel to the tents of the tribunes they begin to encamp the legions [...] Bisecting the above line, they start from this spot and along a line drawn at right angles to the first, they encamp the cavalry [...] The whole camp thus forms a square, and the way in which the streets are laid out and its general arrangement give it the appearance of a town. [...] Given the numbers of cavalry and infantry, whether 4000 or 5000, in each legion, and given likewise the depth, length, and number of the troops and companies, the dimensions of the passages and open spaces and all other details, anyone who gives his mind to it can calculate the area and total perimeter of the camp.[7]

Archaeological research has established that, while nowhere as standardized as Polybius described, early remains of stone-built Roman camps reveal indeed regular geometrical patterns.[8] Polybius also drew out the implications of geometrically-organized space in a comparison of the Roman camp with the Greek one:

> The Romans by thus studying convenience in this matter pursue, it seems to me, a course opposite to that usual among the Greeks. The Greeks in encamping think it of primary importance to adapt the camp to the natural advantages of the ground, first because they shirk the labour of entrenching, and next because they think handmade defences are not equal in value to the fortifications which nature provides unaided on the spot. So that as regards the plan of the camp as a whole they are obliged to adopt all kinds of figures to suit the nature of the ground, and they often have to shift the parts of the army to unsuitable situations, the consequence being that everyone is quite uncertain whereabouts in the camp his own place or the place of his corps is. The Romans on the other hand prefer to submit to the fatigue of entrenching and

> other defensive work for the sake of the convenience of having a single type of camp which never varies and is familiar to all.[9]

The symbolic significance of imposing a geometrical pattern on a territory was evident in another context. As at earlier times, the establishment of a new city came with the apportionment of the territory around it, which needed to be surveyed and divided up. We have evidence of such divisions in, for instance, Larissa (perhaps end of third century BC) and Halieis (perhaps second century BC) in Thessaly, and in some areas conquered by the Romans such as Northern Italy and North Africa (second century BC).[10] The most common pattern was rectangular, while the square shape, so pervasive in Polybius' account, is more commonly associated with Roman foundations and land-divisions. As well as facilitating allocation, which was often carried out through drawing of lots, the fact that the pieces of land were geometrically uniform and consequently of equal size provided at least an appearance of justice in the distribution. Geometrical patterns signified fairness.

We have further evidence about land-surveys from Egypt, where the Ptolemies took over well-established pre-existing administrative structures. The king was the biggest landlord in the country – not only did he lease portions of land to local farmers and new settlers, he also told them what to grow, and established a monopoly on various products such as oil. Thus, no-one but government officials was allowed to sell oil, and the possession of instruments such as presses was regarded as a criminal offence. In order to maintain such a degree of control over properties, tenants and their activities, periodical surveys were necessary. A typical *geometria* (for this is the term commonly used) would have looked like this:

> To the west entering the north along the canals that have been surveyed before, proceeding from the east, the seven-aroura cleruchic holding of Pathebis son of Teephraios, one of Chomenis' soldiers: the remainder
>
> $6\ {}^1/_2\ {}^1/_4$ $\quad 1\ {}^1/_4$ $\quad \frac{1\ {}^1/_8\ {}^1/_{10}\ {}^1/_6}{\text{same}}$ $\quad$ 1 <the area is> $1\ {}^1/_4\ {}^1/_{10}\ {}^1/_6$, (wheat).
>
> To the west proceeding from the south, the seven-aroura cleruchic holding of Besis son of Kollouthes, one of Chomenis' soldiers: $6{}^1/_2$, crown ${}^1/_2\ {}^1/_4\ {}^1/_8\ {}^1/_{10}\ {}^1/_6\ {}^1/_{30}\ {}^1/_2$, <total> $7\ {}^1/_4\ {}^1/_8\ {}^1/_{10}\ {}^1/_6\ {}^1/_{30}\ {}^1/_2$, at $4{}^1/_2$ each
>
> $6\ {}^1/_2\ {}^1/_8\ {}^1/_{10}\ {}^1/_6\ {}^1/_{30}\ {}^1/_2$ $\quad \frac{1\ {}^1/_8\ {}^1/_{10}\ {}^1/_6}{1\ {}^1/_4\ {}^1/_{10}\ {}^1/_6}$ $\quad 6\ {}^1/_8\ {}^1/_{10}\ {}^1/_6\ {}^1/_{30}\ {}^1/_2$
>
> <the area is> 8, excess ${}^1/_2\ {}^1/_{30}\ {}^1/_2$, black cumin, self cultivated. To the north coming from the east along the seven-aroura cleruchic holding that has been surveyed before, ${}^1/_4\ {}^1/_{10}\ {}^1/_6$ schoinoi revenue

from those $3\,^1/_6$.
$6\,^1/_2\,^1/_{30}\,^1/_2\,^1/_4\,^1/_8\,^1/_{10}\,^1/_6\,^1/_{30}\,^1/_2 \quad 6\,^1/_2\,^1/_4$,
$^1/_2\,^1/_{30}\,^1/_2$
<the area is> $3\,^1/_4\,^1/_{10}\,^1/_6$, of which lentils 1 black cumin $2\,^1/_4\,^1/_{10}$ $^1/_6$. To the north of the canal $^1/_{10}\,^1/_6$. To the north proceeding from the west, of Chales son of Pasitos, crown 2 at 1 each
$^1/_2\,^1/_4$ same $^1/_2\,^1/_4\,^1/_8$
$2\,^1/_4$
<the area is> 1 [$^1/_2$] $^1/_4\,^1/_{10}\,^1/_6\,^1/_{30}\,^1/_2$, black cumin. From the east revenue from those $9^1/_2\,^1/_4\,^1/_8$
$6 \quad 1^1/_2\,^1/_4\,^1/_8 \quad 5^1/_2\,^1/_8$
$2\,^1/_4$
<the area is> $11^1/_2\,^1/_4\,^1/_8\,^1/_{10}\,^1/_6\,^1/_{30}\,^1/_2\,^1/_{60}\,^1/_4$. To the east proceeding from the south
$^1/_8\,^1/_{10}\,^1/_6$ [..] $^1/_8$ nothing
1 [.]
<the area is> $^1/_8\,^1/_{30}\,^1/_2$, <total> $9^1/_2\,^1/_4\,^1/_8\,^1/_{10}\,^1/_6\,^1/_{30}\,^1/_2$.[11]

A few words of explanation: although the document is in Greek, the numbers are expressed as sums of series of parts, as was common in ancient Egyptian mathematics. Apart from telling what piece of land was cultivated by whom, and often what was being grown, the survey measured the land, either by simply providing a figure for the area, or by reporting the lengths of the sides of a plot of land, written down with a horizontal line to separate them. If the allotment was quadrangular (the most common case, but there is also a triangular holding, where the writer notes 'nothing' in correspondence of one of the sides), the area was then obtained by adding opposite sides, halving the resulting sums and multiplying them by each other. Tax was calculated on the basis of the area. It has been observed that the method of obtaining surfaces on the basis of the sides, while lacking accuracy, achieved the far more desirable result (in the government's eyes at least) of always overestimating the size of the allotment, thus forcing the tenant to pay more tax.[12]

Surveying instruments were employed for these operations, and remains of some of them have been found. The instrument in Figure 3.2 would have helped with the drawing of straight lines and right angles, and therefore with the laying in place of an orthogonal grid.

We also have evidence (again from Egypt) about the geometers involved: a letter sent *c.* 252 BC by a Polykrates to his father Kleon, whom we know to have been an engineer employed in the public administration, mentions that he was 'proceeding in the study of geometry'.[14] We do not know for sure where Polykrates was writing from: perhaps Alexandria, perhaps

Figure 3.2 A surveying instrument (the plumblines are a modern reconstruction) from the Fayum, dating from Ptolemaic times[13]

Krokodilopolis. Datable to the same period (251 BC) is this letter, from the Arsinoite district:

> Antipater to Pythocles greeting. I append for you a copy of the letter written to me by Phanias. As soon as you receive my letter, inspect and survey (*geometreson*) all the holdings under your superintendence, as Phanias has ordered, and after making a list as accurate as possible according to crops send me the survey (*geometrian*) to submit to Phanias. Do the work scrupulously in the manner of one prepared to sign the royal oath. Goodbye. [...] Phanias to Antipater greeting. [...] As sowing has begun in your

> district, take at once some expert surveyor and inspect all the holdings under your superintendence and after surveying them make a list according to crops, as accurate as possible, of the land sown in each holding, continuing until you have inspected all. Do the work scrupulously in the manner of one prepared to submit the survey to me with a sworn declaration. Take care that you present it to us.[15]

The emphasis on scrupulousness and expertise was not purely rhetorical. Disputes and confusions often arose because the standards of measure varied from place to place, or because equal size did not amount to equivalent quality of land, or because of miscalculations, or because not all officers were as honest as their sworn declarations would have required.[16] A great part of the surviving papyri from this period consists of accounts, receipts of payment and lists (of tax-payers, of tenants). They often state that accounts must be produced regularly and checked carefully. For instance:

> Eukles to Anosis greeting. I learn that you have deposited in the record office the accounting of the pottery without bringing into it even the breakage which has occurred through the donkey drivers, and, in the matter of the pay still due in the case of the potters, that you have entered 8 drachmas per hundred pots instead of the 6 drachmas pay which had been given them; [...] and that, up to the present, you have not handed in to the record office the accounting of the hogs ready for slaughter; and that, on the whole, you have begun to act like a scoundrel [...] I have written regarding these matters to Lykophron and to Apollonios that, if they discover any discrepancies from the accounts handed in by you, they write to me immediately in order that I, being present in person, may have my case judged in relation to you. For it is right that you, who are entering excess charges and are not reporting the totals correctly throughout the accounts, should pay the discrepancies, not I.[17]

We know that Eukles was *epistates* (administrator) of Philadelphia, and therefore quite an influential personage. A much humbler position in society was occupied by the author of this letter, who made similar claims:

> Pemnas to Zenon greeting. About the money owed on the pigs in previous time together with the rent of $37^1/_2$, Herakleides has made a deal with Thoteus, and, without me, they have balanced the accounts and have not allowed me to look anything through until

> now, and then without a justification they have dared to give me the account. And about these things I have complained to Jason many times that it is not right that they make deals together. And Herakleides also has all the documents about the pigs. I have written to you about these things so that you know. Good-bye.[18]

Many documents indicate that the correctness of accounts and, to some extent, of land-surveys, could be guaranteed by *collective* checking – official instructions for administrators stated that the calculations had to be examined by more than one government officer, to make sure that everything was in order. In other words, doing mathematics together was seen as a guarantee of smooth and honest functioning of the state machine. Of course, this view was not confined to the Hellenistic period (see chapter 1), or to Egypt: for instance, we have evidence of collective surveying from Heraclea in Southern Italy. Two bronze tablets, datable to the early third century BC, record a survey of holdings of the temples of Dionysius and Athena. A group of boundary-men (*oristai*), elected by the city assembly, 'measured together' the lands, expressing sizes and setting up boundary stones. The first table was signed by the citizens involved, as well as by Aristodamus, son of Symmachus, as secretary, and by Chaireas son of Damon, of Neapolis, as geometer.[19]

The prevailing impression from the Egyptian documents, however, is that, beyond the rhetoric of fairness that mathematics implemented, checking the sums was not open to anyone – these texts were not public inscriptions, and gaining access to them could be difficult, and perhaps status- and ethnicity-related, as is illustrated by the case of the Egyptian swineherd Pemnas.[20] Numeracy and literacy obviously went along with status and, since getting an education chiefly meant getting a *Hellenic* education, with cultural identity. A few documents from this period which contain simple mathematical exercises, sometimes together with grammatical drills such as declensions of verbs or lists of names, can be situated within the context of basic education. For instance, a third-century BC schoolbook from the Fayum starts with syllables, follows with the numbers up to 25, names of (Greek) gods, quotations from (Greek) literature, including Euripides and Homer, then a table of squares and, finally, a table of subdivisions of the drachma.[21]

Perhaps also linked to an educational context is a third-century BC papyrus from Hermopolis in demotic Egyptian, containing arithmetical and geometrical problems: measurement of land of various shapes, amounts of cloth, size of masts, square roots. The two examples on the facing page are, respectively, an application of the theorem of Pythagoras and a version of the quadrature of the circle.[22]

[Problem 34] A plot of land that <amounts to> 60 square cubits, [that is rec]tangular, the diagonal (being) 13 cubits. Now how many cubits does it make [to a side]? You shall [reckon 13, 13 times: result 169]. You shall reckon 60, 2 times: result 120. You shall [add] it to 1[69]: result 28[9]. Cause that it reduce to its square root: result 17. You shall take the excess of 169 against 120: result 49. Cause that it reduce to its square root: result 7. Subtract it from 1[7]: remainder 10. You shall take to it $^{1}/_{2}$: result 5. It is the width. Subtract 5 from 17: remainder 12. It is the height. You shall say: 'Now the plot of land is 12 cubits by 5 cubits'. <To> cause that you know it. Viz. You shall reckon 12, 12 times: result 144. You shall reckon 5, 5 times: result 25. Result 169. Cause that it reduce to its square root: result 1[3]. It is its diagonal of plot.

[Problem 37] A plot of land [that is round that amounts to 675 square cubits, the diameter being 30,] ... a piece that is square [within] it, that has four corners up to the circumference of the plot of land. Now the piece ... within it makes how many [square] cubits? You shall reckon 30, 30 times: result 900. You shall take to it [half]: result 450. Cause that it reduce to its square root: result 21$^{1}/_{5}$ $^{1}/_{60}$. They are the [meas]urements of the piece. To cause that you know it. ... the plot of land in which is the piece (is) [6]75 square cubits. Its plan. Viz. (see Diagram 3.1a)

Now [the piece] that is squared [... reckons] 21$^{1}/_{5}$ $^{1}/_{60}$ divine-cubits [2]1$^{1}/_{5}$ $^{1}/_{60}$ times: result 450 square cubits. Four segments, [what are] their measurements? Viz. Subtract 21$^{1}/_{5}$ $^{1}/_{60}$ from the

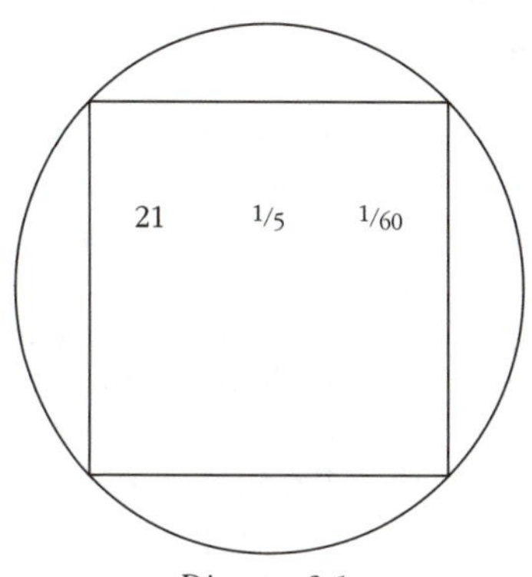

Diagram 3.1a

diameter of the circumference, which (is) 30 cubits: remainder $8^2/_3\ ^1/_{10}\ ^1/_{60}$. Its quantity at 2 amounts to $4^1/_3\ ^1/_{20}\ ^1/_{120}$ at 1. It is the middle height of the segment. Here is its plan. (See Diagram 3.1b.)

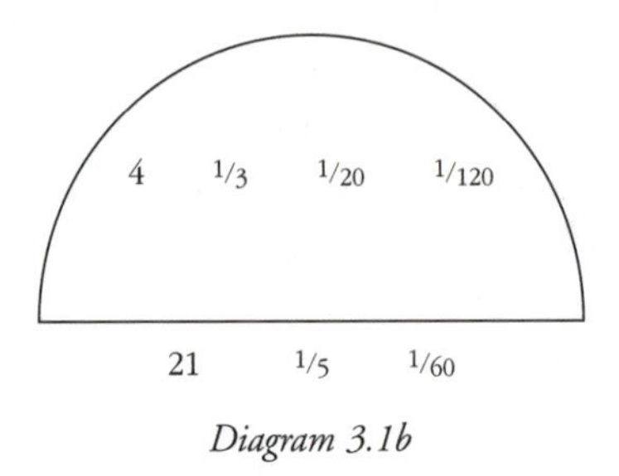

Diagram 3.1b

You shall add $4^1/_3\ ^1/_{20}\ ^1/_{120}$ to $21^1/_5\ ^1/_{60}$: result $25^1/_2\ ^1/_{10}\ ^1/_{20}\ ^1/_{120}$. Take [their] half: result $12^2/_3\ ^1/_{10}\ ^1/_{30}\ ^1/_{240}$. You shall reckon $12^2/_3\ ^1/_{10}\ ^1/_{30}\ ^1/_{240}$, $4^1/_3\ ^1/_{20}\ ^1/_{120}$ times: result $56^1/_4$. You shall reckon $56^1/_4$, 4 times: [result 225]. Result: [6]7[5] cubits again.

The problems in the demotic papyrus are solved not generally, but for specific cases, and, rather than a deductive proof, they contain a verification, or check step, introduced by the expression 'to cause that you know it'. A teaching context is suggested by direct appeals to the reader and by statements such as this:

> When another [add-fraction-to-them] (problem) is stated to you, it will be successful according to the model. If you take the excess of the small (number) against the large (number), you shall put it opposite 1 until it completes.[23]

A group of third or mid-second century BC *ostraka* (pottery fragments) from the island of Elephantine in Southern Egypt have also been found, written over with mathematical texts. Their contents have been identified with Euclid's *Elements* 13.16: the construction of a regular icosahedron inscribed within a sphere, i.e. quite a complex procedure. The Elephantine *ostraka* raise a number of questions. While their contents denote a high level of education, both the humble material and the location (a remote outpost in the heart of 'Egyptian' Egypt) seem to jar with that conclusion. Besides, was the person who produced them acquainted with Euclid's work, or with a different account of the same subject-matter?

The earliest extant papyrological remains of Euclid date from the late second century BC. They formed part of a treatise on geometry by the Epicurean philosopher Demetrius of Laconia. He reported, for polemical purposes, a definition of the circle and statements on the bisection of the angle which correspond to *Elements*, definition 1.15 and propositions 1.3, 1.9 and 1.10.[24] Again, was Demetrius actually referring to Euclid's text, or to something similar that has not come down to us?

Let us defer a discussion of those questions to chapter 4, and move on with our examination of the evidence.

Non-mathematical authors: the rest of the world

For a section about the world (minus philosophers and mathematicians), this is a pretty short survey. The problem is that, unlike the Demosthenes, Aristophanes and Lysias of the previous chapter, not many writers of the Hellenistic period seem to mention mathematics at all. Could that be a reflection of the fact that mathematics played a much more 'public' public role in the Athens of the fifth and fourth century than in the communities of the third and second century, where the majority of Hellenistic literary texts was produced?

One exception, as we have started to see, is Polybius. In line with his geometrized vision of Roman troops, he stated that a knowledge of mathematics was a positive asset for military leaders. In order to be a good general, one needed a combination of experience, observation and methodical knowledge. In particular, a military leader had to be acquainted with astronomy and geometry. Astronomy was necessary to know the variations of day and night, in order to calculate movements of the troops, days of march and so on. Geometry, on the other hand, helped calculate the right length for siege ladders – if one did not know the height of the enemy walls, Polybius reminded them, an easy method (probably via gnomon and similar triangles). was available to those who wanted to exercise themselves in the mathematical studies. Geometry, especially proportion theory, also came in handy when one wanted to change the size of the camp while retaining shape and arrangement.[25]

That not everybody was as knowledgeable in mathematics as they should have been emerged on occasions. Here Polybius criticizes the historian Callisthenes for his report of a battle:

> It is difficult to understand how they posted all these troops in front of the phalanx, considering that the river ran close past the camp, especially in view of their numbers, for, as Callisthenes himself says, there were thirty thousand cavalry and thirty thousand

> mercenaries, and it is easy to calculate how much space was required to hold them. For to be really useful cavalry should not be drawn up more than eight deep, and between each troop there must be a space equal in length to the front of a troop so that there may be no difficulty in wheeling and facing round. Thus a stade will hold eight hundred horse, ten stades eight thousand, and four stades three thousand two hundred, so that eleven thousand two hundred horse would fill a space of fourteen stades. If the whole force of thirty thousand were drawn up the cavalry alone would very nearly suffice to form three such bodies, one placed close behind the other. Where, then, were the mercenaries posted, unless indeed they were drawn up behind the cavalry? This he tells us was not so [...] For such mistakes we can admit no excuse. For when the actual facts show a thing to be impossible we are instantly convinced that it is so. Thus when a writer gives definitely, as in this case, the distance from man to man, the total area of the ground, and the number of men, he is perfectly inexcusable in making false statements.[26]

Using weak demonstrative formulae such as *phaneron oti* and *delon oti* (it is clear that, it is evident that), which were common in mathematical discourse, Polybius proved Callisthenes wrong on the basis of the very figures quoted in his account. In fact, *because* Callisthenes had given a seemingly accurate, mathematical, report, it was all the easier for Polybius to check it through and show that it was incorrect.[27]

Using mathematics in historical reports, even if only to expose someone else's ignorance or misuse of mathematics, was a rhetorical device which accrued accuracy and even an air of objectivity to the report itself. More mathematical mistakes in the following passage:

> Most people judge of the size of cities simply from their circumference. So that when one says that Megalopolis is fifty stades in contour and Sparta forty-eight, but that Sparta is twice as large as Megalopolis, what is said seems unbelievable to them. And when in order to puzzle them even more, one tells them that a city or camp with a circumference of forty stades may be twice as large as one the circumference of which is one hundred stades, what is said seems to them absolutely astounding. The reason of this is that we have forgotten the lessons in geometry we learnt as children.[28]

Clearly, Polybius expected mathematical knowledge to be part of the cultural baggage of a political leader – not too much, but enough to be aware for instance of issues of isoperimetrism, which, it is implied with rhetorical exaggeration, were part of elementary geometry. In fact, an in-depth treatment of isoperimetrism is contained in Euclid's *Elements*, book 13, to which, as we have seen, one could relate the material in the Elephantine *ostraka*.

The references in Polybius suggest that his audience was sensitive to certain aspects or functions of mathematics. The same can be said to explain the numerical imagery in Theocritus, a poet who benefited from the patronage of the Ptolemies:

> From Zeus let us begin, and with Zeus in our poems, Muses, let us make end, for of immortals he is best; but of men let Ptolemy be named, first, last, and in the midst, for of men he is most excellent. [...] Of what am I to make mention first, for beyond myriads to tell are the blessings wherewith heaven has honoured the best of kings? [...] Infinite myriads and myriads of tribes of men with the aid of rain from heaven bring their crops to ripeness, but none is so prolific as are the plains of Egypt [...] Three hundreds of cities are built therein, and three thousands and three times ten thousand therewith, and twice three and three times nine beside; and of all Lord Ptolemy is king.[29]

Theocritus uses numbers in two different ways: very large figures, which are 'accurate' in the sense that they express a definite quantity, and numbers so large that they are beyond counting. Both are meant to convey the greatness of Ptolemy Philadelphus and of his domain: large, very large, so large that numbers cannot suffice to describe it. These rhetorical uses of numbers are not unique to Hellenistic poetry, but they acquire a particular resonance when set against the various acts of official measuring and numbering that took place in Ptolemaic Egypt. At least in intention, accounts and surveys were meant to quantify the domain of the king, say exactly how big, how great, how rich he and his land were. More than that: as we will see in the section on little people on p. 85, geography, at the hands of Ptolemaic employees such as Eratosthenes, became a geometrizing, mathematizing effort. The Earth itself was measured, using Alexandria as reference point. And in chapter 4, we will also see how Archimedes went beyond the indefinite infiniteness of myriads, stretching to extreme limits the capacity of numbers to count things. He estimated the number of grains of sand that could be contained in the entire universe, and expressed their number, dedicating the work to his king, on whom the importance of counting and measuring would certainly not have been lost.

Non-mathematical authors: the philosophers

Many philosophers between the third and the second century BC were concerned with the problem of knowledge, and with mathematics in particular. Getting to know their views is made difficult by two major factors. First, we rarely have first-hand evidence about them – most of what we know comes from later reports. Second, in the wake of Plato and Aristotle, both founders of schools, the Hellenistic period saw the emergence of several philosophical movements, groups and sects: the Epicureans, the Stoics, the Sceptics, the Cynics, plus Academics (Platonists) and Peripatetics (Aristotelians). Although some people can definitely be assigned to one school rather than another (for instance, Epicurus was an Epicurean), great unclarity reigns as to how and whether those labels really corresponded to institutions with unified curricula, to what extent certain theories were distinctive of one school rather than another, and especially, how the intellectual history of those schools can be traced chronologically. Our sources all too often tend to attribute theories to for example 'the Stoics' in general, glossing over developments over time or diversity of opinions among members of the same school. Moreover, in the case of some individuals it is not clear that they subscribed to any philosophical creed exclusively.

A tradition which is even more difficult to profile than the others is Pythagoreanism, which, as we mentioned in chapter 1, had been around since the sixth century BC. Interest in it not only continued, but seems to have positively increased in the Hellenistic period, when many of the texts through which we know the story, or legend, of earlier Pythagoreans were written. In other words, Hellenistic Pythagoreanism to a great extent 'created' earlier Pythagoreanism in the form of books or views, including specific mathematical discoveries, usually in the field of proportion theory, attributed to Archytas, Philolaus and Pythagoras himself. The forgeries or imitations produced in this period can be identified with some reliability because they are only quoted by authors later than the fourth century BC, and also because they often use Aristotelian and/or Platonic philosophical terminology, or anachronisms of a similar kind. Several of these texts deal with mathematical issues, especially with the relation between numbers and the world, and with the ethical or philosophical significance of arithmetic qualities such as even or odd, or of numbers: five or ten and so on. Numbers are often made to carry meanings beyond simple quantity – for instance, according to some of these texts god is an irrational number.[30]

The Peripatetics and Academics also inherited a tradition where mathematics played a fundamental role. Plato's immediate successors had been particularly interested in the possibility that Forms were numbers, and that numbers held the key to universal knowledge.[31] Later Academics concerned themselves with explaining the many mathematical passages of the *Timaeus* or the *Republic*. Names that have been transmitted include Crantor (end of fourth/beginning of third century BC), Theodorus and Clearchus, known also as a Peripatetic, all from Soli. Of the people generally associated with the Aristotelian tradition, Dicearchus was interested in geographical measurements; Aristoxenus, as we will see, wrote about harmonics; the name of Heraclides of Pontus is linked to astronomical theories; Eudemus of Rhodes, as we have seen, wrote histories of geometry, but also of arithmetic, astronomy and music.[32] The Aristotelian *corpus* preserves a work *On Indivisible Lines*, aimed at showing that it is 'neither necessary, nor believable that there are indivisible lines'[33] – a line basically shared by Stoics like Chrysippus.[34]

Indivisibles conflicted with the assumption, common among geometers, that magnitudes could be divided indefinitely. Their existence in nature, on the other hand, was one of the main tenets of Epicurus, who put forth an atomic theory of matter inspired by Democritus. Indeed, as mentioned in the section on material evidence on p. 73, the Epicurean Demetrius set out to attack Euclid's *Elements*, or at least some of the contents of the first book. That, however, does not amount to evidence that the Epicureans spurned mathematics entirely: Demetrius seems to have been knowledgeable in the very subject he set out to criticize; two early third-century Epicureans, Polyaenus and Pythocles, allegedly studied mathematics; the works of mathematicians like Apollonius and Hypsicles mention people (Philonides, and Basilides and Protarchus, respectively), who can be identified with contemporary Epicurean philosophers by the same name.[35]

More general reflections about the nature of demonstration on the part of Epicureans and Stoics are reported in Proclus:

> Up to this point we have been dealing with the principles [definitions, postulates and common notions in Euclid's *Elements* book 1], and it is against them that most critics of geometry have raised objections, endeavoring to show that these parts are not firmly established. Of those in this group whose arguments have become notorious some, such as the Sceptics, would do away with all knowledge [...] whereas others, like the Epicureans, propose only to discredit the principles of geometry. Another group of critics, however, admit the principles but deny that the propositions coming after the principles can be demonstrated unless they grant something that is not contained in the principles. This method of controversy was followed by Zeno of Sidon, who belonged to the school of Epicurus and against whom Posidonius has written a whole book.[36]

Both Zeno and Posidonius of Rhodes (this latter usually labelled a Stoic) lived at the end of the second century BC, thus providing a chronological framework to Proclus' picture. His testimony about the Sceptics, on the other hand, is typical of the vagueness that often surrounds both individual thinkers and what they thought. Diogenes Laertius offers a glimpse of what the father himself of Scepticism, Pyrrho (*c.* 365–275 BC), thought about mathematical certainty. He had distinguished ten modes in which seemingly undisputable questions turned out not to be so:

> The seventh mode has reference to distances, positions, places and the occupants of the places. In this mode things which are thought to be large appear small, square things round; flat things appear to have projections, straight things to be curved.[37]

Later Sceptics added five modes to Pyrrho's ten, focussing more explicitly on demonstration and casting doubt on argumentative structures typical of mathematics:

> If you think, they add, that there are some things which need no demonstration, yours must be a rare intellect, not to see that you must first have demonstration of the very fact that the things you refer to carry convinction in themselves.[38]

Although the Stoics appear to be not as critical of mathematics, they did not assign it a prominent place in their epistemology. In line with the belief that knowledge is ultimately sense-based and should concern itself with the real world, they designated medicine, divination, dialectic and virtue as the four primary sciences. Mathematics, however, played a role in the Stoics' physical theory, which included astronomy and causation, because it shared its subject-matter with those latter.[39] Again, according to Diogenes Laertius, points of contact existed between Stoic physical theory and Euclidean-style preoccupation with the definition of geometrical objects:

> Body is defined by Apollonius in his *Physics* as that which is extended in three dimensions, length, breadth, and depth. This is also called solid body. But surface is the extremity of a solid body, or that which has length and breadth only without depth. [...] A line is the extremity of a surface or length without breadth, or that which has length alone. A point is the extremity of a line, the smallest possible mark.[40]

A parallel, not in contents but in argumentative structure, may also be drawn between mathematics and Stoic logic. As part of their enquiry into the rationality of nature and of discourse, the Stoics analysed language, defining and distinguishing not only its components (propositions, both simple and complex), but also the way those components were combined to form arguments. They studied and classified types of valid argument, including the *modus ponens* and a form of *modus tollens* very similar to the one used in many indirect mathematical proofs.[41] In other words, as well as identifying features of demonstrative discourse that were shared by mathematical demonstrative discourse, the Stoics seemed to be doing for common language what Euclid was doing in the first book of his *Elements*: establishing basic assumptions, notions and constructions, organizing knowledge into a new, rigorous structure.

Little people

If indeed he lived between 360 and 290 BC, the earliest mathematician whose work has come down to us directly is Autolycus of Pitane. We have two treatises by him, on the moving sphere and on the risings and settings of the stars, in which heavenly bodies are reduced to their geometrical shape and the conclusions about them are drawn on the basis of the mathematical properties of those geometrical shapes. Many of the proofs are indirect, by *reductio ad absurdum*; occasional appeal is made to things assumed to be valid earlier in the treatise. All these features characterize Autolycus' account as mathematical astronomy, and are paralleled by other works from the Hellenistic period. Euclid's *Phenomena*, for instance, also treated astronomy in a geometrical style. Basic notions (horizon, meridian, etc.) are defined at the outset, and, although astronomical observations, including the use of a dioptra, are mentioned, the argumentative structure is very similar to that of the *Elements*, with its typical enunciation, setting-out, definition of goal and proof. Along similar lines is Euclid's *Optics*, which starts off with assumptions whereby the phenomenon of vision is geometrized, and continues in deductive mode, results being proved on the basis of previously established propositions.

In late antiquity, Autolycus' *On the Moving Sphere*, Euclid's *Phenomena* and perhaps also Euclid's *Optics* came to be part of the same collection as three more Hellenistic treatises: Aristarchus of Samos' *On the Sizes and Distances of the Sun and Moon* and Theodosius of Tripoli's *Sphaerics* and *On Days and Nights*.[42] Aristarchus (early third century BC), well-known to us for his heliocentric theory of the universe, started his book with hypotheses, i.e. undemonstrated statements, and on the basis of them proved several propositions along deductive lines. It is to be noted that, although he considered heavenly bodies from the point of view of their geometrical properties (sizes and distances), physical characteristics such as movement, illumination and perceptibility were not eliminated from his account.

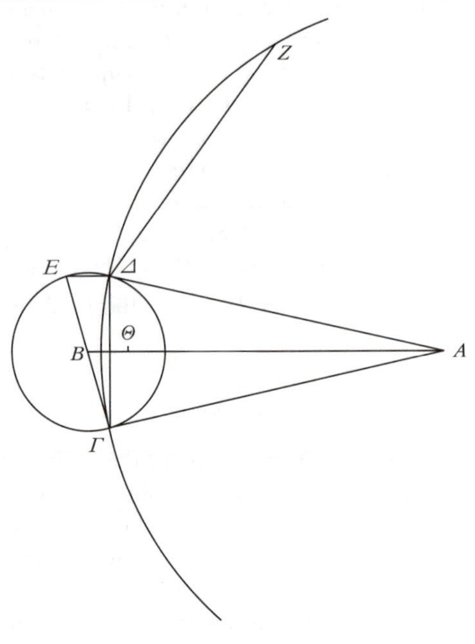

Diagram 3.2

[From Aristarchus' work – an example of approximation via upper and lower boundary (see Diagram 3.2)] The diameter of the moon is less than two 45ths, but greater than one 30th, of the distance that separates the centre of the moon from our eye. For let our eye be at A, and let B be the centre of the moon when the cone surrounding both the sun and the moon has the vertex at our eye. I say that what said in the enunciation takes place. For let AB be joined, and let the plane through AB be prolonged; it will produce as section a circle in the sphere and lines in the cone. Let it then produce the circle $\Gamma E\Delta$ in the sphere, and the lines $A\Delta$, $A\Gamma$ in the cone, and let $B\Gamma$ be joined, and be prolonged to E. It is evident then from what has been proved before that the angle $BA\Gamma$ is the 45th part of half a right angle; and in the same way $B\Gamma$ is less than ΓA by one 45th. Therefore $B\Gamma$ is much less than one 45th part of BA. And ΓE is double $B\Gamma$; therefore ΓE is less than two 45th of AB. And ΓE is the diameter of the moon, while BA is the distance that separates the centre of the moon from our eye; therefore the diameter of the moon is less than two 45ths of the distance that separates the centre of the moon from our eye. I say then that ΓE

> is also greater than one 30th part of BA. For let ΔE and $\Delta\Gamma$ be joined, and with centre A and radius $A\Gamma$, let a circle $\Gamma\Delta Z$ be described, and let ΔZ equal to $A\Gamma$ be fitted in the circle $\Gamma\Delta Z$. And since the right <angle> $E\Delta\Gamma$ is equal to the right <angle> $B\Gamma A$, but the <angle> $BA\Gamma$ is equal to the <angle> $\Theta\Gamma B$, therefore the remaining <angle> $\Delta E\Gamma$ is equal to the remaining <angle> $\Theta B\Gamma$; therefore the equiangular triangle $\Gamma\Delta E$ is equal to the triangle $AB\Gamma$. Therefore it is as BA to $A\Gamma$, so $E\Gamma$ to $\Gamma\Delta$; and alternately as AB to ΓE, so $A\Gamma$ to $\Gamma\Delta$, that is, ΔZ to $\Gamma\Delta$. But since again the angle $\Delta A\Gamma$ is the 45th part of a right <angle>, therefore the circumference $\Gamma\Delta$ is the 180th part of the circle; and the circumference ΔZ is the sixth part of the whole circle; so that the circumference $\Gamma\Delta$ is the 30th part of the circumference ΔZ. And the circumference $\Gamma\Delta$, which is less than the circumference ΔZ, has to the circumference ΔZ itself a ratio less than that of the line $\Gamma\Delta$ to the line $Z\Delta$; therefore the line $\Gamma\Delta$ is greater than the 30th of ΔZ. $Z\Delta$ is then equal to $A\Gamma$; therefore $\Delta\Gamma$ is greater than the 30th of ΓA, so that ΓE is also greater than the 30th of BA. And it has been proved to be also less than two 45ths.[43]

Theodosius (late second/early first century BC) explicitly relied on the tradition of mathematical astronomy before him: he mentioned Euclid (*Elements* and *Phenomena*), as well as our old acquaintance Meton. Book 1 of the *Sphaerics* again starts off with basic definitions, such as that of 'sphere', and continues in a deductive mode towards more and more complex propositions involving, for instance, arcs and tangents to spherical segments. Some of the theorems in book 3 of the *Sphaerics* are divided into sub-cases each with a different diagram, a practice which, as we will see, is often found in Apollonius.[44]

Mathematical astronomy was not the only form of knowledge of the skies: immensely popular since its composition and throughout the Graeco-Roman period was an astronomical poem, the *Phenomena* by Aratus of Soli, written *c.* 276–74 BC at the invitation of King Antigonus II of Macedonia (Aratus later moved to the court of Antiochus I Seleucid). The *Phenomena* provided a wealth of information about constellations, the stars that constitute them, and on how to interpret heavenly signs to forecast the weather. Practical applications to agriculture, navigation and calendar-making were all mentioned, and positive statements were made about the fact that sign-reading was more than random guessing – in fact, when sign confirmed sign, the knowledge derived from them could for Aratus be considered certain.[45] Aratus offers a fascinating example of the relation and contrast between mathematical and non-mathematical accounts of basically the same phenomena, especially if we consider that he apparently sourced his knowledge from the almost certainly *mathematical* treatment of astronomy by Eudoxus of Cnidus. The same issues recur in the only surviving work by Hipparchus (of Nicaea, but he worked mostly in Rhodes around the second half of the second century BC): the *Commentary on the Phenomena of Aratus and Eudoxus*. He is credited, among other things, with the discovery of the precession of the equinoxes. The target of criticism in the *Commentary*, along with the two authors of the title, is an Attalus who had also written a commentary on Aratus. Hipparchus' book does not contain much in the way of mathematics, but it does draw a distinction between what Aratus was doing (poetry, cannot be expected to get things right) and the enterprise Hipparchus himself and Eudoxus are engaged in. A revealing passage contrasts the belief, apparently held by Aratus, that circles like the equator have an actual width, with what 'all the mathematicians think', i.e. that they have none.[46]

A similar contrast between mathematized and non-mathematized knowledge is found, with a rather different outcome, in Aristoxenus, probably a pupil of Aristotle. Although parts of his main work, the *Elements of Harmonics*, are organized as enunciation-like statements followed by proofs, some of them indirect, he made a point of distantiating himself from mathematical treatments of music:

> We try to give these matters demonstrations which conform to the appearances, not in the manner of our predecessors, some of whom used arguments quite extraneous to the subject, dismissing perception as inaccurate and inventing theoretical explanations, and saying that it is in ratios of numbers and relative speeds that the high and the low come about. Their accounts are altogether extraneous, and totally in conflict with the appearances. [...] While it is usual in dealing with geometrical diagrams to say 'Let this be a straight line', we must not be satisfied with similar remarks in relation to intervals. The geometer makes no use of the faculty of perception: he does not train his eyesight to assess the straight or the circular or anything else of that kind either well or badly: it is rather the carpenter, the wood-turner, and some of the other crafts that concern themselves with this. But for the student of music accuracy of perception stands just about first in order of importance, since if he perceives badly it is impossible for him to give a good account of the things which he does not perceive at all.[47]

Aristoxenus' target could be exemplified by the *Section of a Canon* (perhaps by Euclid), whose introduction justified the application of mathematics to the physical phenomenon of musical sound on the basis of the fact that notes are composed of parts and parts have numerical ratios to each other.[48] The mathematization of harmonics thus postulated at the beginning is carried out in full in the body of the treatise, with occasional references to (perhaps) the *Elements* themselves. That the question of mathematics' ability to provide adequate knowledge of nature was widely discussed is also evidenced by Diocles:

> Sometimes people who try to discredit the mathematical scientists and say that they construct their subject on a weak foundation scoff <at this>: for some of them <the mathematicians> assert that the radii of the sphere are known and that each one is greater than the one <next to it> by more than 30 million stades, while others assert <that it is greater> by more than 50 million stades.[49]

Diocles' only extant treatise, *On Burning Mirrors* (*c.* 200 BC) itself presents an interesting interplay of geometrical proofs and recourse to 'the real world'. In his attempt to provide a mathematical description of a paraboloid mirror which can concentrate the sun rays onto a target, setting it on fire, he indicates, for instance, that intensity of burning has to be assessed empirically (31 ff.) and employs material aids in mathematical demonstrations, such as the problem reported in the exmple below, or his solution to the duplication of the cube, which relies on a curved ruler (195). A later tradition credited Archimedes with the use of burning mirrors as weapons, but Diocles suggests rather that they make accurate time-keeping devices, as shadow-less gnomons that burn a trace instead of casting a shadow (16–17). Or they can be employed in temples to light the fire on an altar in a spectacular way (36–7).

> [From Diocles' book on burning mirrors – Do *not* try this at home (see Diagram 3.3)] How do we shape the curvature of the burning-mirror when we want the point at which the burning occurs to be at a given distance from the centre of the surface of the mirror? We draw with a ruler on a given board a line equal to the distance we want: that is line *AB*. We make *AK* twice *BA* and erect *BE* perpendicular and equal to *AB*; we join *EK*. We make *AF* equal to *BE*, and join *EF*: then *ABEF* is a square, and also *EF* is equal to *FK*. We mark on *BA* two points, *G*, *D*, and make *EZ* equal to *BG* and *HE* equal to *BD*. We join *ZG*, *HD*, and produce them on both sides: let them meet *EK* in *M*, *L*. Then if, with *A* as centre

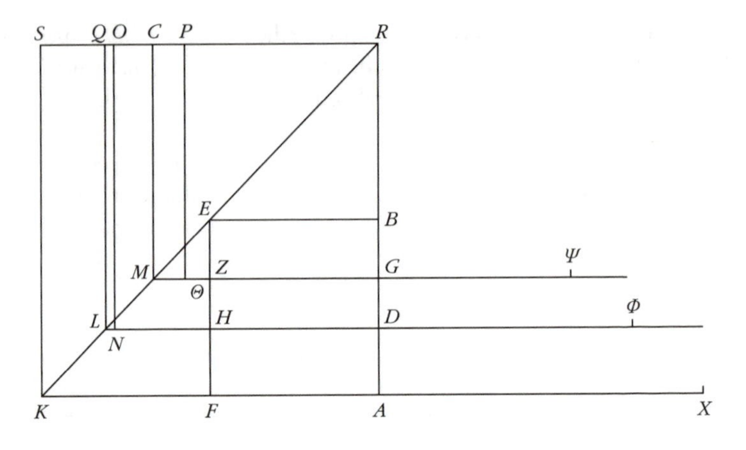

Diagram 3.3

and *GM* as radius, we draw a circle, it cuts *GM*: let it cut in *Q*. Then we continue to draw it in the same way until it cuts it in *Y*. Again, if, with centre *A* and radius *DL*, we draw a circle, it cuts *DL*: let it cut in *N*. Then we continue to draw it about centre *A* until it cuts it again in *P*. Then we draw *AX* as an extension in a straight line of *KA* and make it equal to it. Then points *K, N, Q, B, Y, P, X* lie on a parabola. For we produce *AB* to *R*, letting *BR* equal *AB*; let us draw *RS* perpendicular to *AB* and equal to *KA*, and join *SK*, and draw from points *L, M, Q, N* to line *RS* perpendiculars *LW, MC, NO, QJ*. Then when *KE* is produced in a straight line it passes through *R*. So *WL* is equal to *LD* and *MC* is equal to *MG*, because *KER* is a diagonal of square *AS*. But *LD* is equal to *NA* and *MG* is equal to *QA*, and *LW* is equal to *LD* also and *MC* is equal to *MG*. So *AN* is equal to *LW* is equal to *NO* and *AQ* is equal to *MC* is equal to *QJ*. And *AK* is equal to *KS* and *AR* is bisected at *B*. Since that is so, points *B, Q, N, K* lie on a parabola, as we shall prove subsequently. Similarly points *Y, P, X* also. So if we mark numerous points on *AB*, and draw through them lines parallel to *AK*, and mark on the lines points corresponding to the other points, and bend along the resultant points a ruler made of horn, fastening it so that it cannot move, then draw a line along it and cut the board along that line, then shape the curvature of the figure we wish to make to fit that template, the burning from that surface will occur at point *A*, as was proved in the first proposition.[50]

A further variation on the question of mathematics and the real world is found in the work of Eratosthenes of Cyrene (*c.* 285–194 BC), who was chief librarian at Alexandria. He had apparently studied philosophy at Athens and wrote a *Platonicus*, which, like all of his works, is only known through fragments. In one of them, a passage from the *Timaeus* lends Eratosthenes the opportunity to articulate the difference between *diastema* (distance) and *logos* (ratio). There can be a ratio between things which are equal, but distance is only between two things distinct from each other; ratio is not the same both ways (2:1 is not the same as 1:2), whereas distance is. Thus,

> ratio is a certain kind of relation of two magnitudes to each other [...] ratio is the principle of proportion [...] for every proportion <comes from> ratios, but principle of ratio <is> sameness.[51]

Eratosthenes also discussed proportion in relation to geometrical figures, and Sextus Empiricus confirms his interest in basic definitions of mathematical concepts by reporting that he analyzed the nature of point and line. Apart from the *Platonicus*, Eratosthenes wrote a book *On Means*, which may have contained the original version of a solution to the duplication of the cube now extant in Eutocius and a method, called 'the sieve', to find prime numbers, and a *Geography*. Tackling the subject mathematically, he mapped the surface of the then-known world by means of intersecting lines, while regions were approximated to geometrical figures by having, for instance, Sicily shaped as a triangle and India as a rhomboid. He also measured the obliquity of the ecliptic and the circumference of the Earth, which he put at 252,000 stades. We do not know for sure which unit of measure was used, and his procedure is described in detail only by the late fourth-century AD Cleomedes. According to the latter, Eratosthenes knew that in the city of Syene at noon on the summer solstice the gnomon cast no shadow; he thus assumed that Syene was on the Tropic of Cancer. He also knew that Syene and Alexandria were on the same meridian; finally, the distance between the two cities had been measured by the royal surveyors. Eratosthenes thus measured the shadow cast by a gnomon at noon on the summer solstice at Alexandria and, by means of the properties of similar triangles, estimated that the shadow amounted to $^1/_{25}$ of the hemisphere, and thus to $^1/_{50}$ of the whole circle.[53]

After music, astronomy, geography, we find mathematics applied to a rather different set of phenomena in Biton's treatise on the construction of war instruments, written between the mid-third and the early second century BC and addressed to a king Attalus (one of the Pergamum monarchs). While Biton portrays himself as knowledgeable about the use of the dioptra, and as the author of a book on optics, all the machines in the treatise are explicitly attributed by him to other people, and the place where they were produced is also specified. Thus, Charon of Magnesia built a certain machine at Rhodes, Isidorus of Abydus another machine at Thessalonika, Posidonius the Macedonian devised the siege tower for Alexander the Great, the siege ladder is due to Damis of Colophon. Zopyrus of Tarentum is credited with two types of belly-bow (a type of small catapult), one made at Myletus, one at Cumae. The resulting picture is one of geographical variety, both in the places of origin of these people, and in where they end up working. A community of engineers/machine-makers is evoked, whose members travelled often, changing employer or patron, sharing and transmitting knowledge about materials, proportions in the construction of the machines, anecdotes about their opportune use in actual situations.

Biton describes the shape of the devices and the dimensions of their components, provides suggestions as to their overall structure, and complements his account with illustrations (now lost). His specifications, or lists of dimensions, are in the form of sets of measurements to which the various pieces of a machine have to correspond as accurately as possible. The use of some geometrical instruments is required in making each piece to the given specification with the required accuracy. Biton concludes the account thus:

> Whatever engines we considered most appropriate for you, we have now described. We are convinced that you will be able to discover similar forms through these. Do not be worried by the thought that, because we have used fixed measurements, it will be necessary for you to use the same measurements, too. If you wish to construct larger or smaller instruments, do so; only try to preserve the proportion (*analogia*).[54]

Philo of Byzantium (third century BC) opened *his* book on war engines (the *Belopoiika*) by pointing out that previous treatments were in disagreement both about the proportion (*analogia*) of the various components of the machines and about the guiding element (*stoicheion*) of the construction, i.e. the standard on the basis of which the dimensions of each element were established. Therefore, he decided to ignore the old authors and to draw on those newer methods

> that can achieve what is required in the facts. I understand you do not ignore that the art (*techne*) has something very obscure and incomprehensible to the many; at any rate, many who have undertaken the construction of instruments of equal size and who have used the same arrangement and similar wood and the same metal

> without even changing its weight, have made some with long range and powerful impact and others which fall short of these. Asked why this happened, they were not able to say the cause. Thus the remark made by Polykleitus the sculptor is proper for what I am going to say; for he said that the good is generated little by little through many numbers. Likewise, in this art, since the things that are done are completed through many numbers, those who make a small discrepancy in individual parts produce a large mistake when adding up at the end.[55]

One of the good things generated through many numbers was the realization that, if one wanted to modify the dimensions of a catapult proportionately, they had to use the hole holding the torsion spring as guiding element. Philo describes this discovery as a process happening over time, through repeated experience, trial-and-error and accumulated knowledge. The culmination of this technical development was the rigorous, mathematical, formulation of the duplication of the cube, which was necessary to change the size of the torsion spring hole. Philo's solution to the problem comes at the end of a passage where he offers lists of dimensions for the construction of the machines, and draws up a mathematical table where, with opportune approximations, weight of projectile and diameter of the torsion spring hole are correlated. The table of weights and diameters is obtained as follows (see Diagram 3.4):

> Reduce to units [i.e. drachmae] the weight of the stone for which the machine must be put together. Make the diameter of the hole of as many dactyls as there are units in the cubic root of the number obtained, adding the tenth part of the root found. If the weight does not have an expressible side, take the nearest one; but, if it is above, try to diminish the tenth part in proportion and, if below, to increase the tenth part. [...] It is possible, also, from one number, [...] to put together the remaining diameters instrumentally by doubling the cube [...] Let there be a straight line given of the diameter, of which it is necessary to find the double power, for example A. I put B, which is double that, perpendicular to it, and from the extremity of B I drew another <line> Γ perpendicularly, indefinitely. And I drew a line K from the angle Θ, and cut it in half; let the dividing point be at K. Using K as centre, and $K\Theta$ as radius, I described a semicircle, cutting also through the angle Z, and taking a straight ruler I carefully connected <them>, cutting both lines and keeping one part of the ruler on the angle Z. Let the ruler then be at Z. I moved the ruler around at the same time

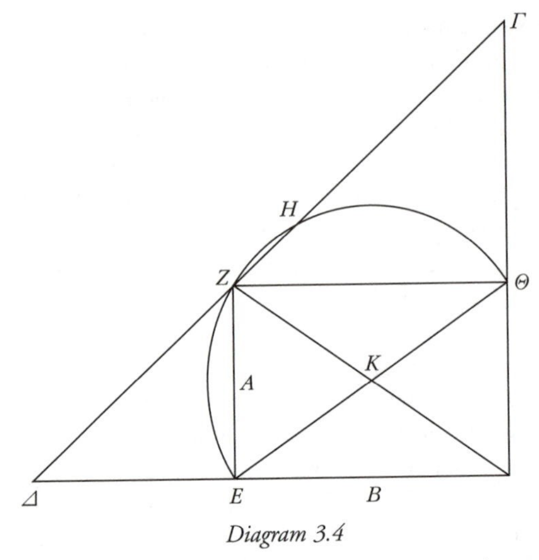

Diagram 3.4

> keeping one part of it touching the angle, and moved around until it happened to me that the part of the ruler from the point of contact Γ, to the point of contact of the circumference H, resulted equal to the <line> from the point of contact Δ, to the angle Z. And ΔE will be the double power of EZ, and $\Theta\Gamma$ of $E\Delta$, and ΘZ of $\Theta\Gamma$.[56]

Philo's is, along with Diocles', the earliest first-hand extant solution to the problem of the duplication of the cube. It, too, employs mechanical aids, in this case a moving ruler, and adds to the conclusion that mathematics was widely practised, benefited from exchanges and accumulation of results over time, and was applied to a number of socially and politically prominent fields.

Euclid

For such a famous mathematician, Euclid remains a very shadowy figure. We do not know with certainty when or where he lived, and even the authorship of his major work, the *Elements*, has come under question. The most educated guess we can make is that Euclid was active in Alexandria, possibly enjoying some form of royal patronage, around the very beginning of the third century BC. His surviving works, apart from the *Elements*, are the *Data*, the *Phenomena*, the *Optics*, the *Section of a Canon*, the *Divisions of Figures*; titles of lost books include *Conics* and *Loci with Respect to a Surface*.[57] In what follows I will concentrate on some features of the *Elements* and of the *Data*.

Contents

The contents of the *Elements* can be briefly summarized thus: book 1 and 2 are on plane rectilineal geometry, book 3 on the circle, book 4 on regular polygons, book 5 on proportion theory, book 6 on plane geometry with an use of proportions (e.g. similar polygons), books 7 to 9 on number theory, book 10 on irrational lines, books 11 and 12 on solid geometry, and, finally, book 13 on regular polyhedra and their relation with the sphere.[58]

Book 1 sets off with a number of basic starting points: definitions, which include that of a point, a line, a circle, the three kinds of angle; postulates, which mostly establish the possibility of elementary constructions, such as

> let it be required to have drawn a straight line from any point to any point [...] and a circle to be drawn with any centre and radius and all right angles to be equal to each other [...][59]

common notions to the effect that, for instance, 'things which are equal to the same <thing> are also equal to one another' and that 'the whole is greater than the part'. Many of the theorems in book 1 were already familiar before Euclid, as we know because they are mentioned by for instance Plato and Aristotle. For example, that in any isosceles triangle the angles at the base are equal to each other (1.5), that the internal angles of a triangle are equal to two right angles (1.32) or that in a right-angled triangle the square on the side subtending the right angle is equal to the squares on the sides containing the right angle, or theorem of Pythagoras (1.47). Some other theorems, such as those on the equality of two triangles, seem new to the *Elements*, even though that may be due to lack of evidence. In general, Euclid inscribes both already known and not previously known results into a systematic demonstrative framework, where the results themselves, the assumptions they use, the possibility of the geometrical constructions they require, and even the geometrical objects they deal with, are all firmly grounded and justified.

Like book 1, books 2, 3 and 4 present a combination of results familiar from elsewhere, results whose simplicity suggests that they were perhaps commonly assumed but not proved in full, and what may have been new results. They are all organized within a deductive structure, with starting points (in the form of definitions) and propositions proved on the basis either of the starting points or of previous propositions. The last item in book 2 is the seminal problem of building a square equal to a given rectilinear figure, or squaring a rectangle, which is equivalent to finding a mean proportional between two straight lines.

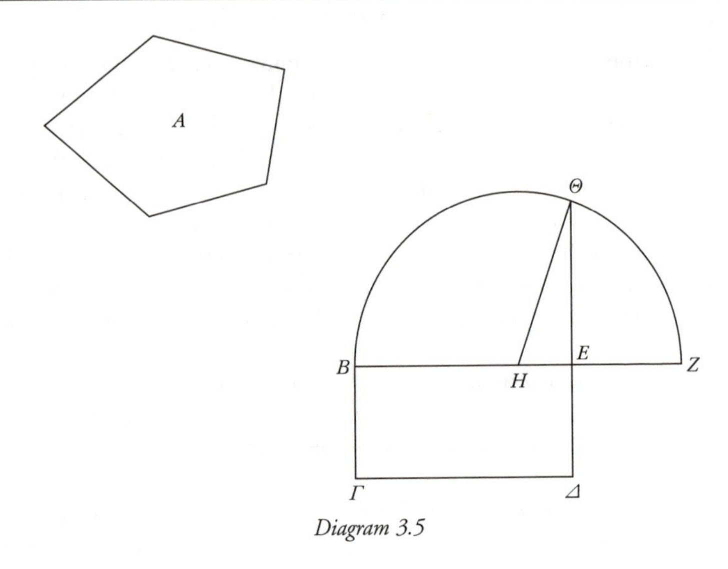

Diagram 3.5

[*Elements* 2.14 (see Diagram 3.5)] To put together a square equal to a given rectilinear figure. Let the given rectilinear figure be A; it is necessary then to put together a square equal to A. Let a right-angled parallelogram $B\Delta$ be put together equal to the rectilinear figure A; if then BE really is equal to $E\Delta$, what has been prescribed would be produced. The square $B\Delta$ equal to the rectilinear figure A has in fact been put together. If instead this is not so, one of the BE, $E\Delta$ is greater. Let the BE be greater, and let it be prolonged until Z, and let EZ be posited equal to $E\Delta$, and let BZ be divided in two in the point H, and with H as centre and one of the HB, HZ as radius let the semicircle $B\Theta Z$ be drawn, and let ΔE be prolonged until Θ, and let $H\Theta$ be conjoined. Since then the line BZ is divided in equal parts by H, but in unequal parts by E, therefore the rectangle formed by BE, EZ plus the square on EH is equal to the square on HZ. For HZ is equal to $H\Theta$. Therefore the <rectangle formed> by BE, EZ plus the <square> on HE is equal to the <square> on $H\Theta$. But the <square> on $H\Theta$ is equal to the squares on ΘE, EH; therefore the <rectangle formed> by BE, EZ plus the <square> on HE is equal to the <squares> on ΘE, EH. Let the common square on HE be subtracted; the remainder therefore, the rectangle formed by BE, EZ is equal to the square on $E\Theta$. But the <rectangle formed> by BE, EZ is $B\Delta$, for EZ is

> equal to $E\Delta$; therefore the parallelogram $B\Delta$ is equal to the square on ΘE. But $B\Delta$ is equal to the rectilinear figure A. Therefore also the rectilinear figure A is equal to the square built on $E\Theta$. Therefore a square has been put together, the <one> built on $E\Theta$ equal to the given rectilinear figure A; which it was necessary to do.

Among the definitions that open book 5, one finds this passage:

> ratio is a sort of relation of two homogeneous magnitudes according to size. It is said that <those> magnitudes have a ratio one with the other, which are capable, multiplied, to exceed one another. It is said that magnitudes are in the same ratio the first with the second and the third with the fourth, if multiplying the same times the first with the third, and multiplying the same times the second with the fourth, in any multiplication whatsoever, the former will either alike exceed, or alike be equal or alike fall short of the latter, respectively taken in the corresponding order. Let magnitudes that have the same ratio be called in proportion.[60]

The starting points of book 5 are stated in a sequence: notice how the concept of ratio is first vaguely explained, then substantiated by a working definition, while the concept of magnitudes in the same ratio is first painstakingly laid out, and then followed by a slimmer synonym. Again, although ratios and proportions were used before Euclid, this is the first extant clear (-ish) statement of what they are.

Book 6 contains, among other things, a definition of similarity for rectilinear figures, a solution to the problem of how to cut a segment into extreme and mean ratio (6.30) and a solution to the problem of finding a mean proportional (see Diagram 3.6):

> To find a mean proportional to two given lines. Let the two given lines be AB, $B\Gamma$; it is necessary then to find a mean proportional to AB, $B\Gamma$. Let them be posited on a line, and let a semicircle $A\Delta\Gamma$ be drawn on $A\Gamma$, and let $B\Delta$ be drawn from the point B perpendicular to the line $A\Gamma$, and let $A\Delta$, $\Delta\Gamma$ be joined. Since <it is> an angle in a semicircle, the angle $A\Delta\Gamma$ is right. And since ΔB is drawn in a right-angled triangle, <namely> $A\Delta\Gamma$, from the right angle to the base perpendicularly, ΔB therefore dividing the base is the mean proportional between AB, $B\Gamma$. Therefore ΔB has been found <to be> mean proportional between the given lines AB, $B\Gamma$; which it was necessary to do (6.13).

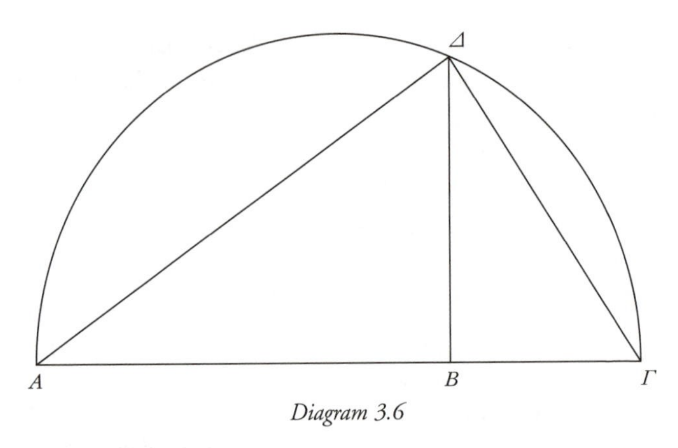

Diagram 3.6

The next three books are usually called 'arithmetical' because they do for basic number theory what book 1 does for geometry. They are in a strict sequence, book 7 beginning with definitions of concepts that are employed throughout books 8 and 9. Euclid says that 'unit' is the thing 'according to which it is said that each of the beings <is> one' (7. definition 1), while a 'number' is 'a multitude put together from units' (7. definition 2). The unit is therefore *not* a number. Other definitions include that of odd and even, of prime number, of multiplication, of plane, square and solid numbers, and (again) of proportionality, this time specifically for numbers: 'numbers are in proportion, if the first of the second and the third of the fourth, are the same times either multiple of, or the same part of, or the same parts of' (7. definition 20).[61]

Results established in the arithmetic books include how to find the greatest common measure of two or three numbers (7.2 and 7.3) and the least common number which two or three numbers measure (7.34 and 7.36), as well as propositions on prime numbers and on numbers in proportion, including means and continuous proportion. In book 8 it is stated that between two square numbers and between two similar plane numbers there is one mean proportional (8.11 and 8.18), while between two cube numbers and between two similar solid numbers there are two mean proportionals (8.12 and 8.19). Proposition 8.12 amounts to a proof that the problem of the duplication of the cube can be reduced to that of finding two mean proportionals between two given lines. Some of the theorems are very general – the one that follows concerns numbers as a whole (see Diagram 3.7):

> The multitude of prime numbers is more than all the multitude of proposed prime numbers. Let the proposed prime numbers be *A*, *B*, *Γ*; I say that prime numbers are more than *A*, *B*, *Γ*. Let the

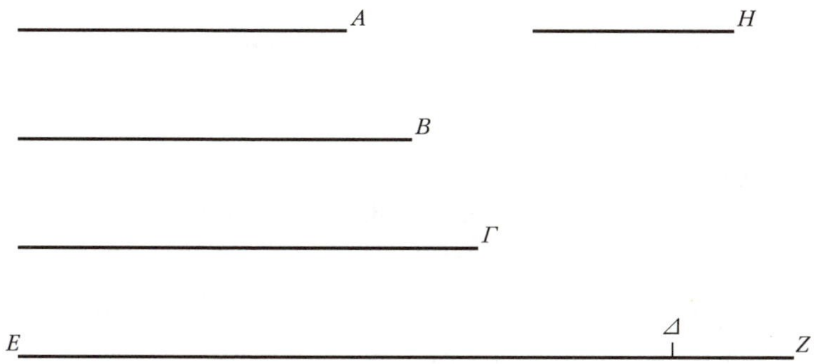

Diagram 3.7

> least dividend of A, B, Γ have been taken and let it be ΔE, and let the unit ΔZ be added to ΔE. EZ then will either be prime or not. Let it first be prime; therefore the found prime numbers A, B, Γ, EZ are more than A, B, Γ. But let EZ be not prime; therefore it will be divided by some prime number. Let it be divided by Z; I say that H is not the same as A, B, Γ. If indeed it is possible, let it be <so>. Then A, B, Γ divide ΔE and therefore H divides ΔE. And it will divide EZ as well. And the number H divides the remaining unit ΔZ; which is absurd. Therefore H will not be the same as A, B, Γ. And it was assumed to be prime. Therefore the prime numbers A, B, Γ, H that have been found are more than the multitude A, B, Γ that had been assumed; which it was necessary to prove (9.20).

Book 10, the longest of the *Elements*, again exhibits a combination of old and new. Its main focus is incommensurability – of magnitudes in general, but more especially of lines, which can be incommensurable in length and in square, as we know from earlier sources including Plato's *Theaetetus*. In fact, book 10 matches the project of thorough classification of incommensurables attributed to Theodorus and/or Theaetetus, as well as extending to magnitudes in general results that had been established for numbers in the arithmetical books, and exploring their further ramifications. For instance, a series of propositions (10.17 onwards) discusses the application of parallelograms to straight lines, a topic already found in books 2 and 6, with the difference that here the notion of incommensurability is brought on board. Thus, cases are distinguished according to whether the segments resulting from the application are incommensurable with each other or with the original line. We also have a method to find the so-called Pythagorean triplets (see Diagram 3.8):

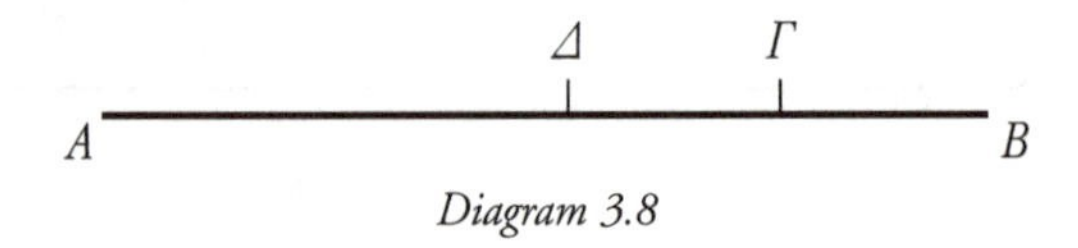

Diagram 3.8

> To find two square numbers, such that their sum will also be a square. Let two numbers be posited, AB, $B\Gamma$, let them be either odd or even. And since, if even is subtracted from even, and if odd from odd, the remainder is even, therefore the remainder $A\Gamma$ is even. Let $A\Gamma$ be divided in two in the point Δ. Let AB, $B\Gamma$ be either similarly plane or square, and these are similarly plane. Therefore the product of AB and $B\Gamma$ plus the square on $\Gamma\Delta$ is equal to the square on $B\Delta$. And the product of AB, $B\Gamma$ will also be square, since it has been proved that if two similarly plane <numbers> multiplied by each other produce something, the result is a square. Therefore the two numbers, the product of AB and $B\Gamma$ and the square on $\Gamma\Delta$, will be found to be square, and those numbers added to each other make the square on $B\Delta$. And it is evident that again the square on $B\Delta$ and the square on $\Gamma\Delta$ will be found to be square numbers, so that the excess of them is the square on AB, $B\Gamma$, if AB, $B\Gamma$ are similarly plane. If instead they are not similarly plane, the square on $B\Delta$ and the square on $\Delta\Gamma$ will be found to be square, then the excess of the product of AB, $B\Gamma$ is not a square; which it was necessary to prove (10. first lemma after 10.28).

Several types of incommensurable lines are introduced and given specific names in the course of the book: medial at 10.21, binomial at 10.36, apotome at 10.73 and a whole group of first, second, up to sixth binomial between propositions 47 and 48, and first, second, up to sixth apotome between propositions 84 and 85.

The following books 11, 12 and 13, on solid geometry, establish, among other things, the relation between pyramid and prism with the same base and same height (12.7) and between cone and cylinder with the same base and same height (12.10). The reader may remember from the section on the problem of later early Greek mathematics p. 50, that those results were by Archimedes attributed to Eudoxus – once again, it would seem that Euclid weaves previous material into the texture of his account.

The main purpose of book 13, the final book of the *Elements*, is the investigation of regular solids with respect to the sphere. Since the faces of regular solids are regular polygons, much space is devoted to those latter, for instance to the relation between the side of a polygon and the radius of the circle within which that polygon is inscribed. Materials from previous books are relevant here, but Euclid also found that in some cases the relation between side and radius, as defined above, or the relation between the side of a polyhedron and the radius of the sphere circumscribing it, could be expressed in terms of incommensurable lines such as had been classified in book 10. He thus brings to fruition several of the topics scattered throughout the work: for instance, results from books 5 and 6 are combined with elements from book 10 to produce 13.6: 'If a rational line is divided into extreme and mean ratio, each of the segments is an irrational, the so-called apotome' or 13.8: 'If lines subtend two successive angles of an equilateral and equiangular pentagon, they cut each other in extreme and mean ratio, and of these the greater segments are equal to the side of the pentagon' (see the example below).

[*Elements* 13.8 (see Diagram 3.9)] If lines subtend two successive angles of an equilateral and equiangular pentagon, they cut each other in extreme and mean ratio, and of these the greater segments are equal to the side of the pentagon. Let the lines $A\Gamma$, BE subtending the two successive angles on A, B of an equilateral and

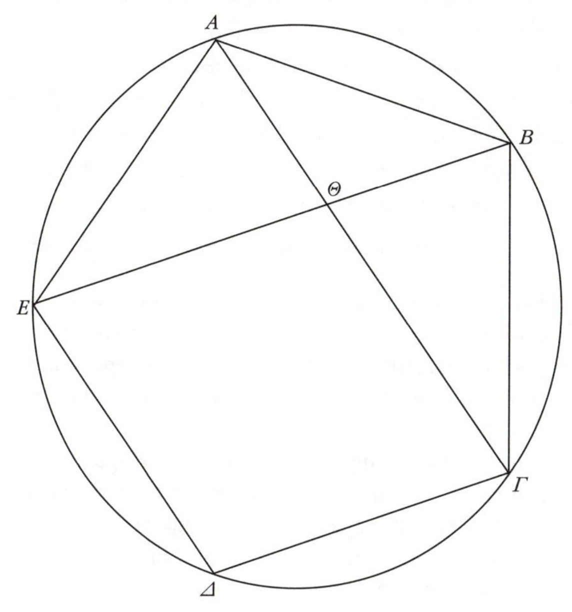

Diagram 3.9

> equiangular pentagon $AB\Gamma\Delta E$, cut each other in the point Θ. I say that each of these is cut in extreme and mean ratio in the point Θ, and of these the greater segments are equal to the side of the pentagon. Let the circle $AB\Gamma\Delta E$ be circumscribed to the pentagon $AB\Gamma\Delta E$. And since the two lines EA, AB are equal to AB, $B\Gamma$ and they surround equal angles, therefore the base BE is equal to $A\Gamma$, and the triangle ABE is equal to the triangle $AB\Gamma$, and the remaining angles will be equal respectively to the remaining angles, <those> by which the equal sides are subtended. Therefore the angle $BA\Gamma$ is equal to ABE; therefore the <angle> $A\Theta E$ is double $BA\Theta$. The <angle> $EA\Gamma$ is double $BA\Gamma$, because the arc $E\Delta\Gamma$ is also double the arc ΓB. Therefore the angle ΘAE is equal to $A\Theta E$, so is the line ΘE to EA, that is, it is equal to AB. And since the line BA is equal to the line AE, the angle ABE is also equal to AEB. But it has been proved that the <angle> ABE is equal to $BA\Theta$; and therefore the <angle> BEA is equal to $BA\Theta$. And the angle ABE is common to the two triangles ABE and $AB\Theta$. Therefore the remaining angle BAE is equal to the remaining angle $A\Theta B$. Therefore the triangle ABE has the same angles as the triangle $AB\Theta$; therefore in proportion as EB is to BA, so AB to $B\Theta$. But BA is equal to $E\Theta$. Therefore as BE to $E\Theta$, so $E\Theta$ to ΘB; but BE is greater than $E\Theta$; therefore $E\Theta$ is also greater than ΘB. Therefore BE is divided in extreme and mean ratio in the point Θ, and the greater segment ΘE is equal to the side of the pentagon. Similarly we prove that $A\Gamma$ is divided into extreme and mean ratio at Θ, and the greater segment of that, $\Gamma\Theta$, is equal to the side of the pentagon; which it was necessary to prove.

The last five propositions of the *Elements* are devoted to the construction of the five regular polyhedra and to their comparison with one another: the icosahedron is the greatest, then the dodecahedron, the octahedron, the cube and the pyramid. An adjunct to prop. 13.18 is as follows:

> I say then that apart from the five mentioned figures no other figure is put together surrounded by equilateral and equiangular <figures> equal to each other. For a solid angle is not put together from two triangles or in general two planes. From three triangles the <solid angle> of the pyramid, from four that of the octahedron, but five than of the icosahedron; from six equilateral and

equiangular triangles put together in one point a solid angle will not <derive>. Since the angle of an equilateral triangle is two thirds of a right <angle> the <angles of the> six <triangles> will be equal to four right <angles>; which is impossible; for any solid angle is surrounded by less than four right angles. Through this, also a solid angle will not be put together by more than six plane angles. The angle of the cube is surrounded by three squares; it is impossible <that it be surrounded> by four; for again they will be four right angles. The <angle> of the dodecahedron by three equilateral and equiangular pentagons; it is impossible by four; since the angle of the equilateral pentagon is a right <angle> and a fifth, the four angles will be greater than four right angles; which is impossible. A solid angle will not be surrounded by other polygons because of this absurdity. Therefore except for the said five figures, no other solid figure will be put together from surrounding equilateral and equiangular <figures>; which it was necessary to prove.

In some aspects as basic a work as the *Elements*, the *Data* also starts off with definitions, which specify what it means that a point, an angle, a rectilinear figure, a circle, a ratio are given in magnitude, position or form. For instance, 'a rectilinear figure is said to be given in form when its angles are given one by one and the ratios of the sides with each other are given'. Areas, lines and angles are given (in magnitude) when 'to them we can find equals'.[62] The definitions are extensively deployed in the proofs, as is clear from the example below.

[*Data* 39 (see Diagram 3.10)] If each of the sides of a triangle is given in magnitude, the triangle is given in form. So, let each of the sides of a triangle $AB\Gamma$ be given in magnitude. I say that the triangle is given in form. So, let a straight line ΔM be given in position, then prolonged towards the point Δ, and without limit towards the remaining point, and let ΔE be posited equal to AB. But AB is given; therefore ΔE is also given; but it is given also in position; and Δ is given; therefore E is also given; and EZ is equal to $B\Gamma$; and $B\Gamma$ is given; therefore $E\Gamma$ is also given; but it is given in position as well; and E is given; therefore Z is also given; and ZH is equal to $A\Gamma$. And $A\Gamma$ is given; therefore ZH is also given. But it is also given in position. And Z is given; therefore H is also given. And with centre the point E and radius $E\Delta$ let the circle

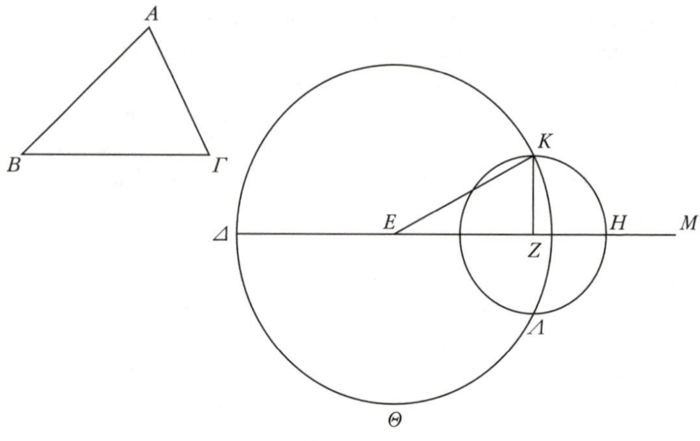

Diagram 3.10

> *ΔKΘ* be drawn; therefore *ΔKΘ* is given in position. Again, with centre the point *Z* and radius *ZH* let the circle *HKΛ* be drawn; therefore *HKΛ* is given in position; and the circle *ΔΘK* is given in position; therefore the point *K* is also given. Let then each of the *E*, *Z* be given; therefore each of the *KE*, *EZ*, *ZK* is also given in position and in magnitude; therefore the triangle *KEZ* is given in form. And it is equal and also similar to *ABΓ*; therefore the triangle *ABΓ* is given in form.

'Being given' means that the mathematical object is determined and identifiable on the basis of its size, position or form. In terms of a diagram, it means that the element is fully identified; when it comes to the solution of a geometrical problem, it means that we can take that element as known and move on in our demonstration. The propositions in the *Data* mostly aim at establishing correlations between givens, so that the reader can recognize how, under certain conditions, the being-given of certain elements entails the being-given of certain other elements. The book seems aimed at a more advanced public than the *Elements*; this is confirmed by its survival in later antiquity as part of a group of texts, the so-called 'treasure of analysis', which were used to acquire and hone problem-solving skills.

Procedure

We have already hinted that propositions in Euclid are proved on the basis either of the starting points or of previous propositions. While this is generally true, several ripples corrugate the smooth logically-ordered surface of the *Elements*. For one, the fifth postulate of book 1:

> if a straight line meeting two straight lines produces the internal angles on the same side less than two right angles, [let it be required that] the two straight lines having been produced indefinitely meet each other on that side on which are the two <angles> less than two right angles

was criticized since antiquity for being not a postulate, but rather a statement requiring proof. Moreover, terms are introduced which have not been previously defined; things are established that are not subsequently used but seem too marginal to have been established just for their own sake; apparently inexplicable detours are not infrequent and there are, as in the case of proportionality, duplicate accounts whose mutual relation is not clear. Some demonstrative procedures are introduced, used a couple of times and then abandoned. For instance, the idea of 'fitting' geometrical figures onto one another is only deployed at 1.4 and 1.8, both about equality of triangles (see Diagram 3.11):

> If two triangles have the two sides equal to the two sides respectively and have the angle surrounded by the equal lines equal to the angle, they also have the base equal to the base, and the triangle will be equal to the triangle, and the remaining angles will be equal to the remaining angles respectively <those> by which the equal sides are subtended. Let the two triangles be $AB\Gamma$, ΔEZ, having the two sides AB, $A\Gamma$ equal to the two sides ΔE, ΔZ respectively, AB to ΔE and $A\Gamma$ to ΔZ and the angle $BA\Gamma$ equal to the angle $E\Delta Z$. I say that also the base $B\Gamma$ is equal to the base EZ, and the triangle $AB\Gamma$ will be equal to the triangle ΔEZ, and the remaining angles will be equal to the remaining angles respectively <those> by which the equal sides are subtended, $AB\Gamma$

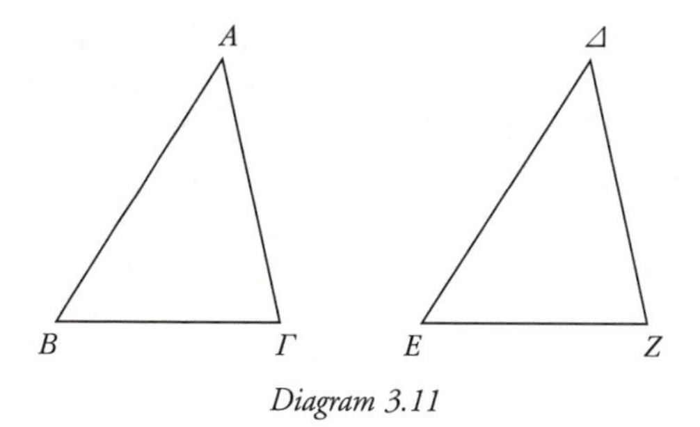

Diagram 3.11

> to ΔEZ, and $A\Gamma B$ to ΔEZ. Indeed the triangle $AB\Gamma$ having been fitted on the triangle ΔEZ and the point A being put on the point Δ and the line AB on the line ΔE, the point B will also fit on E because AB and ΔE are equal. AB having been fitted on ΔE, the line $A\Gamma$ also <fits> on ΔZ because the angles $BA\Gamma$ and $E\Delta Z$ are equal. Therefore the point Γ will also fit on the point Z because again $A\Gamma$ and ΔZ are equal. But also B will fit to E; therefore the base $B\Gamma$ will fit on the base EZ. If then B having been fitted on E, and Γ on Z, the base $B\Gamma$ did not fit on EZ, two straight lines would surround an area, which is impossible. Therefore the base $B\Gamma$ will fit on EZ and will be equal to it. Therefore also the whole triangle $AB\Gamma$ will fit on the whole triangle ΔEZ and will be equal to it, and the remaining angles will be on the remaining angles and will be equal to them, $AB\Gamma$ to ΔEZ and $A\Gamma B$ to ΔZE. If therefore two triangles have the two sides equal to the two sides respectively and have the angle surrounded by the equal lines equal to the angle, they also have the base equal to the base, and the triangle will be equal to the triangle, and the remaining angles will be equal to the remaining angles respectively <those> by which the equal sides are subtended, which it was necessary to prove (1.4).

Some modern interpreters see the 'fitting' method as a relic from the pre-Euclidean past, because its reliance on a certain 'physicality' of geometrical objects would indicate a lower level of abstraction, and therefore an earlier stage of development. I hardly need to point out that the assumption that there is a correlation between level of abstraction, stage of development and relative chronology is founded exclusively on one's subjective views of mathematics and of intellectual history. Because reliable detailed information about early Greek mathematics is so scarce, we are in no position safely to discriminate between earlier or later procedures. Also, greater or lesser development or level of advancement tend to be measured against the yardstick of Archimedes and Apollonius. This may overestimate the degree to which advanced authors affected mathematical practice across the spectrum, and oversimplifies the variety of mathematical traditions which persisted and in fact flourished even after the gathering of consensus around a certain mathematical model embodied by Euclid, Archimedes and Apollonius (our Big Guys). In other words, the introduction and successful wide application of axiomatico-deductive structure, indirect proof and so on did not cause the total extinction of alternative methods, so that the use of a procedure rather than another is no univocal indication of a later or earlier date, or of a supposedly corresponding more or less advanced stage of development.

Many of the propositions in book 1 of the *Elements* are in the form of problems: to construct an equilateral triangle on a given finite straight line (1.1), to bisect a given rectilinear angle (1.10) or more complex tasks, such as 'to apply to a given straight line in a given rectilinear angle a parallelogram equal to a given triangle' (1.44). Problems of this kind are normally solved by means of straight ruler and compass, i.e. nothing more than circles or arcs of circle are required for the construction. Take for instance 1.12 (see Diagram 3.12):

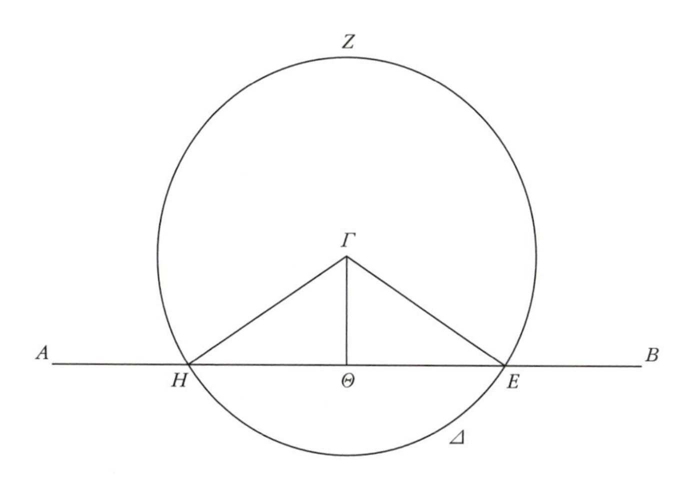

Diagram 3.12

To draw a perpendicular straight line to a given infinite straight line from a given point which is not on the same line. Let then the given infinite straight line <be> AB and the given point which is not on the same line Γ. It is necessary to draw a perpendicular straight line to the given infinite straight line AB from the given point Γ, which is not on the same line. Let then be taken on the other side of the line AB a point whatever Δ, and with centre Γ and radius $\Gamma\Delta$ let the circle EZH be drawn, and let the line EH be divided into two at Θ, and let the lines ΓH, $\Gamma\Theta$, ΓE be joined. I say that $\Gamma\Theta$ has been drawn perpendicular to the given infinite straight line AB from the given point Γ, which is not on the same line. Since $H\Theta$ is equal to ΘE; $\Theta\Gamma$ is common; the two $H\Theta$, $\Theta\Gamma$ are equal to the two $E\Theta$, $\Theta\Gamma$ respectively; and the base ΓH is equal to the base ΓE; therefore the angle $\Gamma\Theta H$ is equal to $E\Theta\Gamma$. And they are one next to the other. If then a line standing onto a line makes the angles one next to the other equal, each of the equal angles is right, and the line that has been erected is called perpendicular to the line on which it has been erected. Therefore, a perpendicular $\Gamma\Theta$ has been drawn to the given infinite straight line AB from the given point Γ which is not on the same line; as it was necessary to do.

The proofs I have quoted so far are direct, but several of the propositions in book 1 are proved indirectly, by showing that the negation of what they affirm is self-contradictory. For instance (see Diagram 3.13),

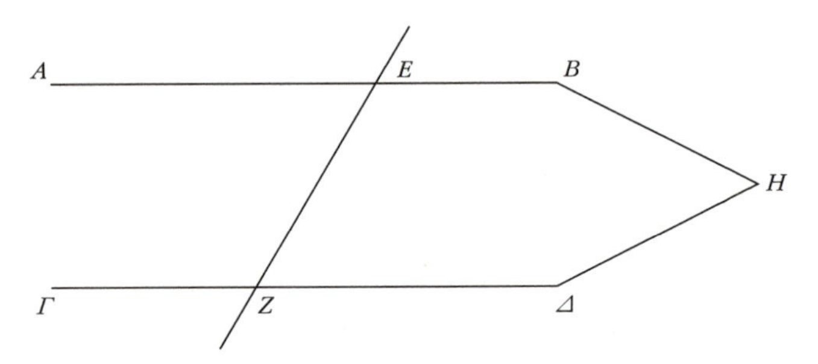

Diagram 3.13

If a line which intersects two lines produces the alternate angles equal to each other, the lines will be parallel to each other. Let indeed a line EZ intersecting two lines AB, $\Gamma\Delta$ produce the alternate angles AEZ, $EZ\Delta$ equal to each other. I say that AB is parallel to $\Gamma\Delta$. If indeed it was not the case, AB, $\Gamma\Delta$ having been prolonged will meet either on the parts B, Δ or on A, Γ. Let them have been prolonged and let them have met on the parts B, Δ in H. But the external angle of the triangle HEZ, AEZ, is equal to the internal and opposite <angle> EZH; which is impossible; therefore AB, $\Gamma\Delta$ prolonged will not meet on the parts B, Γ. Similarly it will be proved that they do not meet on the parts A, Γ. But the <lines> which do not meet on any part are parallel; therefore AB is parallel to $\Gamma\Delta$. If a line which intersects two lines produces the alternate angles equal to each other, the lines will be parallel to each other, which it was necessary to prove (1.27).

A particular type of indirect method, first found in Euclid, is commonly known as the 'method of exhaustion', although no ancient source designates it by any name in particular, and we are not absolutely sure as to what extent it was a standardized method. Used especially in order to determine the relation between circular and rectilinear objects, the method of exhaustion worked by proving that all possible alternatives to the result one was trying to establish were absurd. It relied on manipulations of ratios between figures, and on a lemma, which various geometers stated in various versions. The one in the *Elements* is as follows:

> Two unequal magnitudes having been posited, if from the greater more than the half is subtracted and more than the half from the remainder, and this is done continuously, a magnitude will be left which will be less than the posited lesser magnitude (10.1).[63]

This lemma, which is used by Euclid to prove, among other things, that any pyramid is a third of the prism with the same base and the same height (12.7), corresponds with a lemma reported by Archimedes and traceable to Eudoxus. Once again, it would seem that Euclid was revisiting old ground, not just in terms of contents but also of procedures. A famous example of method of exhaustion is *Elements* 12.2 (see Diagram 3.14):

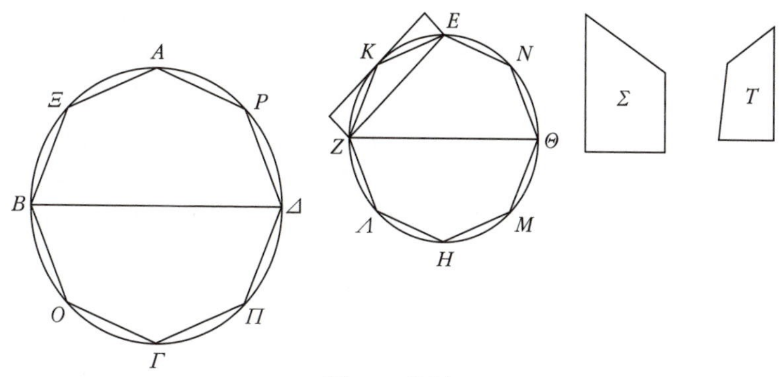

Diagram 3.14

> Circles are to each other as the squares on their diameters. Let $AB\Gamma\Delta$, $EZH\Theta$ be circles, their diameters $B\Delta$, $Z\Theta$. I say that the circle $AB\Gamma\Delta$ is to the circle $EZH\Theta$ as the square on $B\Delta$ is to the square on $Z\Theta$. If indeed it is not <the case that>, as the circle $AB\Gamma\Delta$ to the $EZH\Theta$, so the square on $B\Delta$ to that on $Z\Theta$, it will be as the <square> on $B\Delta$ to that on $Z\Theta$, so the circle $AB\Gamma\Delta$ either to an area less than the circle $EZH\Theta$ or to <an area> greater. Let it be first to the lesser, the <area> Σ. And let the square $EZH\Theta$ be inscribed in the circle $EZH\Theta$; the inscribed square is greater than half the circle $EZH\Theta$, because if tangents to the circle are drawn through the points E, Z, H, Θ, the square $EZH\Theta$ is half the square circumscribed to the circle, and the circle is less than the circumscribed square. So the inscribed square $EZH\Theta$ is greater

than half the circle *EZHΘ*. Let *EZ*, *ZH*, *HΘ*, *ΘE* be divided into two in the points *K*, *Λ*, *M*, *N*, and conjoin *EK*, *KZ*, *ZΛ*, *ΛH*, *HM*, *MΘ*, *ΘN*, *NE*; therefore each of the triangles *EKZ*, *ZΛH*, *HMΘ*, *ΘNE* is greater than the half of its segment of circle, because if tangents to the circle are drawn through the points *K*, *Λ*, *M*, N, and the parallelograms on the lines *EZ*, *ZH*, *HΘ*, *ΘE* are completed, each of the triangles *EKZ*, *ZΛH*, *HMΘ*, *ΘNE* will be half of its parallelogram, but its section <of the circle> is less than the parallelogram. So each of the triangles *EKZ*, *ZΛH*, *HMΘ*, *ΘNE* is greater than the half of its section of circle. Dividing then the remaining arcs in two and conjoining the lines and doing this continuously some parts of circle are left over which will be less than the excess by which the circle *EZHΘ* exceeds the area *Σ*. For it is proved in the first theorem of the tenth book that two unequal magnitudes having been posited, if more than the half is subtracted from the greater and more than the half from the remainder and this is done continuously, a magnitude is left which is less than the posited lesser magnitude. Let it be left then, and let the sections of the circle *EZHΘ* on *EK*, *KZ*, *ZΛ*, *ΛH*, *HM*, *MΘ*, *ΘN*, *NE* be less than the excess by which the circle *EZHΘ* exceeds the area *Σ*. Therefore the leftover polygon *EKZΛHMΘN* is greater than the area *Σ*. Let the polygon *AΞBOΓΠΔP* similar to the polygon *EKZΛHMΘN* be inscribed in the circle *ABΓΔ*. Therefore as the square on *BΔ* is to the square on *ZΘ*, so the polygon *AΞBOΓΠΔP* is to the polygon *EKZΛHMΘN*. But also as the square on *BΔ* is to that on *ZΘ*, so the circle *ABΓΔ* is to the area *Σ*. And therefore as the circle *ABΓΔ* is to the area Σ, so the polygon *AΞBOΓΠΔP* is to the polygon *EKZΛHMΘN*. Conversely therefore as the circle *ABΓΔ* is to the polygon inscribed in it, so the area *Σ* is to the polygon *EKZΛHMΘN*. But the circle *ABΓΔ* is greater than the polygon in it. Therefore the area *Σ* as well <is greater> than the polygon *EKZΛHMΘN*. But it is less; which is impossible. Therefore it is not <the case that> as the square on *BΔ* is to the square on *ZΘ*, so the circle *ABΓΔ* to some area less than the circle *EZHΘ*. Similarly we prove that it is not <the case that> as the square on *ZΘ* to the square on *BΔ*, so the circle *EZHΘ* to some area less than the *ABΓΔ*. I say then that it is not <the case that> as the square on *BΔ* is to the <square> on *ZΘ*, so the circle *ABΓΔ* is to some area greater than the circle *EZHΘ*. If indeed it is possible, let it be to a greater area *Σ*. And again therefore as the square on *ZΘ* is to that on *BΔ*, so the area *Σ* to the circle *ABΓΔ*, but as the area *Σ* to the circle *ABΓΔ*, so the circle *EZHΘ* to some area less

> than the circle $AB\Gamma\Delta$. And therefore as the <square> on $Z\Theta$ to that on $B\Delta$, so the circle $EZH\Theta$ to some area less than the circle $AB\Gamma\Delta$; which it has been proved to be impossible. Therefore it is not <the case that> as the square on $B\Delta$ is to that on $Z\Theta$, so the circle $AB\Gamma\Delta$ to some area greater than the circle $EZH\Theta$. It has been proved that it is not to a lesser <area>. Therefore it is as the square on $B\Delta$ to that on $Z\Theta$, so the circle $AB\Gamma\Delta$ to the circle $EZH\Theta$. Circles are to each other as the squares on their diameters; which it was necessary to prove.

In the fragment attributed to Hippocrates of Chios and reported on p. 57 we read:

> He made a starting-point, and posited as first thing useful for these, that similar segments of circles have the same ratio to each other as the squares on their bases. And this he proved by showing that the squares on the diameters have the same ratio as the circles.

Let us ignore for a moment the problems related to Simplicius' testimony, and let us say that Hippocrates in the early fourth century BC already knew the result on the ratio of circles. What he may not have known, however, was how to prove his result rigorously, in a way that was immune from the criticisms of self-contradiction or circularity or inappropriateness variously levelled by Aristotle. Eudoxus then formulated a lemma which allowed him to talk rigorously about what happens when processes of subtraction, addition or multiplication are carried out indefinitely. The combination of Hippocrates' result and Eudoxus' lemma is seen in Euclid's *Elements*, whether he reproduced it from an extant source or put the two things together himself. The old and the less old are yet again combined to powerful effect.

Archimedes

Archimedes was one of the many victims of the many wars fought in the Hellenistic period. He was killed in 212 BC when his city Syracuse was taken by the Romans after two years of siege, as part of a larger conflict between Rome and Carthage. Syracuse, as we mentioned, had excellent fortifications in which Archimedes may have been involved. He also played a major role in the devising of military machines. Polybius says that his engines were capable of throwing projectiles both at long and close range, and describes an iron hand attached to a chain which would clutch a ship, lift it up in the air and then drop it into the sea with disastrous consequences.[64] There are also reports of machines Archimedes built in peace time – a weight-lifting instrument with which he launched an enormous ship, and a sphere that imitated the heavens.[65]

Apart from his adventurous life and wondrous inventions, Archimedes is famous as the author of several mathematical works: *Sphere and Cylinder*, in two parts (=*SC*); *Quadrature of the Parabola* (*QP*); *Measurement of the Circle* (*MC*); *Spirals* (*SP*); *Equilibrium of Planes* in two parts (*EP*); *Sand-Reckoner* (*AR*); *Conoids and Spheroids* (*CS*); *Floating Bodies* in two parts (*FB*); *Method to Eratosthenes*; *The Cattle Problem*. Various shorter works of more uncertain attribution are also extant (*Stomachion*; *Lemmas; On Circles Tangent to Each Other*), as are the titles of several lost works, one or more of them on mechanics. Nearly each work is introduced by a letter, where Archimedes explains and summarizes the contents, and occasionally provides a history of the issues at hand. His addressees are Dositheus, Eratosthenes and, for the *Sand-Reckoner*, Gelon, who succeeded Hiero as king of Syracuse. Archimedes mentions that a former correspondent, Conon, has recently died, and that he had sent arithmetical studies to a Zeuxippus. Like Eratosthenes, both Dositheus and Conon were probably based at Alexandria, and many historians think that Archimedes himself spent some time there.[66]

Contents

Archimedes' name is linked to some discoveries in plane and solid geometry, particularly the determination of the area and volume of curvilinear figures. Most famously, he established the equivalence between a circle and a rectilinear figure, thus finally squaring the circle (see Diagram 3.15):

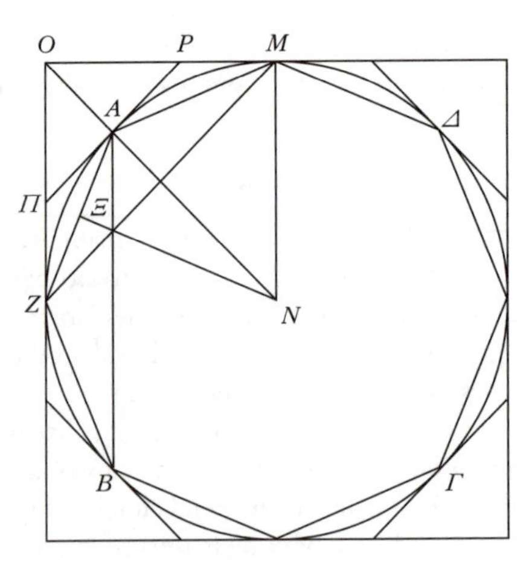

Diagram 3.15

> Any circle is equal to a right-angled triangle, whose radius is equal to one of the <sides> around the right angle, while the perimeter <is equal> to the base. Let the circle *ABΓΔ* have to the triangle *E* <the relation> as assumed; I say that it is equal. If possible, let the circle be greater, and let the square *AΓ* be inscribed, and let the arcs be divided in half, and let the segments already be less than the excess by which the circle exceeds the triangle; therefore the rectilinear figure is even greater than the triangle. Let a centre *N* be taken and a perpendicular *NΞ*; therefore *NΞ* <is> less than the side of the triangle. Then the perimeter of the rectilinear figure is also less than the remainder, since <it is also less> than the perimeter of the circle. Therefore the rectilinear figure is less than the triangle *E*; which is absurd. Let then the circle, if possible, <be> less than the triangle *E*, and let the square be circumscribed, and the arcs be divided in half, and tangents throught the points drawn. Therefore the <angle> *OAP* is right. Therefore *OP* is greater than *MP*; for *PM* is equal to *PA*; and the triangle *POΠ* therefore is greater than half of the figure *OZAM*. Let the segments similar to *ΠZA* then have been left less than the excess by which *E* exceeds the circle *ABΓΔ*; therefore the circumscribed rectilinear figure <is> even less than *E*; which is absurd; it is in fact greater, since *NA* is equal to the perpendicular of the triangle, and the perimeter is greater than the base of the triangle. Therefore the circle is equal to the triangle *E*.[67]

While the theorem expresses the area of a circle in terms of such-and-such a triangle, another proposition from the same book provided a numerical expression of the same problem, by stating that 'The circumference of any circle is three times the diameter plus an amount which is less than the seventh part of the diameter, and more than ten seventy-one parts <of the diameter>' (*MC* 3). Archimedes also determined the area and volume of a sphere, and proved several results about spherical segments.[68] His interest in the relationship between rectilinear and curvilinear figures may have been behind his enquiries into geometrical objects which are not in Euclid's *Elements*. For instance, the conics, produced by cutting a cone with a plane (especially the parabola, which Archimedes calls 'section of a right-angled cone') or conoids and spheroids, obtained by rotating conics, or the spiral, which originates from a segment rotating around one of its extremities with a point moving along it at the same time.[69] As with more traditional geometrical objects, his efforts were directed at determining the area or volume of these figures.

Archimedes also tackled topics which belong to what we would call mathematical physics: centres of gravity of plane figures, the problem of determining conditions of equilibrium and the problem of determining whether a body will sink or float when immersed in a liquid. He stated and proved the so-called law of the lever, according to which magnitudes, whether commensurable (*EP* 1.6) or incommensurable (*EP* 1.7), are in equilibrium when their distance from the fulcrum of a balance from which they are suspended is inversely proportional to their weight. Unfortunately, none of these works comes with a prefatory letter, and none of them provides a definition of centre of gravity. Archimedes does refer, however, to other mechanical books, which may have introduced basic notions.

Procedure

Like Euclid, Archimedes employed both direct and indirect proofs. He often used the method of exhaustion, for instance to prove the theorem about the area of the circle reported above. Another procedure worth noting is exemplified as follows (see Diagram 3.16):

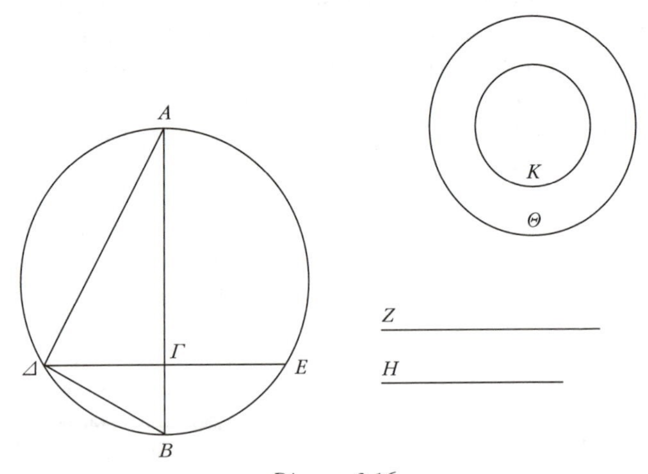

Diagram 3.16

> To cut the given sphere with a plane, so that the surfaces of the segments have to each other the same ratio as a given one. Let it be done, and let the greatest circle of the sphere be *AΔBE*, a diameter of that be *AB*, and let a perpendicular plane have been dropped towards *AB*, and let the plane produce a section *ΔE* in the circle *AΔBE*, and let *AΔ*, *BΔ* be conjoined. Since now the ratio of the surface of the segment *ΔAE* to the surface of the

> segment ΔBE is <given>, but the surface of the segment ΔAE is equal to the circle in which the radius is equal to ΔB, as the said circles to each other, so the <square> on $A\Delta$ to that on ΔB, that is $A\Gamma$ to ΓB, therefore the ratio of $A\Gamma$ to ΓB has been given; so that the point Γ is given. And ΔE is perpendicular to AB; therefore the plane across ΔE is also given in position. Let it be put together like this; let there be a sphere $AB\Delta E$, in which <there is> a greatest circle, and a diameter AB, and the given ratio that of Z to H, and let AB be cut in Γ, so that it is as $A\Gamma$ to $B\Gamma$, so Z to H, and let the sphere be cut across the plane Γ perpendicularly to the line AB, and let ΔE be the common section, and let $A\Delta$, ΔB be conjoined, and let two circles Θ, K, be posited, and Θ having the radius equal to $A\Delta$, while K has the radius equal to ΔB; therefore the circle Θ is equal to the surface of the segment ΔAE, while the <circle> K <is equal> to the segment ΔBE; this indeed has been already proved in the first book. And since the <angle> $A\Delta B$ is right and $\Gamma\Delta$ is perpendicular, it is as $A\Gamma$ to ΓB, that is Z to H, <so> the <square> on $A\Delta$ to that on ΔB, that is the <square> on the radius of the circle Θ to that of the radius of the circle K, that is the circle Θ to the circle K, that is the surface of the segment ΔAE to the surface of the segment ΔBE of the sphere (*SC* 2.3).

In the solution to this problem, Archimedes first assumes that the problem is already solved ('Let it be done'), then works from this assumption backwards, so to speak, until every element of the problem is accounted for and given. This part is called analysis, a 'breaking down' the construction into its elements. The second part of the process, the synthesis or 'putting together', consists of a normal proof, where every step is basically repeated, in the confidence that it can be justified. Analysis-and-synthesis procedures, which had been described by Aristotle and were not confined to mathematics, allow a glimpse into the way ancient geometers attained their results before organizing them in a demonstrative structure – what is called their heuristics (process of discovery). Most of the proofs one finds are in the form of synthesis; Archimedes is pretty unusual in appending analyses as well.

Further unparalleled glimpses into the heuristics of Greek mathematics are afforded by the *Method* – in the introductory letter, Archimedes explains to Eratosthenes that he wants to make public the way in which many of his results had occurred to him, so that people can benefit from it and discover even more theorems. The procedure combines geometry and mechanics, and can be applied when one needs to establish the equivalence between a certain object (say, a sphere) and another object whose characteristics are better known (say, a cone). The two objects are imagined to be at the two ends of a balance – if the conditions for their *equilibrium* can be established, then so can the conditions for their *equivalence*. This is achieved using on the one hand the results about centres of gravity and equilibrium that Archimedes had formulated in *EP*, and, on the other hand, by considering each figure to be made up by an infinite number of lines, for each of which the equilibrium conditions are then proved to apply. In other words, mechanics can be applied to geometry because a connection can be made between geometrical being, extension, which determines area or volume, and mechanical or physical being, which has to do with equilibrium and centre of gravity.

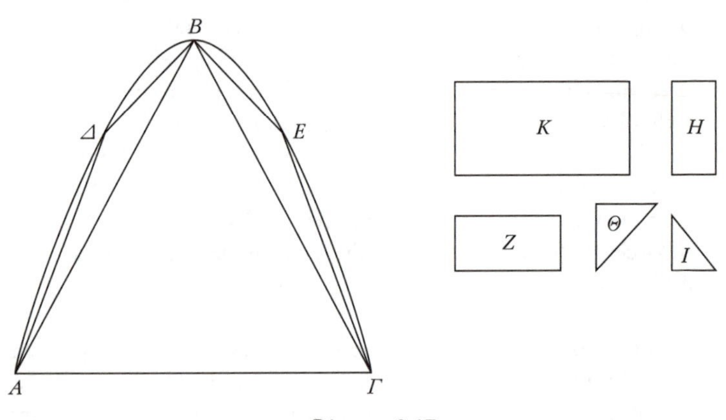

Diagram 3.17

An example of a result proved 'normally': [*QP* 24 (see Diagram 3.17)] Any segment surrounded by a line and a section of right-angled cone [a parabola] is four thirds of the triangle having the same base and equal height. Let *AΔBEΓ* be a segment surrounded by a line and a section of right-angled cone, let *ABΓ* be a triangle having the same base and equal height, let the area *K* be four thirds of the triangle *ABΓ*. It is to be proved that <the area *K*> is equal to the segment *AΔBEΓ*. If in fact it is not equal, it is either greater or lesser. First, if possible, let the segment *AΔBEΓ* be greater than the area *K*. I have inscribed the triangles *AΔB*, *BEΓ*, as it is said, I have inscribed also in the remaining segment all the triangles having the same base as the segments and the same height, and always in the next segments let two triangles be inscribed having the same base as the segments and the same height; the segments which are left over will be less than the excess by which the segment

AΔBEΓ exceeds the area *K*. So that the inscribed polygon will be greater than *K*, which is impossible. Since then there are posited areas one next to the other in a quadruplicate ratio, first the triangle *ABΓ* is four times the triangles *AΔB*, *BEΓ*, since these same <triangles> are four times those inscribed in the said segments and always like that, it is clear that all the areas together are less than a third of the greatest, while K is four thirds of the greatest area. Therefore the segment *AΔBEΓ* is not greater than the area *K*. Let it then, if possible, be lesser. Let the triangle *ABΓ* be posited equal to *Z*, *H* <being> a fourth of *Z*, and *Θ* similar to *H*, and let it be set always one next to the other, until the last comes out less than the excess by which the area *K* exceeds the segment, and let *I* be less; then the areas *Z*, *H*, *Θ*, *I* and the third of *I* are four thirds of *Z*. *K* is also the four thirds of *Z*; therefore *K* is equal to *Z*, *H*, *Θ*, *I* and a third part of *I*. Since now the area *K* exceeds *Z*, *H*, *Θ*, *I* by less than *I*, but the segment by more than *I*, it is clear that even more the areas *Z*, *H*, *Θ*, *I* are greater than the segment; which is impossible; it has been proved in fact that if the area taken next to it in quadruple ratio is as large, the greatest is equal to the triangle inscribed in the segment, the areas all together will be lesser than the segment. Therefore the segment *AΔBEΓ* is not lesser than the area *K*. It has been proved that it is not greater either; therefore it is equal to *K*. The area *K* is then four thirds of the triangle *ABΓ*; and therefore the segment *AΔBEΓ* is four thirds of the triangle *ABΓ*.

The same result established through a combination of geometry, mechanics and infinitesimals: [*Method* 1 (see Diagram 3.18)] Let the segment *ABΓ* be surrounded by the line *AΓ* and by the section of right-angled cone *ABΓ*, and let *AΓ* be divided in half at *Δ*, and let *ΔBE* be drawn <parallel to> the diameter, and let *AB*, *BΓ* be conjoined. I say that the segment *ABΓ* is four-thirds of the triangle *ABΓ*. Let from the points *A*, *Γ* be drawn *AZ* <parallel> to *ΔBE*, and *ΓZ* tangent to the section, and let <*ΓB*> be prolonged <to *K*, and let *KΘ* be posited equal to *ΓK*>. Let the balance *ΓΘ* be

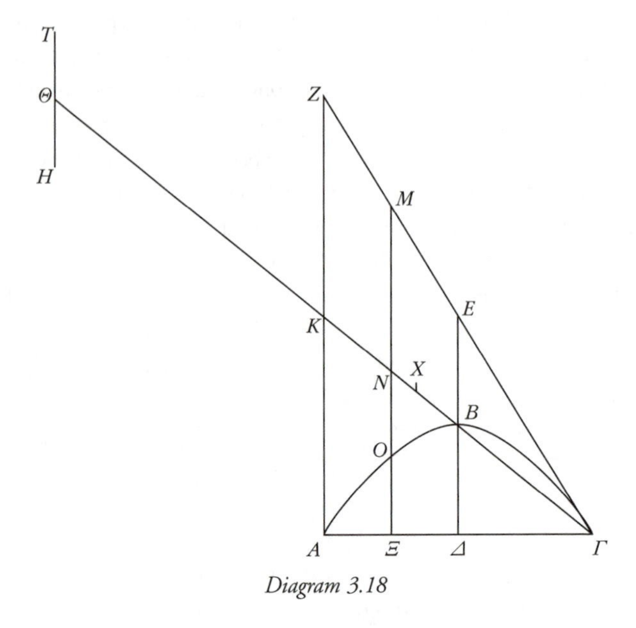

Diagram 3.18

imagined and its centre K and a parallel whatever to $E\Delta$, $M\Xi$. Since now ΓBA is a parabola, and ΓZ is a tangent and $\Gamma\Delta$ is an ordinate, EB is equal to $B\Delta$; this in fact has been proved in the elements; because of this, and because ZA, $M\Xi$ are parallel to $E\Delta$; MN is also equal to $N\Xi$, and ZK to KA. And since it is as ΓA to $A\Xi$, so $M\Xi$ to ΞO, on the other hand as ΓA to $A\Xi$, so ΓK to KN, and ΓK is equal to $K\Theta$, therefore <it is> as ΘK to KN, so $M\Xi$ to ΞO. And since the point N is centre of gravity of the line $M\Xi$, because MN is equal to $N\Xi$, if therefore we set TH equal to ΞO and Θ as its centre of gravity, so that $T\Theta$ is equal to ΘH, $T\Theta H$ is in equilibrium with $M\Xi$, this latter remaining <in place>, because ΘN is divided inversely with respect to the weights TH, $M\Xi$, and as ΘK to KN, so $M\Xi$ to HT; so that the centre of gravity of both is K. Similarly also if one draws all the parallels to $E\Delta$ in the triangle $ZA\Gamma$, they are in equilibrium, remaining <in place>, with the <lines> that they cut off from the section <of right-angled cone>, transported to the <point Θ, so that> the centre of gravity <of both is> K. And since the triangle GZA is made up from the <lines> in the triangle ΓZA, the segment $AB\Gamma$ is made up of the <lines> taken in the section <of right-angled cone> similarly to ΞO, therefore the triangle $ZA\Gamma$

will be in equilibrium remaining <in place> with the segment of section <of right-angled cone> having been posited around the centre of gravity Θ in the point K, so that the centre of gravity of both is K. Let ΓK be divided in X, so that ΓK will be three times KX; therefore the point C will be the centre of gravity of the triangle $AZ\Gamma$; it has been proved in fact in the *Equilibria*. Since then the triangle $ZA\Gamma$ remaining <in place> is in equilibrium with the segment $BA\Gamma$ staying in K around the centre of gravity Θ, and the centre of gravity of the triangle $ZA\Gamma$ is X, therefore it is as the triangle AZG to the segment ABG posited around the centre Θ, so ΘK to XK. But ΘK is three times KX; therefore the triangle $AZ\Gamma$ will also be three times the segment $AB\Gamma$. The triangle $ZA\Gamma$ is four times the triangle $AB\Gamma$ because ZK is equal to KA and AD to the $\Delta\Gamma$; therefore the segment $AB\Gamma$ is four thirds of the triangle $AB\Gamma$.

Archimedes made it quite clear that his mechanical method did not amount to a rigorous proof, either because he was wary of using mechanics to yield geometrical results, or because he had a problem with infinitesimals, or because of both. He seemed aware of a distinction between the context of discovery and that of justification, and of the need to package a discovery in a way that was acceptable. Discovering a result could turn into nothing if the supporting proof did not work, or could not be formulated. Archimedes also intimated that a result may be felt to be correct but it could still take a long time for a valid proof to be found.

'Acceptable', 'valid' are all terms that imply the existence of an accepting or rejecting public on the one hand, and of an agreed-upon set of rules on the other hand. Both seem to have been present at Archimedes' time – his addressees, both the ones he names and the many unnamed ones, constituted a sort of community, where people knew each other, exchanged work, shared interests, a language and, at least to some extent, some criteria of what constituted 'good' and 'bad' mathematics.

Apollonius

Apollonius, according to Eutocius, was born at Perga in Pamphylia during the reign of Ptolemy III (246–21 BC). He lived between Alexandria and Pergamum, and mentioned a stay at Ephesus. The only surviving work by him is the *Conics*, originally in eight books, of which four are extant in Greek and three in Arabic only. There were other works, of which the titles and occasional excerpts are preserved: on rules for multiplication and/or a system to express large numbers, on cutting off a ratio, an area and a determinate section, on tangencies, plane loci, *neuseis*, on a comparison of the dodecahedron with the icosahedron. Apollonius is also credited with studies on the spiral, on astronomy and on optics. Like Archimedes, he prefaced his work with letters: the first two parts of the *Conics* are addressed to Eudemus of Pergamum, the last four to an Attalus, who might be king Attalus I of Pergamum. Apollonius also mentioned the geometers Naucratis, Philonides, Conon of Samos and Nicoteles of Cyrene, as well as Euclid.[70]

Contents

Apollonius himself summarized the contents of the *Conics* in his introduction to the first book:

> At the time when I was with you [Eudemus] in Pergamum, I saw that you were eager to get a copy of the Conics which I had worked out. So I send you the first book, which I have corrected, and the remaining ones will be sent off when I am satisfied with them. For I think that you do not forget that you heard from me how I undertook the composition of this matter at the request of the geometer Naucrates, at the time when he was relaxing with us after he arrived in Alexandria; and how, having elaborated it in eight books, I immediately gave copies to him in a hurry, without revising them, since he was on the point of sailing away: instead I put down everything as it occurred to me, with the intention of coming back to it in the end. Hence, having now got the opportunity, I am publishing each part as it gets its revision. Now since it so happened that some others of those who came into contact with me got copies of the first and second books before they were corrected, do not be surprised if you come across these in a different version. Of the eight books, the first four constitute an elementary introduction. The first contains the methods of generating the three sections and the opposite branches [of the hyperbola], and their basic properties, developed more fully and more generally than in the works of the other [writers on conics]; the second contains the properties of the diameters and axes of the sections, the asymptotes, and other matters which have typical and essential applications in *diorismoi* [...] The third contains many surprising theorems useful for the synthesis of solid loci and for diorisms; of these the greater part and the most beautiful are new. It was the discovery of these that made me aware that Euclid has not worked out the whole of the locus for three and four lines, but only a

> fortuitous part of it, and that not very successfully; for it was not possible to complete the synthesis without additional discoveries. The fourth deals with how many times the conic sections may intersect each other and the circumference of a circle, and other matters in addition; neither of these two problems have been written about by our predecessors, namely in how many points a conic section or circumference of a circle <can intersect [opposite branches] and in how many points opposite branches> can intersect <opposite branches>. The remaining [books] are more particular: one deals somewhat fully with *minima* and *maxima*, another with equal and similar conic sections, another with theorems concerning *diorismoi*, another with determinate conic problems.[71]

Conics were not a new topic, but it is evident that, as well as distancing himself from some of his predecessors by declaring his account fuller, clearer and more general than theirs, Apollonius was keen to add new things and redefine concepts. For instance, he re-christened as parabola, hyperbola and ellipsis what were already known as sections of a right-angled, obtuse-angled and acute-angled cone, respectively. He showed how all three curves could be produced by cutting the same cone, rather than three different cones, and redefined them on the basis of the relation between some of their elements. For instance (see Diagram 3.19),

> If a cone is cut by a plane through the axis, and it is cut also by another plane cutting the base of the cone along a line perpendicular to the base of the triangle through the axis, and the diameter of the section prolonged meets one side of the triangle through the axis outside the vertex of the cone, any <line> which is drawn parallel from the section to the common section of the cutting plane and the base of the cone until the diameter of the section, is equal in square to some area applied to some line, to which <line> the <line> which is the prolonging of the diameter of the section, and which subtends the angle outside the triangle, has a ratio which <is> the square on the <line> drawn from the vertex of the cone parallel to the diameter of the section until the base of the triangle to the <rectangle> formed by the bases of the section, as produced by a <line> drawn, having as latitude the <line> cut off by this from the diameter to the vertex of the section, exceeding by a shape similar and similarly posited the <rectangle> formed by the <line> subtending the external angle of the triangle and by the parameter; let then this section be called hyperbola [...] (1.12).[73]

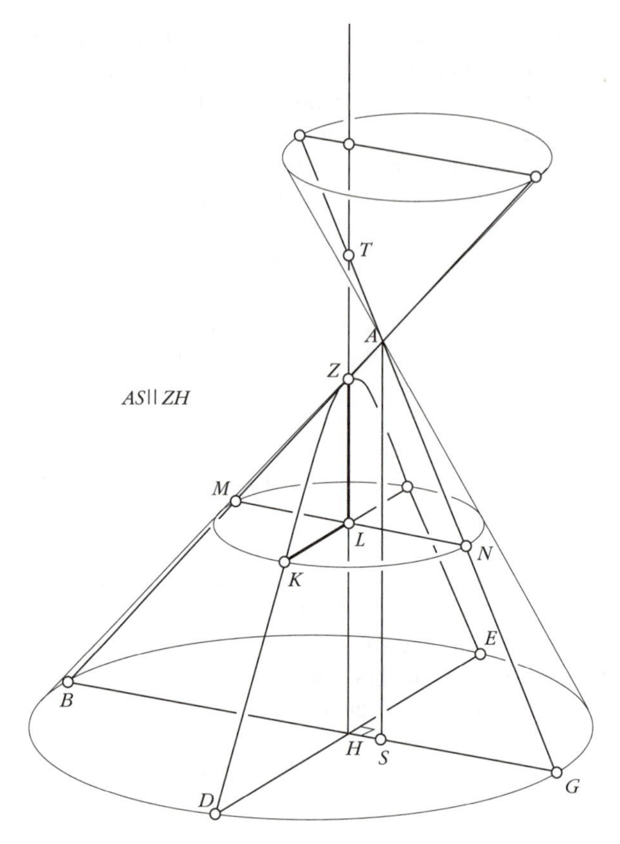

Diagram 3.19

Also, for each type of conic, Apollonius extended the properties defined for one particular diameter, i.e. the axis, to all diameters, and the properties defined for ordinates perpendicular to the diameter to all ordinates (1.50–51). This allowed him to solve some problems of construction – for instance, how to find a certain conic given certain data at the outset. Above all, he can be said to have streamlined the treatment of conics; by defining them on the basis of a common origin and by generalizing their properties, he constituted them as fully-determined, rigorously-described, geometrical objects.

Procedure

Apollonius used both direct and indirect methods; a proof of the conditions of equality for two parabolas remarkably echoes Euclid's superposition or 'fitting' procedure (see Diagram 3.20):

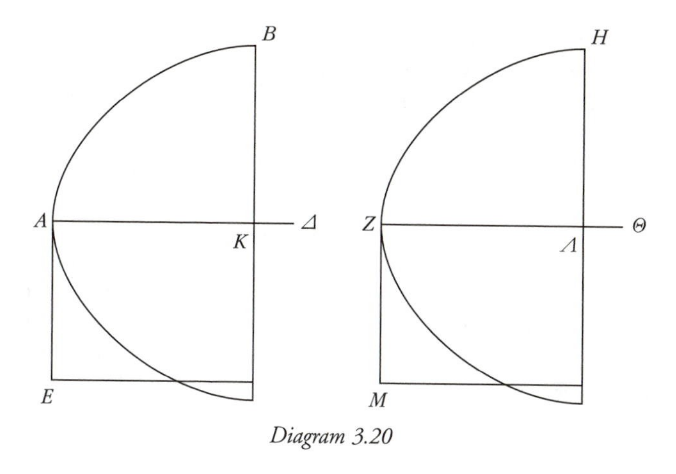

Diagram 3.20

Parabolas in which the parameters of the perpendiculars to the axes are equal are themselves equal, and if parabolas are equal, their parameters are equal. Let there be two parabolas, with axes $A\Delta$, $Z\Theta$, and equal parameters, AE, ZM. Then I say that these sections are equal. When we apply axis $A\Delta$ to axis $Z\Theta$, then the section will coincide with the section so as to fit on it. For if it does not fit on it, let there be a part of section AB which does not fit on section ZH. We mark point B on the part of it which does not coincide with ZH, and draw from it [to the axis] perpendicular BK, and complete rectangle KE. We make $Z\Lambda$ equal to AK, and draw from point Λ a perpendicular to the axis [meeting the section in H], ΛH, and complete rectangle ΛM. Then lines KA, AE are equal to lines ΛZ, ZM, each to its correspondent; therefore the rectangle from KA, AE is equal to the rectangle from ΛM, ZM. And the square on KB is equal to the rectangle EK, as is proven in proposition 11 of the first book. And similarly too the square on ΛH is equal to the rectangle ΛM; therefore KB is equal to ΛH. So when the axis [of one section] is applied to the axis [of the other], line AK will coincide with line $Z\Lambda$, and line KB will coincide with line ΛH, and point B will coincide with point H. But it was supposed not to fall on section ZH: that is absurd. So it is impossible for the section not to be equal to the section. Furthermore, we make the section equal to the section, and make AK equal to line $Z\Lambda$, and draw the perpendiculars [to the axis] from points K, Λ, and complete rectangles EK, $M\Lambda$: then section AB will coincide with section ZH, and therefore axis AK will coincide with axis $Z\Lambda$. For if it does not coincide with it, parabola

> ZH has two axes, which is impossible. So let it coincide with it. Then point K will coincide with point Λ, because AK is equal to $Z\Lambda$. And point B will coincide with point H. Therefore BK is equal to ΛH; therefore the rectangle EK is equal to the rectangle ΛM. And AK is equal to $Z\Lambda$, therefore AE is equal to ZM.[74]

In fact, Apollonius explicitly mentions Euclid in the introduction to the first book of the *Conics*, as quoted before, and claims to have improved on him. There are also interesting parallels between lost works by Apollonius, and subjects which Archimedes deals with: both studied the spiral, both produced notation systems to express very large numbers, both calculated the ratio between diameter and circumference of a circle. Eutocius even reports that Apollonius had been accused of plagiarizing Archimedes in his treatment of conics.[75] Be that as it may, Apollonius' work clearly presupposes an accumulation of mathematical results: his style is deductive in the mould of Euclid and Archimedes, but also very conscious of the need to systematize and order his material, and of the necessity carefully to distinguish between similar results proved for cognate, but non-identical objects. See for instance 1.26, where the case of a parabola and a hyperbola are considered separately (see Diagram 3.21):

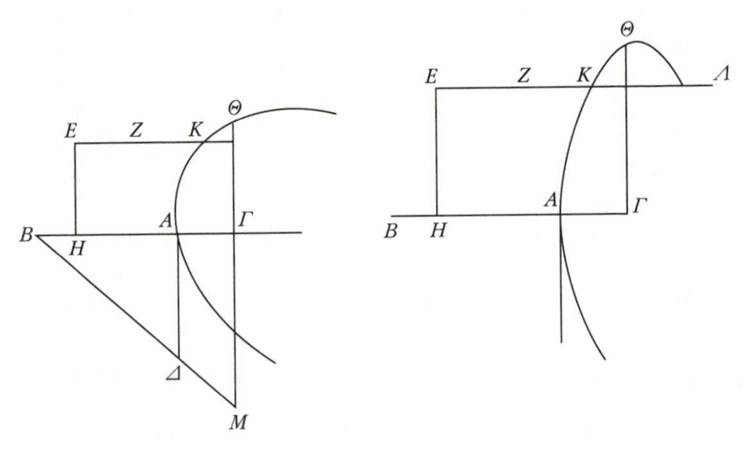

Diagram 3.21

> If in a parabola or a hyperbola a line is drawn to the diameter of the section, it meets the section at only one point. Let there be a parabola first, with diameter $AB\Gamma$, and parameter $A\Delta$, and let EZ be drawn parallel to AB. I say that EZ prolonged will meet the section. Let a point E be taken on EZ, and from E let EH be drawn parallel ordinately, and let the <rectangle> formed by $\Delta A\Gamma$ be

> greater than the <square> on HE, and from Γ let $\Gamma\Theta$ be erected ordinately; therefore the <square> on $\Theta\Gamma$ is equal to the <rectangle> formed by $\Delta A\Gamma$. And the <rectangle> formed by $\Delta A\Gamma$ is greater than the <square> on EH; therefore the <square> on $\Theta\Gamma$ will be greater than that on EH as well; and therefore $\Theta\Gamma$ will be greater than EH as well. And they are parallel; therefore EZ prolonged cuts $\Theta\Gamma$; so that it will also meet the section. Let it meet <the section> in K. I say that it will meet it in only one point K. If in fact possible, let it meet <the section> in Λ as well. Since then a line cuts a parabola in two points, prolonged it will meet the diameter of the section, which is absurd; for it is assumed to be parallel. Therefore EZ prolonged meets the section in only one point. Let the section be a hyperbola, with transverse side AB and parameter $A\Delta$, and let ΔB be conjoined and prolonged. Having constructed these things let ΓM be drawn from Γ parallel to $A\Delta$. Since the <rectangle> formed by $M\Gamma A$ is greater than that formed by $\Delta A\Gamma$, and the <square> on $\Gamma\Theta$ is equal to the <rectangle> formed by $M\Gamma A$, and the <rectangle> formed by $\Delta A\Gamma$ is greater than the <square> on HE, therefore the <square> on $\Gamma\Theta$ will be greater than that on EH as well. So that $\Gamma\Theta$ is greater than EH as well, and the same things will happen as before.[76]

The proof combines direct and indirect methods, is divided into sub-cases, it takes advantage of Apollonius' new definitions, and it builds on previous results. Sub-cases consider variations in the characteristics of the geometrical figures under examination, and are already found in the *Elements*, but they are particularly frequent in the *Conics*, where sometimes we have more than one diagram for the same proposition (e.g. 4.56, 4.57).

Apollonius provides ample evidence that the objects of geometrical inquiry were becoming increasingly complex: they no longer had simple, univocal properties and a limited field of variation ranging e.g. from acute to right-angled to obtuse. Conics allowed much more room for manoeuvre; the task of providing some grasp of their manifold configurations could be articulated as a *diorismos*. The proposition below is an example: Apollonius set the question and then considered how the solution varied with the variation of elements in Diagram 3.22.

> If there is a hyperbola, and the transverse diameter of the figure constructed on its axis is not less than its parameter, then the parameter of the figure constructed on the axis is less than the parameter of [any of] the figures constructed on the other diameters of the section, and the parameter of [any of] the figures constructed on

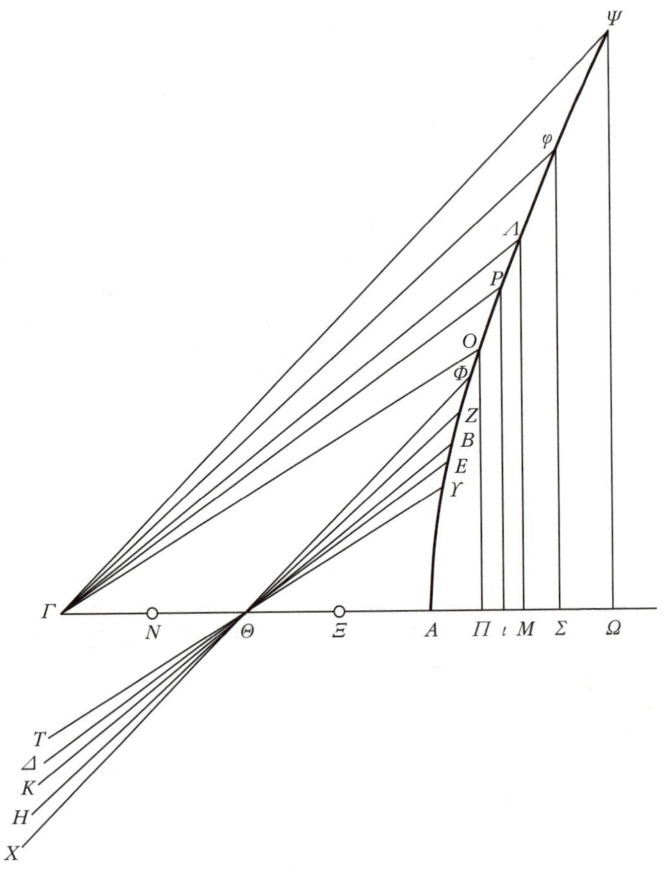

Diagram 3.22

diameters closer to the axis is less than the parameter of the figures constructed on [diameters] farther from the axis. Let there be a hyperbola with axis *ΑΓ* and center *Θ*, and with two of its diameters *ΚΒ*, *ΥΤ*. Then I say that the parameter of the figure of the section constructed on *ΑΓ* is less than the parameter of the figure of the section constructed on *ΚΒ*, and that the parameter of the figure of the section constructed on *ΚΒ* is less than the parameter of the figure of the section constructed on *ΥΤ*. First, we make axis *ΑΓ* equal to the parameter of the figure constructed on it. [...] Furthermore, we make axis *ΑΓ* greater than the parameter of the figure of the section constructed on it [...] Furthermore, we make line *ΑΓ* less than the parameter of the figure constructed on it, but not less than half the parameter of the figure constructed on it:

> then I say that, again, the parameter of the figure constructed on *AΓ* is less than the parameter of the figure constructed on *KB*, and that the parameter of the figure constructed on *KB* is less than the parameter of the figure constructed on *TΥ*. [...] Furthermore, we make *AΓ* less than half the parameter of the figure of the section constructed on it: then I say that there are two diameters, [one] on either side of this axis, such that the parameter of the figure constructed on each of them is twice that [diameter]; and that [parameter] is less than the parameter of the figure constructed on any other of the diameters on that side [of the axis]; and the parameter of figures constructed on the diameters closer to those two diameters is less than the parameter of a figure constructed on a [diameter] farther [from them] [...].[77]

In sum, Apollonius' use of subcases, his interest in generalizations and in determination of the conditions of validity or solubility of a proposition, his (explicit) references to Euclid and (implicit?) to Archimedes, as well as the complexity itself of his topics, suggest the presence of cumulative mathematical knowledge, without which some operations (organizing, redefining, being 'clearer', adding new things) would lack significance. His works hint that a mathematical tradition was in the making, and that written culture (see the awareness he shows of editing, correcting, of having different copies of his work circulating) played a giant role in the formation of that tradition.

Notes

1 Archimedes, *Sand-Reckoner* (Mugler 134–5), my translation.
2 Thucydides, *The Peloponnesian War* 1.102.2; 2.18.1; 2.58.1; 2.75 ff.
3 On Philo and fortifications, see Winter (1971), esp. 118 f.; Garlan (1974); Lawrence (1979); McNicoll and Milner (1997). Some of the elements of Fort Euryalus are so sophisticated that more than one scholar is inclined to suggest Archimedes as the engineer behind them, see Lawrence (1946); Winter (1963). Cf. also Ober (1992).
4 Philo, *Poliorketika* 1.3–7; 1.39; 1.64; 1.87; 2.19 (quotation). The *Poliorketika* and the *Belopoiika* were part of a larger work, entitled *Mechanical Syntaxis*, see Ferrari (1984).
5 References in Garlan (1974), 207 ff.
6 Castagnoli (1956); Owens (1991). We have references to surveying and town planning in Callimachus, *Aetia* 1.24 (a ten-foot pole used both as a goad for oxen and a measure for land), 2.fragment 43 (foundation rites for a city) and Plautus, *Poenulus* 46–9 (the person explaining the plot of the play compares himself to a surveyor determining the boundaries of a territory).
7 Polybius, *Histories* 6.27–32, Loeb translation with modification.
8 Pamment Salvatore (1996) – mid-second century BC evidence from various sites in Spain.
9 Polybius, *Histories* 6.42, Loeb translation with modifications.
10 Larissa in Salviat and Vatin (1974); Halieis and Chersonesos (fourth century BC) in Boyd and Jameson (1981). For a Hellenistic dating of the Chersonesos land-division, see Dufková and Pecírka (1970); Wasowicz (1972). For Roman cases, see e.g. Dilke (1971),

Gabba (1984), Salmon (1985), Moatti (1993). For redistributions, see e.g. Austin (1981), documents 180 (*c.* 275 BC) from Asia Minor, 235 (259/8 BC) and 240 (257 BC) from Egypt, 271 (240 BC) from Lycia.

11 *P. Tebt.* 87 (late second century BC, probably from Berenikis Thesmophorou in the Egyptian Fayum), 46–62, partial translation in Thompson/Crawford (1971), 13, with modifications. For land-surveys in Egypt see also Déléage (1933), Cuvigny (1985), who indicates that some of the papyri had diagrams or maps, 88. A survey like this was not the only type available: we have at least one example of what was called simply a 'determination of boundaries' (*periorismos*): see document 185 in Austin (1981), which is an inscription from Didyma in Asia Minor, dating 254/3 BC. The main difference of this latter to a *geometria* was that no size was indicated.

12 A similar procedure in *P. Mich.* 3245 (probably second century BC, perhaps from the Arsinoite nome), in Bruins *et al.* (1988).

13 Reproduced from Lyon (1927).

14 *P. Flind. Petr.* 2.11.2, cf. Lewis (1986), 42.

15 *P. Freib.* 7, in Hunt (1934), number 412, Loeb translation with modifications.

16 See e.g. *P. Cairo Zen.* 59132 (256 BC), *P. Cairo Zen.* 59188 (255 BC); *P. Lond.* 2027 (not dated, but also from Zenon's archive); *P. Col. Zen.* 2.87.1–22 (244 BC), in Westermann and Sayre Hasenoehrl (1934); *P. Ent.* 66 (218 BC), mentioned in Lewis (1986), 65 f.; *P. Tebt.* 24 (117 BC), in Grenfell (1902).

17 *P. Col. Zen.* 88 (243 BC), translation in Westermann and Sayre Hasenoehrl (1934). Cf. also e.g. *P. Tebt.* 24 (see note above); *P. Col. Zen.* 54 (250 BC); *P. Cairo Zen.* 59355 (243 BC), and see Lewis (1986), 44, 53.

18 *P. Cairo Zen.* 59330, my translation. In 59331, which is more fragmentary, Pemnas manifests to Zenon the suspicion that Herakleides 'has entered to his debit a larger quantity than he really owes'. Both documents are dated 30 June, 248 BC, and are in Edgar (1925), 52–4 (the quotation is from his introduction to 59331).

19 The documents in Uguzzoni and Ghinatti (1968).

20 Official instructions in *P. Rev. Laws* (259/8 BC). For Hellenistic examples of public accounts, cf. Burford (1969); Austin (1981), document 97, from Olbia on the Black Sea (late third–early second century BC), and document 194, from the temple of Apollo at Delos (third century BC); Rhodes and Lewis (1997).

21 *P. Cairo* 65445, in Guéraud and Jouguet (1938). Cf. also *P. Berol.* 21296, second century BC, which has a table of parts, in Ioannidou (1996), 202. Fowler (1988) and (1995) list other examples (multiplication and tables of parts, three from the second century BC, one from either the second or the first century BC); see also Morgan (1998).

22 *P. Cairo* J.E. 89127–30, 89137–43 translation and numbering in Parker (1972).

23 *P. Cairo* J.E. 89127–30, 89137–43, problem 6, translation and numbering as above. The reference is to problems about series of parts where one has to find the next element in the series.

24 The ostraka in Mau and Müller (1987); the papyri in Angeli and Dorandi (1987) and Dorandi (1994). See also Fowler (1999), 209 ff.

25 Polybius, *Histories* 9.14 ff.

26 Polybius, *Histories* 12.17–22.

27 According to Strabo, *Geography* 2.4.1–3, Polybius expressed similar criticisms about the mathematical data reported by some geographers.

28 Polybius, *Histories* 9.26a, Loeb translation with modifications.

29 Theocritus, *Idylls* 17.1–85, tr. A.S.F. Gow, Cambridge 1952, with modifications.

30 The main collection is Thesleff (1965), with introductory material in Thesleff (1961). Also important Burkert (1972).

31 See e.g. Napolitano Valditara (1988).

32 The Academics in Dörrie (1987), ch. 5-8; the Peripatetics in Wehrli (1959).
33 [Aristotle], *On Indivisible Lines* 969b, my translation. The date of the treatise is uncertain, but there is a reference to apotome at 968b which might provide a date *post quem* since the earliest extant definition of apotome is in Euclid's *Elements* book 10. Unfortunately, we do not know Euclid's date with accuracy.
34 As reported by Plutarch, *On Common Conceptions Against the Stoics* 1079d–f.
35 Apollonius, *Conics* preface to book 2; Hypsicles, *Book 14 of the Elements* preface. See Sedley (1976); Mueller (1982), Appendix; Angeli and Dorandi (1987); Mansfeld (1998), 36.
36 Proclus, *Commentary on the First Book of Euclid's Elements* 199, translation as above.
37 Diogenes Laertius, *Lives of the Philosophers* 9.85.
38 Diogenes Laertius, *ibid.* 9.90–91.
39 Diogenes Laertius, *ibid.* 7.132–133.
40 Diogenes Laertius, *ibid.* 7.135. Apollonius is believed to have lived in the second century BC.
41 Diogenes Laertius, *ibid.* 7.81.
42 Mansfeld (1998), 14 ff.
43 Aristarchus, *On the Sizes and Distances of the Sun and Moon* 11, my translation.
44 References to Euclid's *Elements* in the *Sphaerics* too numerous to list; the *Phenomena* at *Days and Nights* 2.10, 126.33; Meton at *ibid.* 2.18, 152.1; the theorems with subcases at *Sphaerics* 3.9, 3.10.
45 Aratus, *Phenomena* 1140 ff.
46 Hipparchus, *Commentary on Aratus and Eudoxus' Phenomena* 90.20–4.
47 Aristoxenus, *Elements of harmonics* 2.32–3, tr. A. Barker, Cambridge 1989.
48 [Euclid], *Section of a Canon* 148–9.
49 Diocles, *On Burning Mirrors* 4, tr. G.J. Toomer, Springer 1976, with modifications, reproduced by kind permission of Spinger-Verlag, Inc. The text is only extant in an Arabic translation.
50 Diocles, *On Burning Mirrors* 97–111, translation as above.
51 In Theon of Smyrna, *Account of Mathematics Useful to Reading Plato* 81.17 ff., my translation. Eratosthenes' definition of ratio is similar to that in Euclid's *Elements* book 5.
52 Sextus Empiricus, *Against the Geometers* 28.
53 Cleomedes, *On Circular Motion* 1.10; cf. Dicks (1971), 390.
54 Biton, *Construction of War Instruments* 67–8, translation Marsden (1971), with modifications.
55 Philo, *Construction of Catapults* 49.12–50.9, translation Marsden (1971), with modifications.
56 Philo, *ibid.* 51.15–52.17, translation as above.
57 See Heath (1926); Murdoch (1971); Caveing (1990).
58 The summary is adapted from Mueller (1981), viii–ix.
59 Euclid, *Elements* 1. Postulates 1–3, translations of Euclid are mine unless stated otherwise.
60 Euclid, *Elements* 5. definitions 4–6.
61 Multiplication, which here takes place between numbers, had already been assumed in the definitions of ratios between magnitudes at the beginning of book 5. This and the double definition of proportionality have led some scholars to conclude that for Euclid numbers are not magnitudes, see Mueller (1981), 121 ff., 144 ff.
62 Euclid, *Data* definitions 1 and 2.
63 Alternative versions concern addition or multiplication, rather than subtraction, of magnitudes.
64 Polybius, *Histories* 8.3 ff.; among other sources are Livy, *From the Founding of the City* 24.34, 25.31; Cicero, *On the Greatest Good and Bad* 5.19.50, *Tusculan Disputations* 1.25.63,

5.23.64, 5.32.64 ff., *Against Verres* 2.58.131, *On the Commonwealth* 1.21.14; Diodorus, *Historical Library* 26.18–19.

65 References in Dijksterhuis (1956).

66 Dositheus in *SC*, *CS*, *SP*, and *QP*; he is also mentioned (if it is the same person) in Diocles, *On Burning Mirrors* 6. Eratosthenes in *Method* and *Cattle Problem*. Conon (of Samos, if again it is the same person) is also mentioned in Apollonius, *Conics* 1.Preface and Diocles, *On Burning Mirrors* 3. Zeuxippus in *AR* Introduction.

67 Archimedes, *MC* 1. Translations of Archimedes are mine unless stated otherwise.

68 Archimedes, *SC* 1.33, 1.34 and e.g. 1.35, 1.37, 1.38, 2.2 (volume of a spherical segment).

69 The conics were probably known to Euclid, who is credited with a treatise on them, and perhaps already to fourth-century BC geometers like Menaechmus. Conoids and spheroids seem to have been Archimedes' own invention. As for the spiral, Pappus explains that it was *proposed* by Conon of Samos and *proved* by Archimedes (*Mathematical Collection* 234.1–4).

70 For biographical data, see Huxley (1963); Toomer (1970).

71 Apollonius, *Conics* 1.Preface, tr. G.J. Toomer (1990), xiv–xv, with modifications, reproduced by kind permission of Springer-Verlag, Inc.

72 Apollonius, *Conics* 1.11, 1.12 and 1.13, respectively.

73 My translation. The diagram here reproduced is from Toomer (1990) 667, by kind permission of Springer-Verlag Inc.

74 Apollonius, *Conics* 6.1, translation as above.

75 Eutocius, *Commentary on Apollonius' Conics* 1.5 ff.

76 My translation. The expression 'lines drawn ordinately' corresponds to what we call 'ordinates', i.e. a line bisected by the diameter of the conic, while an 'abscissa' is the segment of diameter cut off by an ordinate.

77 Apollonius, *Conics* 7.33–5, translation as above. The diagram is also reproduced from Toomer (1990), 822, by kind permission of Springer-Verlag Inc.

4

HELLENISTIC MATHEMATICS: THE QUESTIONS

The previous chapter depicts mathematics as a collective enterprise: Ptolemaic officers were told to check the accounts in groups; the surveyors at Heraclea measured the land together; the central element of a catapult was identified thanks to accumulated experience. The prefaces to several mathematical texts recount of networks of people, who sometimes communicated by letter, sometimes physically met and talked to each other, maybe pored over a diagram together. Some other times, they communed across the boundaries of time, in a dialogue with past works – take Archimedes and Eudoxus, or Apollonius and Euclid. One of the two questions raised in this chapter will be about communities of mathematicians. I will try better to describe them, and to relate them to the wider context of Hellenistic culture. But before I do that, I will tackle another question.

If from the group picture sketched above we change focus to a close-up of one individual – Euclid – we get a very blurry image. As we have said, we know nothing about his life – he is, simply, the author of the *Elements*. Or is he? Archimedes' testimony on Eudoxus implies that substantial parts of book 10 and possibly of book 12 may not be by Euclid. Still, if the *Elements* contain pre-Euclidean material, at least we can source it for precious information about early Greek mathematics. Or can we? The following three lines by the fourth-century AD mathematician Theon of Alexandria have, in Maurice Caveing's words, dominated the history of the text of the *Elements*:

> But that within equal circles sectors are to each other as the angles on which they insist, we have proved in our edition of the elements at the end of book six.[1]

Now, the great majority of manuscripts on which we depend for our text of Euclid unselfconsciously contain the proposition described above, as if it just belonged to the text. Had Theon not mentioned it, we may never have known that the corollary of *Elements* 6.33 is not Euclid's at all. How do we know that the same is not true for other parts of the text? How do we go about searching for the real Euclid? This is what we shall try to find out next.

The problem of the real Euclid

To get an idea of the story behind our text of the *Elements*, take a look at Figure 4.1. And if that looks complicated, you can imagine a similar, if simpler one, attached at the top of the archetypal *G*, to signify the pre-Euclidean sources incorporated into the *Elements*. Given the rather desolate scenario of the evidence about early Greek mathematics, it comes as no surprise that historians have clung to Euclid with near-obsessive interest. The quest for the real Euclid has thus acquired a double purpose: to access the original work for its own sake, and because an 'unadulterated' version would show more transparently traces of the past.

Because the problem is one of transmission of texts, it is useful to turn to the actual manuscripts through which a work is known to us. Many of them are arranged thus:

with the main body of writing in the middle and smaller pieces of writing on the side, sometimes even on both sides

such as this, which could go on for an entire paragraph

or this here

The notes made at the margins of the main text in order to explain or complement it, are called scholia. They became widespread especially from the third century AD onwards, as papyrus scrolls were gradually replaced by parchment or vellum codexes as the main writing medium. A codex had large stackable pages and adding writing to an already extant body of text was made easier by its format (you did not have to unroll it and it stayed open by itself). It is thought that sometimes scholia got incorporated into the main text due to copying accidents – many interpolations (passages contained in a work which are not original to it) perhaps started their lives that way.

In parallel with the huge number of Euclidean manuscripts, an enormous quantity of scholia to the *Elements* have come down to us, and they are only partially edited. Insofar as it has been possible to date them, mostly on paleographical grounds, the earliest group may have been produced around the sixth century AD, and it draws heavily on Proclus' and almost certainly Pappus' commentaries to the *Elements*. Scholia often offer clues about the original authorship of one piece or another of Euclid's text. For instance,

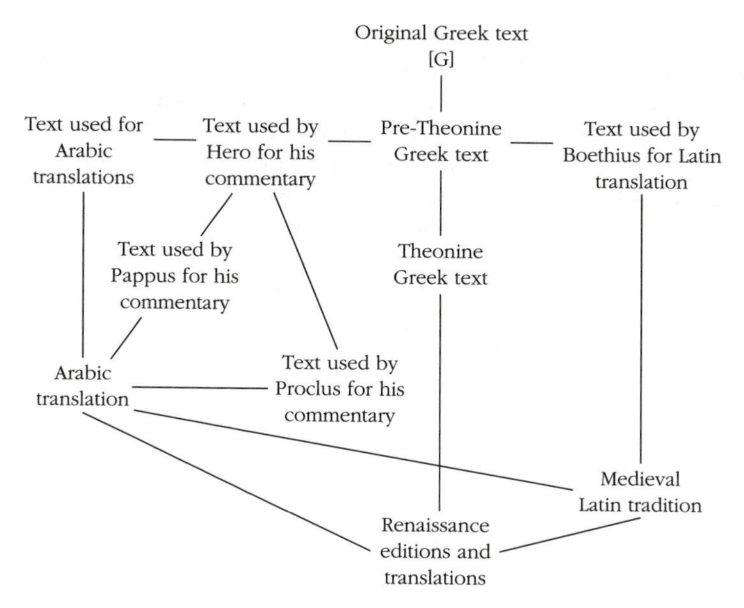

Figure 4.1 The medieval and Renaissance tradition of Euclid's *Elements* (adapted from Murdoch (1971))

they attribute proposition 1.26 (on the conditions of equality for triangles) to Thales, on the authority of Eudemus; 1.47 (the so-called theorem of Pythagoras) to (unsurprisingly) Pythagoras; 10.9 (on incommensurables in square and length) to Theaetetus. Their reliability as sources, however, should be assessed in the same way as that of any other source: they are late, and, like modern historians, they may be unduly stretching the vague information found in Plato or other earlier authors.

The identification of pre-Euclidean material more often relies on conceptions of logical coherence or of how mathematics proceeds. On this rather risky basis, it has been shown that parts of the *Elements* do not fit with the rest, and consequently must be pieces from other, presumably previous, works. One of the most famous examples of such reconstructions is the last part of book 9, which was by Oskar Becker ascribed to the Pythagoreans. The subject of the book is arithmetic, in particular prime numbers and divisibility into factors, but after theorem 9.20 the text switches to a different topic, and the next fourteen items (9.21 to 9.34) deal with odd and even numbers (e.g. 9.26: if from an odd number an odd number be subtracted, the remainder will be even). Only the two final propositions seem to return to the main path, both in content and procedure. Now, odd and even are topics that we know to have been studied by the Pythagoreans, and Becker was able to recast the simple demonstrations of propositions

9.21 to 9.34 in terms of so-called pebble arithmetic, a technique which had also been linked to the Pythagoreans and which consists in arranging pebbles in strings, squares or gnomons. Thus, an even number is represented by a double string where all the pebbles are in pairs. Prop. 9.21, that the sum of even numbers is even, can for instance be shown to be correct via this method, because any string of even numbers/pebbles remains double when any even number of pebbles is added to it.[2] Becker concluded that this part of the *Elements* was a relic from a former mathematical age, framed uneasily within the greater sophistication of later arithmetical theory.

Another interesting case of 'missing link' is the so-called proposition 10.117, as follows (see Diagram 4.1):

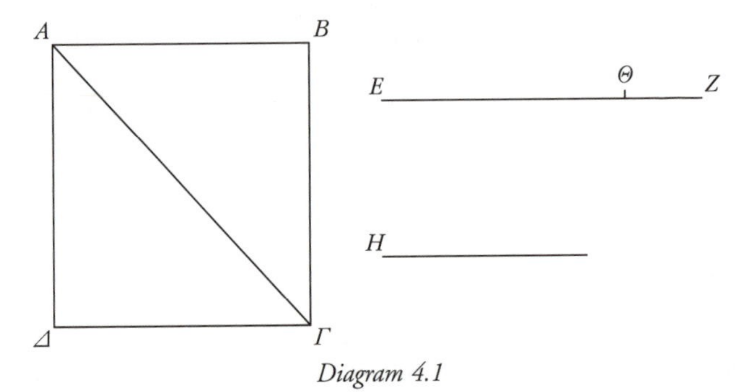

Diagram 4.1

> Let it be proposed by us to prove that the diagonal of square figures is incommensurable in length with the side. Let $AB\Gamma\Delta$ be a square, its diagonal $A\Gamma$; I say that ΓA is incommensurable with the length AB. If in fact possible, let it be commensurable; I say that it will happen that the same number is even and also odd. Since it is evident that the <square> on $A\Gamma$ is double that on AB. And since ΓA is commensurable with AB, therefore ΓA has a ratio to AB as of a number to a number. Let it have the ratio of EZ to H, and let EZ, H be the least of those having the same ratio as them; therefore EZ is not a unit. If in fact EZ was a unit, it would have the <same> ratio to H as $A\Gamma$ has to AB, and $A\Gamma$ is greater than AB, therefore EZ would also be greater than the number H. Which is absurd. Therefore EZ is not a unit; therefore it is a number. And since it is as ΓA to AB, so EZ to H, and therefore as the <square> on ΓA to that on AB, so that on EZ to that on H. The <square> on ΓA is double that on AB; therefore the <square> on EZ is also double that on H; therefore the <square> on EZ is even; so that EZ is also even. If in fact it was odd, the square on it would also be odd because if a certain number of odd numbers are added to each

> other and their quantity is odd, the whole is odd; therefore *EZ* is even. Let it be divided in half along *Θ*. And since *EZ*, *H* are the least of those having the same ratio, they are prime with respect to each other. And *EZ* is even; therefore *H* is odd. If in fact it was even, two would measure the *EZ*, *H*; for a whole even has a half part; <and they> are prime with respect to each other; which is impossible. Therefore *H* is not even; therefore it is odd. And since *EZ* is double *EΘ*, therefore the <square> on *EZ* is four times that on *EΘ*. The <square> on *EZ* is double that on *H*; therefore the <square> on *H* is double that on *EΘ*; therefore the <square> on *H* is even. Therefore *H*, through what has been said, is even; but it is also odd; which is impossible. Therefore *ΓA* is not commensurable with *AB* in length; which is what it was necessary to prove.[3]

This is a proof of the incommensurability of the side and diagonal of the square, which uses *reductio ad absurdum* to show that, if diagonal and side were commensurable, the same number would be both odd and even. But *Elements* book 10 contains another, more general, proof to the same effect, that 'squares which do not have to one another the ratio of a square number to a square number have their sides incommensurable in length' (10.9, attributed by a scholion to Theaetetus). Proposition 10.117, on the other hand, matches the description of a proof mentioned by Aristotle:

> In the latter [*sc.* arguments which are brought to a conclusion *per impossibile*], even if no preliminary agreement has been made, men still accept the reasoning, because the falsity is patent, e.g. the falsity of what follows from the assumption that the diagonal is commensurate, viz. that then odd numbers are equal to evens.[4]

According to one line of interpretation, if 10.9 was discovered by Theaetetus, then the proof by Aristotle, which is less general, must have been discovered earlier, but by whom? The best candidates are the Pythagoreans, who seem to have been interested both in incommensurability and in odd and even. On this reading, 10.117 would be another relic from a former mathematical era, tacked on more sophisticated material. A different, and now widely accepted, interpretation was given by Wilbur Knorr, who observed that the third-century AD author Alexander of Aphrodisias, commenting on that very same passage from the *Prior Analytics*, quoted material from the *Elements* but did not appear to know 10.117. Knorr argued that 10.117 was produced at some point between Alexander of Aphrodisias and Theon, whose version of the *Elements* already contained it. The 'pre-Euclidean' proof was 'a result of the continuing activity of later Aristotelian commentators'.

These two cases illustrate the difficulties facing the historian who wants to use the *Elements* as a source for earlier material. The crucial question is Euclid's attitude to past mathematics: did he incorporate entire treatises into his text with the seams still showing, or did he drastically intervene, changing them beyond recognition? And how can we really tell what kind of editor Euclid was, if first, we do not know his sources – they are all lost to us, and, second, even the Euclid we do have is not the real Euclid anyway? I retreat to a safe and cosy agnostic position with respect to the retrievability of pre-Euclidean mathematics from the body of the *Elements*. When Plato, Aristotle or Archimedes testify that a result was already known, or the general features of a procedure already established, then we have some ground, but hardly a lot of detailed ground, to run on. Suspending judgement with respect to pre-Euclidean material, however, still leaves the other horn of the problem unattacked. Leaving aside the (virtual) tree at the top of G in Figure 4.1, we still have the many ramifications underneath it, and first of all the Theonine question.

In 1808 the French scholar François Peyrard observed that an Euclidean manuscript in the Vatican library in Rome (known as *Vat. gr.* 190, or P, tenth century) showed no trace of the infamous corollary to proposition 6.33 and was in other respects different from the majority of other manuscripts of the *Elements*. He proceeded to conjecture that P was pre-Theonine, unscathed by Theon's intervention, and therefore closer to Euclid's original. J.L. Heiberg, editor of what is still widely considered the standard edition of the Greek text of the *Elements*, accepted Peyrard's conclusions. While he used manuscripts from the so-called Theonine family, he held P 'to represent the authentic Euclidean text'.[5]

But not even P could be said to contain the real Euclid. In an article entitled 'The Wrong Text of Euclid', Knorr questioned the primacy of P. He revived a previous hypothesis by M. Klamroth, a contemporary of Heiberg, to the effect that the Arabic tradition of the *Elements*, which also lacked the corollary to 6.33, may have derived from a better text, or texts, of Euclid than P. Indeed, the Arabic *Elements* differ from Heiberg's on several counts, such as shorter proofs or a number of missing corollaries. Heiberg had entertained but rejected, on the basis of several factors, including incompleteness, the idea that the Arabic tradition may be truer to the real Euclid. Yet, as Knorr pointed out, 'incompleteness' is a subjective factor. But there is more. Already Heiberg had occasionally chosen the lesson of Theonine manuscripts over that of P wherever he suspected scribal error. One manuscript from the Theonine family, called b (from Bologna, eleventh century) is unique in that it agrees with the rest of the group except for the last part of book 11 and the whole of book 12. Knorr showed that the divergent parts of b aligned with the Arabic tradition, and could then be used further to support his thesis. He concluded his article thus:

> We have never had a 'genuine' text of Euclid, and we never will have one [...] But we can do much better than has been done so far.[6]

One of the ways to go would be to get a clear picture of the extent of Theon's intervention on the text. Comparison with what he did in his commentary on Ptolemy could be useful to identify operations such as correction or improvement where he believed Euclid was mistaken or confused, or additions where he thought that the text was difficult to understand. We have to keep in mind that the *Elements* were seen first and foremost as a reservoir of results, not necessarily as a work that should be preserved in its linguistic or methodological integrity. Although the notion of preserving ancient texts did exist – Hellenistic editions of Homer, say, were produced which respected archaic language – it seems to have been applied only to literary works. No philological concern seems to have prevented later mathematicians[7] from correcting, simplifying, updating previous material. In fact, Theon was not the first. In the centuries between the early third BC and late fourth AD, other people like him had read, used and reworked the *Elements*. We know, to begin with, of at least two commentaries on Euclid prior to Theon: Hero's and Pappus', and Proclus' commentary mentions many unspecified others. It is possible that these commentators edited parts of the *Elements*, or that they made copies, perhaps for personal use, with additions or modifications which later came to be incorporated into the text.

Unique access to pre-Theonine Euclidean material is given by three papyri: one, Hellenistic, mentioned in the section on material evidence in chapter 3, contains the definition of a circle (1. definition 15) and propositions 1.9 and 1.10; a second- or third-century AD papyrus from the Fayum has the equivalent of 1.39 and 1.41 and a third- or fourth-century AD papyrus fragment from Oxyrhynchus has the equivalent of 2.5. The following example will allow the reader to compare the versions in the papyri with those in P.

> [*P. Herc.* 1061 – 1. definition 15 (see Diagram 4.2)] ... circle is a plane figure surrounded by one line, all the lines falling on it from one point of those situated within the circle are equal; for in this way the geometers define it [...] [1.9] let, they say, A be the given angle surrounded by the side AB and by the <side> $A\Gamma$ and a point of the side AB <is> Δ, it is necessary to subtract a length $A\Delta$ sufficient ... since in fact $A\Delta$ is equal to AE and AZ is common, therefore the two $A\Delta$ and AZ are equal to the two AZ and AE and the base ΔZ to EZ; for the triangle has been put together equilateral

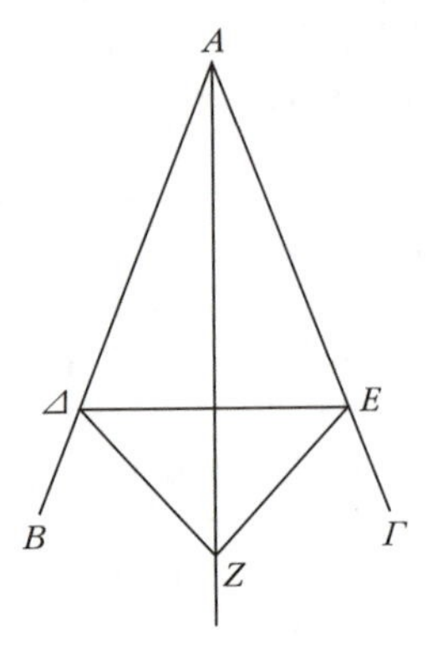

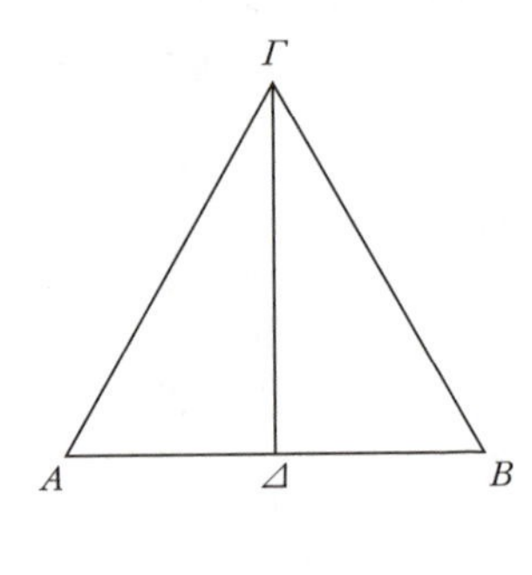

Diagram 4.2

... [1.10] on the given line MN, they say, let it be put together the equilateral triangle ... divide the given line into two ... for if it is <possible?> to divide into two the given line ... if MP and PN are equal and the given line MN is divided into two, the same will be also ... ΞM is equal to ΞN the angle O to P and MN will be divided into two, so that ...

[*Vat. gr.* 190 – 1. definition 15 (see Diagram 4.2)] circle is a plane figure surrounded by one line, which is called circumference, all the lines falling on it, on the circumference of the circle, from one point of those situated within the figure are equal to each other. [1.9] [...] for let a point whatever Δ be taken on AB, and let AE equal to $A\Delta$ be subtracted from $A\Gamma$ [...] since in fact $A\Delta$ is equal to AE and AZ is common, the two ΔA, AZ are equal to the two EA, AZ respectively; and the base ΔZ is equal to EZ; therefore the angle ΔAZ is equal to EAZ [...] [1.10] [...] Let an equilateral triangle $AB\Gamma$ be put together on it [...] I say that the line AB will be divided into two along the point Δ. Since in fact $A\Gamma$ is equal to ΓB, $\Gamma\Delta$ is common, the two $A\Gamma$, $\Gamma\Delta$ are equal to the two $B\Gamma$, $\Gamma\Delta$ respectively; and the angle $A\Gamma\Delta$ is equal to $B\Gamma\Delta$ [...] therefore the given delimited line AB will be divided into two along Δ [...]

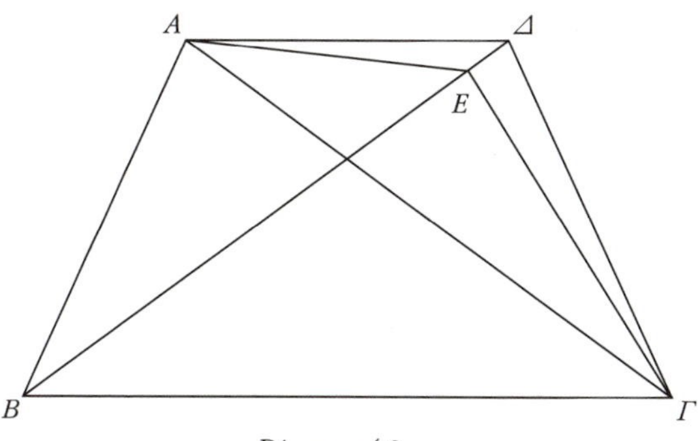

Diagram 4.3

[*P. Fayum* 9 – 1.39 (see Diagram 4.3)] are ... on the same side ... parallels ... on the same base ... let *AΔ* be conjoined ... is to *BΓ* ... parallel to *BΓ* ... *ABΓ* ... on the same base ... parallels ... is equal to *BΔΓ* ... the greater to the lesser ... *AE* is parallel to *BΓ* ... we have proved that it is not other ... therefore *AΔ* is parallel to *BΓ* ... [1.41] if a parallelogram as a triangle ... the same and in the same ... the parallelogram will be ... parallelogram ... the base ... double ... let it be conjoined ... to the triangle *EBΓ* ... of *BΓ* and ... to *BΓ AE* but ... parallelogram ... and of *EBΓ* [...]

[*Vat. gr.* 190 – 1.39] Equal triangles which are on the same base and on the same side are also in the same parallels. [...] For let *AΔ* be conjoined; I say that *AΔ* is parallel to *BΓ*. [...] and the triangle *ABΓ* is equal to the triangle *EBC*; for it is on the same base [...] and in the same parallels. But *ABΓ* is equal to *BΔΓ*; and *ΔBΓ* is equal to *EBΓ* the greater to the lesser [...] neither therefore is *AE* parallel to *BΓ* [...] [1.41] If a parallelogram has the same base as a triangle [etc.]

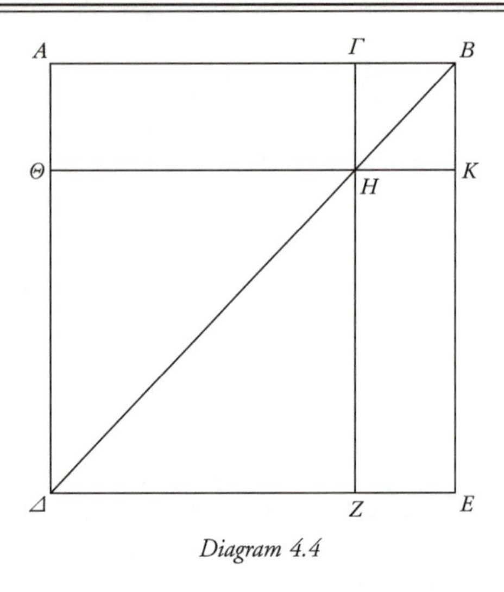

Diagram 4.4

[*P. Oxy.* 29 – 2.4 (see Diagram 4.4)] to the surrounded rectangle ... [2.5] If a straight line is divided into equal and unequal <segments>, the whole rectangle formed by the unequal segments plus the square on the sections in between is equal to the square on the half.

[*Vat. gr.* 190 – 2.4] to the rectangle; which it was necessary to prove. Corollary. From this it is evident that in square areas the parallelograms formed around the diagonal are squares. If a straight line is divided into equal and unequal <segments>, the whole rectangle formed by the unequal segments plus the square on the sections in between is equal to the square on the half.

The Euclidean papyri offer a good sample of ways in which a mathematical text could undergo modifications. The first account was probably polemical, and distanced the author from the 'geometers' through use of 'they say' to mark the two propositions, which in their turn differ quite significantly from P. Yet, the correspondence as far as the definition of circle goes is remarkable. The Fayum papyrus has preserved only the inner core of columns of text, and P seems to fill it in rather nicely – but the papyrus goes from what we now call 1.39 straight to 1.41. That means that the *Elements* known to the person who wrote the Fayum papyrus did not contain our present 1.40. As for the third papyrus, along with good parallelism in the enunciation of 2.5 and the end of 2.4, there is a major element of divergence in that the papyrus does not have the corollary (which Heiberg himself considered an interpolation).

In sum, the quest for the real Euclid seems fraught with serious, if not completely insurmountable, difficulties. There is in it a lesson for the historian: collecting, reworking, explaining, systematizing previous texts became one of the main features of ancient mathematical practices at least since Euclid himself, thus basically since our very first 'big' work. At least in part, this is the result of the growing presence of the written medium, so impressively represented by the hundreds and hundreds of scrolls which are said to have been stacked on the shelves of the library in Alexandria. We need better to understand those operations of collecting and reworking, if we are to understand ancient mathematics itself. Moreover, the relation between the *Elements* and the other surviving works by Euclid, whose editorial history is marginally less complicated, could fruitfully be explored. Finally, it bears reflection that the people who wrote the Fayum and Oxyrhynchus papyri are representative of a large, unnamed public for mathematics, who may have contributed no new result to it, but may have nonetheless shaped its practice by the mere act of reproducing and transmitting texts.

The problem of the birth of a mathematical community

A rather large, if not entirely unnamed, public for mathematics emerges also from the works of Archimedes and Apollonius, but also of lesser figures like Philo or Diocles. At the very beginning of his account on burning mirrors, the latter mentions a number of other people who were interested in the same topic as him: Zenodorus, Pythion from Thasus who suggested a certain problem to Conon, and Dositheus, who gave a practical solution to it.[8] A similar scenario is evoked by Hypsicles (mid-second century BC), who addressed the so-called book 14 of the *Elements* to a Protarchus and mentioned a Basilides of Tyre. Basilides shared a passion for mathematics with Hypsicles' own father and, together with him, had written corrections to one of Apollonius' texts. Hypsicles in his turn had been so intrigued by a proof of Apollonius' that he decided further to examine the issue, and to send Protarchus the results, so that they could be assessed by those who 'are proficient, thanks to their experience, in all studies and in particular in geometry'.[9]

In a recent study, Reviel Netz has examined the language and argumentative structure of a cross-section of mathematical texts, primarily Euclid, Archimedes and Apollonius. He has established that they shared a method of proof based on deduction and on the use of the lettered diagram. He has also convincingly demonstrated that the lexicon used by these texts was small, often organized formulaically, and that they relied on a small system of background knowledge, a set of mathematical results which were taken for granted. Acquaintance with this toolbox on the part of the reader was also taken for granted, and amounted to the expertise required for a full comprehension of the text.[10] The practices described by Netz (agreement on procedures, compartmentalized language, notion of expertise) pinpoint a group, among whose members we can count the usual suspects – Euclid, Archimedes and Apollonius – and their immediate audience, i.e. the people mentioned in their works. So, the internal mathematical structure itself of at least a selection of our texts points to the existence of a community of mathematicians who, even across time, shared a specialized language, discursive conventions, criteria of validity and rigour.

How did this community originate? A partial answer can be found in the cultural policies of the Hellenistic kings; moreover, we could talk about multiple births, of mathematical communities and of similar groups in other fields. The most famous cultural space of the Hellenistic period was the Museum of Alexandria, founded and financed by Ptolemy I. Although many details of the functioning of this institution remain obscure, we know that it was home to a diverse group of scholars, mostly Greek-speaking but not limited to people of Greek descent; that there was a massive library; and that there was some degree of communal life, perhaps with shared meals and in-house accommodation. People associated with the Museum included poets, grammarians, historians, philosophers, doctors and, crucially for us here, natural philosophers, geographers, machine-builders, astronomers, geometers. Other cities such as Pergamum and Rhodes were also active cultural centres, complete with libraries and sundry scholars.

Given the obvious expense entailed by patronage on such a large scale, a natural question could be, what was in it for the Ptolemies (and the Attalids, and the Seleucids)? In some cases, promoting science had immediate practical benefits. It is rather obvious why Hellenistic monarchs would want to support military engineers like Biton or Philo. Geography and astronomy could be used for map-making and time-keeping. But there was more than simply utility in sight for the royal patrons of the arts and sciences: rulers could now present and represent themselves as well-educated in the Greek way, and could flaunt knowledge as yet another jewel in their crown. In this sense, the scientists and scholars of the Museum constituted yet another collection, along with that of books and of rare and exotic animals. Indeed, they were described accordingly by a contemporary: 'many are kept to graze in populous Egypt, well-fed bookworms, who quarrel without end in the Muses' bird-cage'.[11]

Another second-hand testimony also conveys significant aspects of the Ptolemaic rule:

> [...] the Grand Procession [...] was led through the city stadium. First of all marched the sectional procession of the Morning Star, because the Grand Procession began at the time when that aforementioned star appeared. [...] After them marched the poet Philikos, who was the priest of Dionysius, and all the guild of the artists of Dionysius [...] After them a four-wheeled cart was led along by sixty men ... twelve feet wide, on which there was a seated statue of Nysa twelve feet tall, wearing a yellow chiton woven with golden thread [...] This statue stood up mechanically without anyone laying a hand on it, and it sat back down again after pouring a libation of milk from a gold phiale. [...] The figure was crowned with golden ivy leaves and with grapes made of very precious jewels. [...] Next, another four-wheeled cart, thirty feet long by twenty-four feet wide, was pulled by three hundred men, on which there was set up a wine-press thirty-six feet long by twenty-two and a half feet wide, full of ripe grapes. [...] Next there came a four-wheeled cart, thirty-seven and a half feet long by twenty-one feet wide, which was pulled by six hundred men. On it was an askos made of leopard skins which held three thousand measures. As the wine was released little by little, it also flowed over the whole street. [...] At the very end, the infantry and cavalry forces marched in procession, all of them fully armed in a marvellous fashion. [...] Besides the armour worn by all these troops, there were also many other panoplies kept in reserve, whose number is not easy to record, but Kallixeinos gave the full count.[12]

Kallixeinos' detailed recounting of this public celebration provides us with a full picture of power itself: the gold, the abundance of food and drink, the connection with the gods, the more explicit embodiment of might marching at the end, and, crucially for us, the display of knowledge. The timing itself of the procession (which included both morning and evening star floats) seems to have been based on astronomical calculations. A poet walked along. The statue of Nysa (the mythical wetnurse of Dionysius), gleaming with gold, was an *automaton*, a self-moving object, whose functioning was to be later explained, mathematically, by Hero. The *automaton* represented an important element of the self-image the Ptolemies wished to convey to the public (we know that delegates from many other states had been invited to the festivities), namely the capacity to create wonder and amazement. The fact that common observers were not able to explain the apparently impossible things they were experiencing – an enormous statue moving 'without anyone laying a hand on it' – reflected back positively on those who knew what was going on, or indeed could *produce* such phenomena: the machine-maker, but above all the patron. Even straightforwardly useful research such as geography compounded effectiveness with wondrousness; Philo was keen that his catapults should look as terrifying as they would prove when actually used; Diocles was very well aware of the potentialities of burning mirrors as spectacle.[13]

Having to operate within such an environment shaped cultural practices in several ways. First, thrown together by the service of a common patron, people were exposed to other forms of knowledge, and cross-pollination occurred frequently. According to recent reconstructions, for instance, Hellenistic medical treatises used mechanical terms that we otherwise only find in Philo of Byzantium. The heroes of Apollonius of Rhodes' *Argonautica* travel through a mythical world whose features seem to have been updated following the latest geographical treatises.[14] Euclid wrote on optics, astronomy and harmonics as well as mathematics. Eratosthenes was interested both in philosophy and in mathematical novelties; he was chief librarian at Alexandria as well as measuring the circumference of the Earth. Archimedes was a machine-maker, and Livy reports that he was well known for his observations of the skies.[15] Philo and Diocles offered solutions to the problem of the two mean proportionals within the context of machine-building, but Archimedes took the same problem for granted (it evidently was part of the geometers' toolbox) in his *Sphere and Cylinder*.[16] Second, reference to the past and collection and accumulation of cultural tokens, especially books, were crucial factors in the construction of the self-image of Hellenistic kings, and they were, or they became, crucial factors in the construction of the self-image of groups of intellectuals as well. The Hellenistic period, to an extent not seen before, saw the emerging of philosophical, medical and philological or grammatical traditions or schools.

This has been amply documented, so I need not rehearse well-known evidence here. Let me just point out a few aspects shared by some of these schools or traditions: above all, the 'canonizations' of some texts or authors. Already Lycurgus, public treasurer at Athens between 337 and 325/4 BC, had elevated the triad of Aeschylus, Sophocles and Euripides to the status of official classics, promoting the establishment of an *authentic* text of their works. Indeed, the Hellenistic age was a great period for fakes, imitations and works written in the style of a chosen authorial model: just think of the mass of Pythagorean texts we mentioned in the section on philosophers in chapter 3, or of *faux*-Platonic or Aristotelian works like the *Epinomis* and *On Indivisible Lines*. This not only indicates that there was a market for books, but also that some names had become prestigious enough to inspire imitation or downright forgery. Thus, more manuscripts of the same work were collected, in an attempt to establish a canonical or 'best' version, which involved selecting between different manuscripts of the same work, or different readings of the same phrase, or different explanations of the same passage. The choice had to do with claiming to oneself the authority to decide what was the real Sophocles, or what Plato really meant, or what Euclid missed out. In the case of Homer, Alexandrine scholars preserved versions that they did not consider authentic, but even so they were still advocating to themselves the expertise to discriminate between genuine Homer and later imitations.[17]

It was in connection with this process of canonization that grammar was born. Codification of texts implied attention to the language employed and formulation of criteria to distinguish between authentic and not authentic. Now that Greek was being learnt by more and more non-native Greek speakers, some criterion had to be articulated systematically to teach correct from incorrect. Grammar thus was a codification of the language, whereby some speakers, taking into account common usage on the one hand and the usage of the canonical texts on the other hand, decided what was good Greek and what was not quite the right Greek. Grammarians defined their pursuit as a *techne*, in the same league as medicine and mathematics. Parallels can indeed be run: both the grammar and the mathematics of this period, as exemplified chiefly by Euclid's *Elements*, operate on already existing material and order it according to criteria of validity. Although people may speak and write correctly even if they ignore grammar, grammar is an explicitation of concepts already in use that by its very act establishes those concepts on firmer foundations. Thus, notions such as point, line, triangle, had been unproblematically used long before Euclid posited them as rigorously defined starting points. The creation of grammar also created the grammatical expert, a person who can claim a more rigorous knowledge than the person in the street, through the firm grasp he has acquired of the canonical texts, combined with observation and participation in common usage and practices. This description fits with what Netz has identified as the features of the mathematical expert implied by the works of Euclid, Archimedes and Apollonius.

Unfortunately, not much survives of the grammatical work done in Alexandria between the third and the first century BC: the earliest treatise we have is the *Grammatical art* by Dionysius of Thracia (late second century BC). Its authorship has sometimes been questioned, but enough scholars deem it authentic for me to include it here. The work is organized systematically: it opens with a definition of grammar as methodic knowledge based on the poets and on prevalent usage, and then proceeds to define, define,

define each and every element of speech, from vowels to consonants, to modes to tenses to conjunctions, usually specifying how many parts are subsumed under each element. If we take these definitions as first principles or starting points, the foundation work here can again be compared to that in the first book of Euclid's *Elements*. Everybody in a sense knows what a line or a circle are, and everybody in a sense knows what a vowel or consonant are – Euclid's and Dionysius' action in defining them are parallel.

On the other hand, the accumulation of cultural resources, both human and papyraceous, seems to have gone hand in hand with a desire to distinguish and define oneself and one's specific pursuit. There was a need for self-identification and distinction, which might have derived from an environment which was both multicultural and slanted towards one particular culture, or could be related to competition for patronage. The hostility, as well as the exchanges, between different schools or sects within the philosophical and medical fields are very well documented. Also, remember the philosophical attacks on mathematics, or the dismissive remarks about mathematicians reported by Aristoxenus and Diocles, or, on the other hand, Hipparchus' not-so-high opinion of Aratus. Or, and this is one of the richest texts available to us for an exploration of the complexities of Hellenistic mathematical practice, let us consider Archimedes' *Sand-Reckoner*. In it, arithmetical and astronomical interests are combined, and Archimedes both put forth a new system of notation for very large numbers and launched into astronomical enquiries. His claim is that, even if one filled the whole universe with sand, it would still be possible to express that multitude of grains of sand with a number. Astronomy comes in when he has to give an estimate for the size of the universe, in order to assess precisely how many grains of sand it could contain. The final number would not have been expressible in ordinary Greek (Milesian) notation, but is 'a thousand myriads of eighth numbers' according to Archimedes' new system – 10^{24} in modern figures.

The text is addressed to a king, probably a provider of patronage;[18] it presents the author both in relation to his colleagues and to previous authors, astronomers and mathematicians past and present; it plays up the wondrousness factor. The anecdotal evidence abounds in tall stories about Archimedes' machines and their marvellous effects, but here we have the man himself, in his own words. Imagine, he says to the king, being able to count not just the sand on the local beach of Syracuse, nor even the sand of the whole of Sicily, but the sand of the entire universe, if the universe were filled with sand. Archimedes reassures Gelon that the tenor of the demonstrations is not too difficult, and that he will be able to follow them, and, after navigating an assured course between astronomical observations and arithmetical calculations, concludes the treatise thus:

> I understand, king Gelon, that these things will seem not believable to the many who do not share in the studies, but to those who instead have taken part [in them] and have reflected on the distances and the sizes of the earth and the sun and the moon and the whole universe, they will be persuasive because of the proof; which is why I thought that it was not inappropriate for you to consider these things.[19]

King Gelon need not know, or notice, that the *Sand-Reckoner* is not structured like Archimedes' other treatises, presumably reserved for his mathematical peers: it has a narrative mode of argumentation, but no starting points and no deductive chains. The king, on the other hand, may appreciate being made a honorary member of the restricted community of people who know, as opposed to 'the many who do not share in the studies'.

As we have already noticed, astronomy was a popular subject, especially with patrons, as testified by Aratus and by poems such as Callimachus' celebration of the lock of Berenice, a constellation named after the hair of a Ptolemaic queen:

> Having examined all the charted sky, and where the stars move ... Conon saw me also in the air, the lock of Berenice, which she dedicated to all the gods.[20]

Conon (the same as in Archimedes' and Diocles' works?) must have cleverly combined science and ego-boosting by inscribing the Ptolemies in the heavens; but it is interesting that Callimachus made the episode his own by celebrating it in style. Aratus' poetic appropriation of Eudoxus, as described by a none-too-pleased Hipparchus, may have been along similar lines. Think also of Eratosthenes' little poem, as reported by Eutocius (and if it is authentic), with which he celebrates his solution to the problem of the two mean proportionals, praises Ptolemy and criticizes a couple of mathematical predecessors. More than direct competition, one could describe this as a dialectic relationship between mathematicians/astronomers and poets. Was Archimedes' *Sand-Reckoner* directly answering Theocritus' evocation of a countless universe? Was he proclaiming the centrality of mathematics to the understanding and managing of an enlarged world? And was he, the best representative of a small and extremely sophisticated mathematical community, also recognizing the necessity of patronage, and the inescapability of the real world?

Notes

1 Theon, *Commentary on Ptolemy's Syntaxis* 1.10 492.7-8. Theon's edition is mentioned in the scholia to the *Elements*, 1.2; 4.4; references in Mansfeld (1998), 25. See Caveing (1990), 45 ff., on which I rely extensively.

2 Becker (1933).

3 Euclid, *Elements* 10. appendix 27 (contained in most manuscripts, including P).

4 Aristotle, *Prior Analytics* 50a, quoted from Fowler (1999), 292, and see also Knorr (1975), 228 ff.

5 Which is not to say that P is the oldest extant manuscript: B, in Oxford, can be dated with accuracy to AD 888. Cf. Knorr (1996), 212.

6 Knorr (1996), 261.

7 Or, as G.E.R. Lloyd points out to me, later medicine writers.

8 Diocles, *On Burning Mirrors* 4, 3 and 6, respectively.

9 Hypsicles, *Book 14 of the Elements* preface.

10 Netz (1999a).

11 The words are attributed to Timon of Phleisus (third century BC) by Athenaeus, *Deipnosophistae* 22d, Loeb translation with modifications.

12 Kallixeinos of Rhodes *ap.* Athenaeus, *ibid.* 197c–203a, translation in Rice (1983). The procession described is thought to have taken place ca. 280–75 BC.

13 Philo, *Mechanica IV (Construction of Catapults)* 61.29–62 ff.; 66.17 ff.

14 Cf. von Staden (1998), Hurst (1998), respectively.

15 Livy, *From the Foundation of the City* 24.34.2.

16 Archimedes, *SC* 2.1.

17 See Irigoin (1998) and, on grammar, Montanari (1993).

18 Archimedes is defined in a source *sungenes*, 'related', to the king of Syracuse – the term has often been taken literally, but in fact could be used as an honorific title to denote a royal familiar, cf. *LS s.v.*

19 Archimedes, *AR* (Mugler 156–7).

20 Callimachus, *Aetia* fragment 110.

5

GRAECO-ROMAN MATHEMATICS: THE EVIDENCE

> That certain persons have studied, and have dared
> to publish, the dimensions [of the universe] is mere madness
> [...] as if indeed the measure of anything could be
> taken by him that knows not the measure of himself.[1]

By the end of the first century BC, Rome had turned from a republic into an empire, and, in the course of the first two centuries AD, it secured control over most of Europe, North Africa, Egypt and the Near East. The empire is a constant presence in our evidence from this period, and it enters mathematical discourse in several ways. Managing an army, collecting taxes, keeping a census on such a vast scale implied centralized administrative practices (accounts, tax rolls, land surveys). Mathematics was also used to articulate views about politics, society and morals. It would be impossible to describe our period in a few words: let us just say that the world had become even larger than after Alexander's expedition, exchanges of all types increased; and the textual past kept accumulating in the form of books and libraries. Turning to the evidence, apart from the usual survey of material sources, there are individual sections on Vitruvius and Hero. For the rest, authors have been assigned to the two sections 'Other Greeks' and 'Other Romans' on the basis of the language they worked in – geographically, they come from all over the place and they all belonged to the same Empire. A more earnest exploration of the Greek/Roman divide will be taken up in chapter 6.

Material evidence

The great majority of papyri from this period comes, as usual, from Egypt. They include the earliest extant philosophical commentary, on Plato's *Theaetetus*. The author, who mentions other commentaries, including one to the *Timaeus*, has not been identified. Given the contents of the original dialogue, this papyrus is quite rich in mathematical passages, explaining for instance the construction of a square on a given line (reference is made to Plato's *Meno*), the difference between square and rectangular numbers, between incommensurability in square and in length, and between 'wedge', 'brick' and 'beam' numbers. I append a sample in the following section (and see Diagram 5.1).

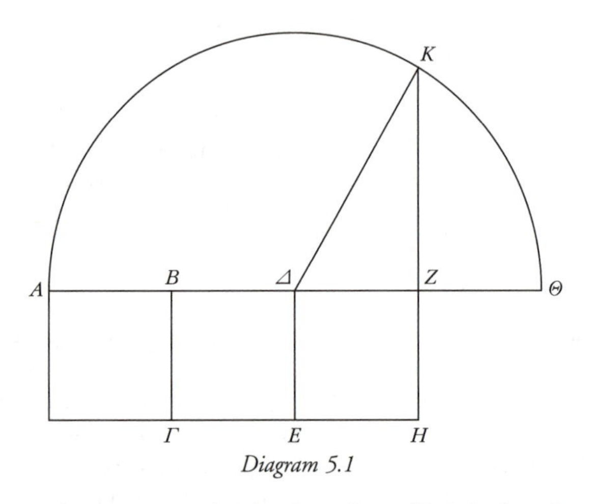

Diagram 5.1

Let $AB\Gamma$ be a square with side of one foot, AB. It is clear that the <square> on this side will be one square foot; for one by one is one. And let a line be prolonged from the line AB and let $B\Delta$ equal to AB be cut on it, and let the square $B\Gamma\Delta E$ be drawn on $B\Delta$; the <square> on $B\Delta$ will be equal to that on AB; the whole AE is not a square but a parallelogram. Again, let a line be prolonged from the line $A\Delta$ and let ΔZ equal to $B\Delta$ be cut on it, and let the square ΔZEH be drawn on it. The square ΔZEH is equal to either of the squares set out before, and the whole area AH is a parallelogram. Again let a line be prolonged from the line AZ, and let $Z\Theta$ equal to ΔZ be cut, and let $A\Theta$ be divided into half by the point Δ. And with centre Δ and radius ΔA, let a semicircle $AK\Theta$ be drawn, and let ZK be drawn perpendicular to HE, and let $K\Delta$ be joined. Since the line $A\Theta$ is divided into equal parts by the point Δ and in unequal parts by Z, the <rectangle> formed by AZ, $Z\Theta$ plus the <square> on the <line> between the sections, ΔZ, is equal to the <square> on $\Delta\Theta$. But ΔK is equal to $\Delta\Theta$. Therefore the <square> on ΔK is equal to the <rectangle> formed by AZ, $Z\Theta$ and the

<square> on ΔZ. But the <squares> on ΔZ, ZK are equal to that on ΔK. Therefore the <squares> on ΔZ, ZK are equal to the <rectangle> formed by AZ, $Z\Theta$ and the <square> on ΔZ. Let the <square> on ΔZ which is common be subtracted; therefore the <square> on ZK which is left is equal to the <rectangle> formed by AZ, ΘZ which is left. The <rectangle> formed by AZ, $Z\Theta$ is the <rectangle> formed by AZ, ZH. For $Z\Theta$ is equal to ZH; therefore the <square> on ZK is equal to the parallelogram AH. The <square> on ZK is then incommensurable to the parallelogram AH, which surrounds inside itself three squares of one foot equal to each other.[2]

Whereas the proposition above follows very closely Euclid's *Elements* 2.14, some other parts of the text (e.g. the 'wedge' and 'brick' numbers) do not. If the commentary was meant for the general public (its tone is quite simple, and interest in Plato was quite widespread), their mathematical knowledge would have been compounded of more than one tradition, both Euclidean and 'other'.

Evidence for the diffusion of the Euclidean tradition is provided by three papyri whose contents are traceable to the *Elements*. All seem to have been written in good, if at times hasty, hands, i.e. they seem to have been produced by educated adults. In fact, two of them, which only have the enunciations and diagrams without proofs, look as if they were written for personal perusal, as if the author wanted to work through the demonstration for himself (herself?).[3] As for the 'other' tradition, it is represented by at least one papyrus with problems such the following (see Diagram 5.2):

If another incomplete cone is given, which has the vertex 2, the base 10, and the inclinations each 5, subtract the 2 of the vertex from the 10 of the base = 8, of which $^1/_2$ = 4; by itself = 16; and the 5 of the inclination by itself = 25. From these subtract the 16 = 9, of which the root = 3. This is the height. And add the vertex to the base, the 2 to the 10 = 12, of which $^1/_2$ = 6. Describe a circle whose diameter is 6. Multiply this by itself = 36, of which $^1/_4$ = 9. Subtract 9 <from 36> = 27. And subtract the 2 of the vertex from the 10 = 8, of which $^1/_2$ = 4. Describe a circle whose diameter is 4, the area is 12, of which $^1/_3$ = 4. Add to the 27 = 31. This multiplied by the 3 of the height = 93. The stone will be of as many feet.[4]

The reference to the truncated cone as a 'stone' ties in with several other clues in this text: some of the geometrical objects are given names of archi-

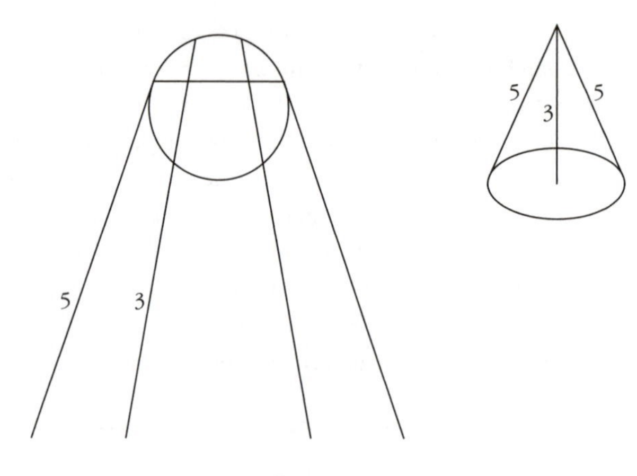

Diagram 5.2

tectural elements, and one of the problems, now completely illegible, is illustrated by a cylindrical column, with a well-defined capital. The papyrus was probably aimed at architects for training or reference, and provided information that could be extended from the particular cases, each involving a solid with such-and-such dimensions, to problems with similar features. In fact, it often specifies that the procedure can be applied in analogous circumstances: one of the problems does not even append a solution, leaving the reader to fill it in, and there is a sequence of four propositions about the volume of a cylinder with a full calculation in the first two and an abbreviated version in the latter two, as if they were, basically, a drill.

We have numerous other papyri containing tables of division, addition and multiplication; financial documents and metrological texts, where instructions are given to convert one unit of measure (of land, of money) into another. One papyrus starts off with a table of parts, and follows with a list of arithmetical problems concerning conversion between different coinages and calculation of freight charges. The problems are mostly about specific examples, but two of them refer to a 'proof' (*apodeixis*) – essentially a brief counter-check:

> The width of a field is [$2^{1}/_{2}$] schoenia; what will be the length so that it makes 200 arouras? As is necessary, reduce the [$2^{1}/_{2}$] schoenia to] $^{1}/_{2}$, = 5; and the 20 arouras to $^{1}/_{2}$ = 40 of which the 5th part = 8; the length will be of as many <units> as that; proof; multiply the $2^{1}/_{2}$ schoenia of the width by the 8 of the length = the above-said 20 arouras.[5]

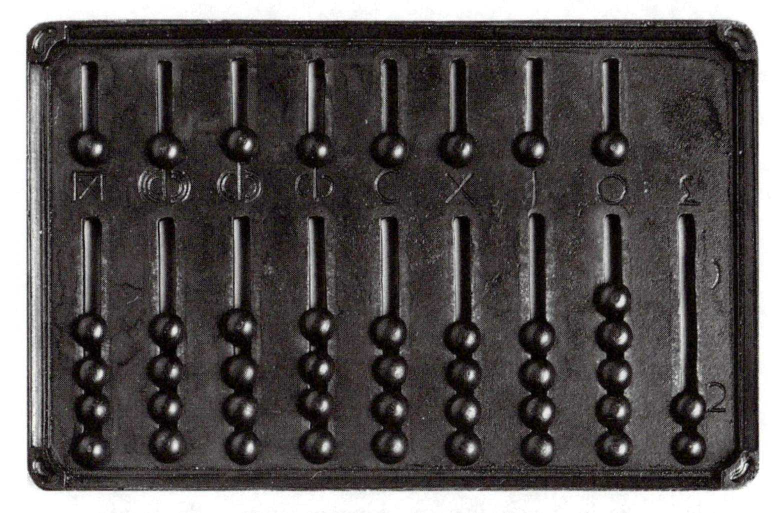

Figure 5.1 Roman abacus (replica)
(© copyright Science Museum/Science and Society Picture Library, London)

Another papyrus has two problems about the distribution of money into unequal parts. Two long calculations are arranged in columns, so that the procedure, as well as explained, is visualized, as if it was being carried out on an abacus. We have at least three concrete examples of abacus from this period, all quite small and in bronze (see Figure 5.1). The most recent find is from a late first-century AD grave in Aosta.

Like earlier Greek ones, these abaci had signs and columns both for numbers and for sums of money. Compared to our extant Greek examples, however, which are slabs of stone, the bronze Roman abaci seem to have been more easily transportable. The fact that the counters are grafted on, rather than loose, must also have facilitated use, and suggests that, although only a few actual items survive, abaci were widely used.

Facility of calculation was also the rationale behind a rather numerous group of papyri: planetary tables. They were used mainly in connection with astrological practice; how they were produced is not entirely clear, but most scholars think that their authors drew on Mesopotamian observations and data. Knowing how to use a planetary table required expertise, and some of our papyri provide instructions for calculations requiring arithmetical skills.[7] Once the position of the planets at a certain point in time was determined, one could compile horoscopes. Most of the ones we have tend to be simple, listing what 'house' each planet is in and little else. The astrologers who made them need not have known very much about astronomical observations or celestial models: interpreting the chart was of course a different matter, and that was where for most people the real skill lay. But, as the public for astrology was vast and diverse, the corresponding practice was just as varied. Along the simple horoscopes we find more sophisticated examples:

> The Egyptian men of old who had faithfully studied the heavenly bodies and had learned the motions of the seven gods, compiled and arranged everything in perpetual tables and generously left to us their knowledge of these things. From these I have accurately calculated and arranged for each one (of the seven gods) according to degree and minute, aspect and phase, and, simply, not to waste time in enumerating each item, whatever concerns its investigation. For thus the way of astrological prediction is made straight, unambiguous, that is, consistent. [...] And Phosphoros, the star of Venus, had completed in Pisces 16 degrees and four minutes, which is the fifteenth part of a degree; in the sign of Jupiter; in its own exaltation; rising at dawn; at the Southern Fish; like crystal; in the terms of Mercury; distant two lunar diameters from the Star in the Connecting Cords. [...] Titus Pitenius computed it as is set forth. [...] Computed in Hermopolis, where the horizon has the ratio seven to five. The time of pregnancy: 276 days. With good fortune.[8]

This planetary picture is quite complex; the astrologer prizes accuracy and emphasizes that the calculations are his own, the result of personal input. By referring to the eternal planetary tables handed down from the distant past, he inscribes his practice within a valuable tradition. In sum, astrology catered for a diverse public, with diverse pockets. While the use of planetary tables was a constant, calculating skills, the ability to make one's own astronomical observations and the resulting greater accuracy made the difference between a luxury horoscope and an ordinary one.

Our papyrological evidence includes of course a great many accounts, receipts, and lists for tax purposes, whose format shows remarkable continuity with earlier periods. The notion that accounts can be tampered with, and that accountability goes hand in hand with political honesty, are also already familiar.[9] Since the time of the Republic, Roman magistrates had had to give accounts at the end of their term of office, and deposit a copy with the central archive in Rome; the charters of newly-founded cities, or of cities which had newly entered the Roman sphere of dominance, often mention the obligation to give accounts.[10] These practices are known from previous chapters; the difference is in whom one is accountable *to*. An inscription from Messene (AD 35–44) puts accounts in the context of a Greek city faced with Roman visitors:

> [Aristocles], secretary [of the councillors], introduced the assessment of the eight obol tax into the council, with an account of the [revenues accruing from] it and how each has been spent on the purposes ordained, showed that he had used it for nothing [other than these purposes], and made his statement of the sums of money still owing from the tax in the theatre before non-members of the council and in the presence of Vibius the praetorian legate [...] In this connection also the councillors approved his care and integrity and with Vibius the praetorian legate were eager to do him the honour of a bronze statue, while Vibius the praetorian legate personally gave him in the presence of all the citizens the right to wear a gold ring [...] Whereas Aristocles the son of Callicrates [...] gave his attention to the clear daily writing up on the wall of all the financial transactions of the city by those responsible for handling any business of the city, setting a beneficial example to worthy men of integrity and justice in the conduct of office [...] in entertaining governors and numerous other Romans too he devotes the expenditure of his own money to the advantage of the city. [...] And on account of the merits inscribed above, Memmius the proconsul and Vibius the praetorian legate have each in recognition of his conduct given him the right to wear the gold ring, as has the council also [...] [Cresphon]tis <tribe>; one hundred and twenty-two talents, thirty minae, a stater, eight obols, a half obol; [Daiphonti]s <tribe>; one hundred twenty-two talents, fifty-six minae, five staters, eight obols [etc.].[11]

The writing on the wall and its corollary of transparency in the use of resources may be seen as a throwback to the golden times of the Greek *polis* (which Messene, founded in the fourth century BC, never experienced anyway), yet everything happens under the eyes of Vibius the praetorian legate, whose presence gives a whole new meaning to the Greek officers' accountability.

Another figure familiar from earlier times is that of the public accountant, in Latin *numerarius* or *tabularius*. He generally worked for upper financial officers called *rationales* or *a rationibus*; moreover, secretaries or *scribae*, whose duties were less specific, were usually numerate as well as literate. To these one should add the accountants and financial secretaries working in big private households, who tended to be slaves or freedmen. In the case of imperial slaves or freedmen, the boundary between 'state' and 'private' accountants becomes very blurred. Our abundant epigraphical evidence indicates that, especially in the second half of the first century AD, many of the *numerarii* and *tabularii* were freedmen, when not slaves, while from the mid-second century onwards we find a greater proportion of free-borns. A connected group whose features are not entirely clear is that of the *calculatores*, or *ratiocinatores*. They do not occur in inscriptions very frequently: the examples I could find were a *doctor* (teacher) of calculation in a provincial town, an Italian *sevir Augustalis* (member of a civic order made up of rich freedmen), and a slave child-prodigy from Ostia, who died at the age of thirteen and was fondly remembered by his teacher as a wondrous *calculator* who had written commentaries on the art.[12] Finally, accountants, secretaries and perhaps architects were included, among others, in a category of public clerks, the *apparitores*, which was only open to men of free or freedman status. We know that several *apparitores*, including the poet Horace, improved their status – he went from son of a freedman to member of the equestrian order, so it would seem that belonging to the category was a good avenue for social mobility.[13]

With the obvious exception of loans from the state to private citizens and taxes and duties,[14] public accounts in the Roman Empire were not usually displayed in the form of inscriptions. What we do find instead is for instance this:

> Gaius Appuleius Diocles, charioteer of the Red Stable, a Lusitanian Spaniard by birth, aged 42 years, 7 months, 23 days. He drove his first chariot in the White Stable, in the consulship of Acilius Aviola and Corellius Pansa. [...] He won his first victory in the Red Stable in the consulship of Laenas Pontianus and Antonius Rufinus. Grand totals: He drove chariots for 24 years, ran 4,257 starts, and won 1,462 victories, 110 in opening races. In single-entry races he won 1,064 victories, winning 92 major purses, 32 of them (including 3 with six-horse teams) at 30,000 sesterces, 28 (including 2 with six-horse teams) at 40,000 sesterces, 29 (including 1 with a seven-horse team) at 50,000 sesterces [...] He won a total of 35,863,120 sesterces.[15]

A charioteer's achievements are celebrated, and enumerated and added up, the better to emphasize his extraordinary career. On the principle that spending money for the community was virtuous and politically a good investment, there are also inscriptions advertising the amounts lavished on public works by private individuals. For instance, the long bilingual document known as *Res Gestae*:

> Below is a copy of the achievements of the deified Augustus, whereby he brought the whole world under the rule of the Roman people, and of the sums which he expended on the Republic and people of

> Rome. [...] I held a *lustrum* after an interval of forty-two years. At this *lustrum*, four million sixty-three thousand Roman citizens were registered. I performed a second *lustrum* [...] At this *lustrum* four million two hundred and thirty-three thousand Roman citizens were registered. [...] To every man of the common people of Rome I paid three hundred sesterces in accordance with my father's will; and in my own name I gave four hundred sesterces from the spoils of war [...] again in my tenth consulship I gave every man a gratuity of four hundred sesterces from my own patrimony [...] These gratuities of mine reached never fewer than two hundred and fifty thousand persons. [...] On four occasions I assisted the treasury with my own funds, paying over a hundred and fifty million sesterces to those in charge of the treasury. [...] I provided the public spectacle of a naval battle on the other side of the Tiber [...] having excavated an area a thousand and eight hundred feet long by a thousand and two hundred feet wide, where thirty beaked ships [...] joined in battle. [...] Of those who fought at that time under my standards, more than 700 were senators. 83 of these have become consuls [...] and approximately 170 became priests. [...] Italy possesses 28 colonies established under my authority [...] At the time of writing, I am in my seventy-sixth year.[16]

Close to the perceived end of his life, Augustus takes comprehensive stock, quantifying and counting not only sums of money he has spent, but also people, territories of the empire, members of the élite and the extent of their participation to the highest offices; down to his own age. A whole life and a whole piece of history are set down in numbers.

Augustus had other ways of making mathematics work for him: a monumental sun-dial was set up in Rome between 10 and 9 BC.[17] Its gnomon was an Egyptian obelisk whose shadow indicated hours, days and months and whose pediment reminded the public of Augustus' victory over Antony and Cleopatra. The obelisk is still extant, albeit in reconstructed form and in a different position, and archaeological excavations have unearthed bronze-inlaid time lines, dating from Domitian's period, which may have been a restoration of the earlier sun-dial.[18] Pliny the Elder (AD 23–79) declared the working of Augustus' massive clock a 'thing worthy of being known' and named the mathematician Facundus Novius as its maker:

> [Facundus] was said to have understood [its] principle from <the shadow cast by?> a human head. The readings have been out of line for about 30 years now, either because the course of the sun itself is out of tune and has been altered by some change in the

> way the heavens work, or because the whole earth has shifted slightly from its central position.[19]

The production of sundials was indeed an important branch of mathematics, discussed by authors such as Vitruvius, Hero and Ptolemy. Interestingly, Pliny was prepared to believe that the whole universe was out of synch rather than doubt the accuracy of Facundus' creation. We have another, earlier, spectacular example of time-keeping object in the so-called Antikythera device, retrieved from a shipwreck off the Greek island by the same name (see Figure 5.2).

A small bronze mechanism, datable to *c.* 87 BC, it consists of some thirty toothed wheels of different sizes, connected by pinions, encased in a box. Although its precise function is not entirely clear, it seems that various aspects

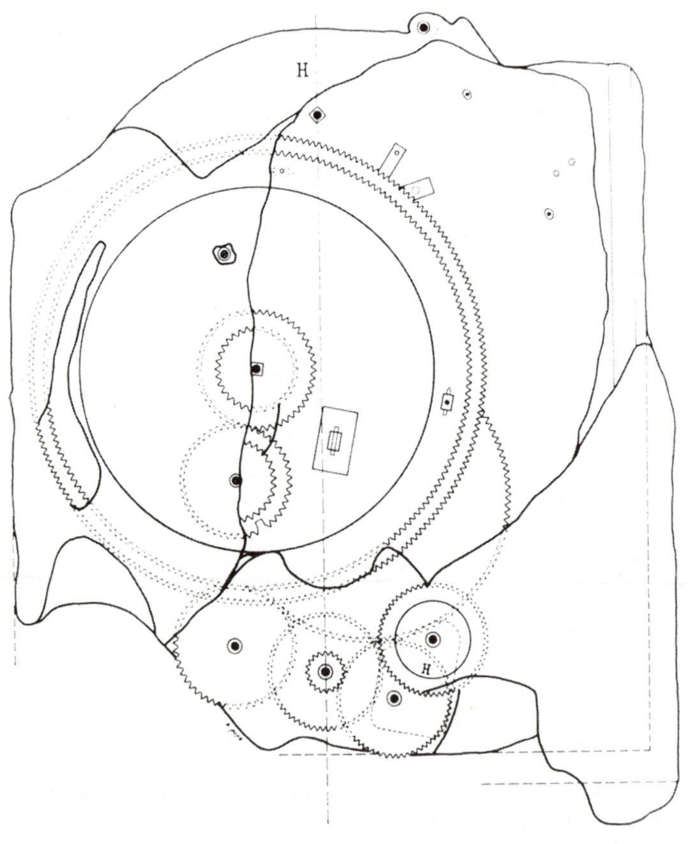

Figure 5.2 Gearwork from the Antykythera object (reproduced with permission of the American Philosophical society from de Solla Price (1974), fig. 14 p. 24)

of mathematical knowledge went into its making: it appears to be based on calendrical cycles, and its gear mechanism required accurate calibration of the diameter and number of teeth of the various wheels, in ways later discussed by Hero and Pappus. The main dial is inscribed with the names of the signs of the Zodiac and those of the months in the Egyptian calendar (widely used since Hellenistic times). Two longer inscriptions, unfortunately fragmentary, are a *parapegma* and some user's instructions. As in the case of planetary tables, the person producing the mathematical instrument is not necessarily the one using it, but mediation or (partial) transmission of expertise is provided in the form of instructions or explanations. The Antikythera device was a precious, prestige object: the craftsmanship involved is remarkable; also, the ship cargo included other luxury items such as statues, some of them in bronze, and amphorae. It is further testimony to the interest members of the upper classes had in astronomy and time-keeping.

Along with these two extraordinary examples, 'normal' sundials were common in the Graeco-Roman period, although it is often difficult to date them with accuracy. They have been found in private houses and working establishments, as well as in squares and markets, in cities such as Pompeii, Athens, Carthage, and Palmyra. Usually rather small, they were often made of stone with a metallic gnomon and with the time-lines engraved and painted in. Most of them only indicated hours, others days, months, solstices and equinoxes; also, their shapes varied from hemispherical to conical to cylindrical, and some dials were more complicated to make than others. We find, in short, a situation similar to that of horoscopes: the product was differentiated on the basis of its customers and their financial means. Besides, differentiation was achieved through a greater or lesser import of mathematical knowledge, as well as through decoration, scale and quality of material employed. To give an example, the sundial in Figure 5.3, found in Pompeii in the so-called *granario* (a sort of corn exchange), shows 'that the hour lines were constructed using arbitrary parallel circles by makers who were not particularly concerned with exact seasonal markings'.[20] On the other hand, other sundials were more accurate and their making must have required some knowledge of conic sections. Examples from Pompeii include some found in wealthy private houses and one found in a shop whose owner was called Verus:[21] it is of small dimensions and encased in a box, hence transportable, and it is made of ivory and carefully carved, so it must have been expensive. Verus' establishment has turned out other objects, mostly in bronze (candelabra, vessels), graffiti that identify him as a *faber* and, above all, writing implements, a ruler, compasses and the pieces of a surveying instrument, which has been reconstructed as in Figure 5.4

The reliability of this reconstruction is confirmed by a similar find in Bavaria, and by engravings on tombstones. Roman land-surveying has left

Figure 5.3 Sundial found in Pompeii (reproduced with permission from Gibbs (1976), plate 4 p. 137, © copyright Yale University Press)

many archaeological remains; the most spectacular are large-scale grids, still visible in many parts of France, Spain, the former Yugoslavia and especially Italy and North Africa. Figure 5.5 is an example.

These patterns are recoverable by various methods: digging, aerial photography, electronic remote sensing or field survey archaeology. On the ground, the lines are walls or roads, or remains of ditches which produce surface irregularities, in their turn detectable from the air or by instruments. Often, boundary stones are excavated in correspondence with these patterns; their function was to indicate the ownership or lease of a plot of land or its position with respect to some reference points. The laying-out of a grid like the one pictured in Figure 5.5 usually started from two designated main perpendicular axes (mostly roads or paths), generally orientated in the directions of the four cardinal points, the *decumanus* east–west and the *cardo* north–south. All the other lines were then laid in place parallel and perpendicular and at regular distances from the main axes. The *groma* and other sighting instruments were used to keep the lines straight; measuring rods, the extremities of which have also been recovered from Verus' shop and from Enns in Austria,[23] were used to ascertain distances; and sun-dials, probably not unlike the one again found in Verus' shop, helped lay the *cardo* and *decumanus* in the right directions. The whole operation was known

Figure 5.4 Reconstruction of a *groma* from Pompeii, *c.* first century AD (© copyright Science Museum/Science and Society Picture Library, London)

as 'centuriation' (the territory as 'centuriated') from *centuria*, a unit of measure corresponding to two hundred *iugera*, equivalent to two-thirds of an acre. Physical or man-made obstacles, such as rivers or irregular borders or temples, often got in the way of the surveyor, but, on the whole, the centuriated territory became in effect a geometrical landscape.

Many inscriptions document land-surveying activities. Especially for emperors like Vespasian or Nerva, who both came to power after very turbulent periods, the division and apportionment of land had strong political and financial motivations: political, because they were presented as the restoration of order and justice; financial, because remeasurement and redistribution of land often meant the establishment or reassertion of tax demands. The emperor often figured as the author of the survey, even when this had been in fact carried out by *mensores*. Clearly, surveying the land was a mark of power; it was up to Caesar to divide, parcel out and assign. One of the most famous monuments of land restitution is the so-called Orange cadaster (see Figure 5.6).

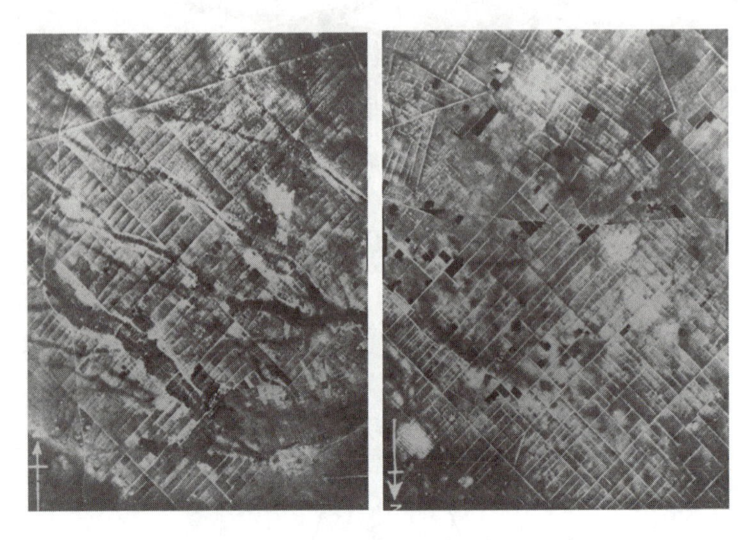

Figure 5.5 Centuriation from Africa Proconsularis (from Paul MacKendrick, *The North African Stones Speak*, p. 31, fig. 2.2, © copyright 1980 by the University of North Carolina Press. Used by permission of the publisher)

Figure 5.6 Fragment 7 of Cadaster Orange A, *c.* AD 77 (reproduced with permission of the Musée d'Orange, Vaucluse, France)

Now in a fragmentary state, it was a large-scale marble map of the region around the city of Arausius (Orange) in France. As one can see from the picture, space had been geometrized and divided into equal squares, and then inscribed with some landscape features (a river) and letters to indicate who occupied the land and, in some cases, how much tax they were supposed to pay.[24] The Orange cadaster was accompanied by an inscription, dated AD 77, with which Vespasian announced that he was returning public land to the state. What that in fact meant was that he was claming the right to tax land that had until then been in private use. Although no surveyors are mentioned, it was they, the 'silent technicians', who measured out the land and produced, or led to the production of, the map. It was their mathematical outlook that made both the act of measurement and the representation of accomplished measurement (the map) possible.[25]

In other cases the surveyor is present in the picture, as an expert called upon to give the benefit of his specialized skill. We have several examples of this in inscriptions about boundary disputes, which occurred very frequently not only between private individuals (in which cases the state was unlikely to intervene), but also between neighbouring communities:

> Decreed by the proconsul Quintus Gellius Sentius Augurinus, read out from the tablets on the kalends of March. Since the good and greatest emperor Trajan Hadrian Augustus had written to me in order that I, having employed surveyors and having investigated the case of the boundary controversy between the people of Lamia and those of Hypata, mark off the boundaries, and since I have been to the place in question rather often and for successive days, and have investigated in the presence of the representatives of either city, having employed the surveyor, Julius Victor, veteran of Augustus, it is resolved that the beginning of the boundaries is from the place where I found out that Siden was, which is below the precinct dedicated to Neptune, and from that place going down one keeps a straight line until the Dercynna spring [...].[26]

Although the emphasis is squarely on Augurinus, in his full proconsular judicial capacity (he inspects the sites concerned, listens to the parties involved, walks along the boundaries), a *mensor* is mentioned and even named. We know from other inscriptions of this kind that the expert testimony of the land-surveyor was only one of the various types of evidence used in a boundary dispute. Those were often sensitive cases: for instance, Greek communities, which in the past had resolved these matters through their own boundary-men and their own arbitrations, now were told what to do by a Roman officer and by surveyors sent from Rome. It is not surprising that decisions were taken after listening to local witnesses, looking at precedent adjudications and searching the terrain for old boundary markers. The expertise of the land-surveyor in some cases fades into the background: the role of the governmental officer as arbiter is brought to the fore, and accurate measuring becomes just one of the many guarantees of a fair decision.

The fact that many inscriptions about land-surveying give the point of view of the person *using* the services of the surveyor does not mean of course that the surveyor did not have a point of view. Had the inscription above been set up by Julius Victor, we may have heard a different version, of how the dispute was settled mostly thanks to his expert intervention. We have one such different point of view in a document found in Algeria:

> Both the most splendid city of Saldae and I, together with the people of Saldae, beg you, o lord, to exhort the surveyor Nonius Datus, veteran of the third Augustan legion, to come to Saldae, to complete the works. I set out on the journey, and was attacked by brigands; naked and wounded my men and I managed to escape; I arrived at Saldae; I met Clemens the provincial governor. He took me to the mountain, where they were uncertain and weeping about the tunnel, on the point of giving up the whole thing, because the tunnellers had covered a distance greater than that from side to side of the mountain. It turned out that the cavities diverged from the straight line, to the point that the upper end of the tunnel was leaning to the right southwards, and analogously the lower end of the tunnel was leaning to its right northwards: so the two parts were diverging, deviating from the straight line. But the straight line had been marked off with stakes on the top of the mountain, from east to west. [...] When I assigned the work, to make them understand how to do the tunnelling, I set a competition between the team from the navy and the team from the javelin division, and in this way they met in the middle of the mountain. [...] Having completed the work, and released the water, the provincial governor Varius Clemens inaugurated it. 5 modii of capacity [...].[27]

Nonius Datus, a retired military surveyor, describes his adventurous trip in his own voice. Unlike the Eupalinus tunnel (see chapter 1), the two arms of the underground water conduit in Saldae had not managed to meet in the middle: their course was divergent, and the people were thrown into despair. Datus went (literally) to set things straight: his intervention is compounded of managerial skills (motivating and organizing the workforce) and surveying expertise. The method of marking the path off with stakes required some simple geometry and accurate measurements, and is described in at least one almost contemporary land-surveying treatise.[28]

We have many other, shorter, inscriptions commemorating land-surveyors themselves.[29] Many of them, like Datus, were or had been in the military; several were freedmen. I think it is indicative of a certain pride in the profession that we have surveying instruments engraved on the tombstones of at least two *mensores*, in both cases freedmen.

Although it has not been possible to reconstruct their training in any detail, the combined evidence of inscriptions and technical treatises goes to show that the surveyors had indeed a sense of belonging to something like a professional group, with shared knowledge, work ethics and mathematical reference points such as Euclid and, to a lesser extent, perhaps Archimedes. At the same time, again both epigraphical and literary evidence also show that there were alternative, when not competing, views about surveying and land division on the part of non-technical administrators and, as we will see, of some members of the educated general public.

Vitruvius

A contemporary of both Julius Caesar and Octavian Augustus, Vitruvius worked as an architect and military engineer and may have been an *apparitor*.[30] His only work, the *Architecture*, deals with a variety of topics: book 10, for instance, is entirely devoted to machines, including catapults, and there is a long section on sun-dials. Mathematics is a ubiquitous presence: Vitruvius uses it to lay the groundplan of a building; to track down the directions of the winds; to calculate the proportions of the various elements of a temple starting from a module or standard element; to build sounding vessels that amplify voices in the theatre according to the principles of harmonics; to construct an analemma, the scheme on which dials are based. He provides lists of ready-made measurements for catapults, calculated on the basis of the weight of their intended projectile, because those who are not 'familiar with the numbers and multiplications through geometrical procedures' may find themselves at a loss for time if they need that crucial information during a siege.[31]

At the very beginning of the *Architecture*, after dedicating it to the emperor, mentioning his service to the emperor's excellent father, his good relations with the emperor's sister, and praising the emperor's own building achievements, Vitruvius states that the architect

> should be a man of letters, an expert draughtsman, know geometry very well, he should have learnt many histories, have listened diligently to philosophers, known music, not be ignorant of medicine,

> have learnt the decisions of jurists and be familiar with astrology and the principles of the sky. The reasons why this should be so are these. [...] Geometry furnishes many resources to architecture. It teaches the use of rule and compass which greatly facilitate the laying out of buildings on their sites and the arrangement of set-squares, levels and lines. [...] By arithmetic, the cost of building is summed up, the procedures of mensuration are explained, and difficult questions of symmetry are solved with geometrical procedures and methods. [...] But those individuals on whom nature has bestowed so much activeness, acumen, memory that they can know geometry, astronomy, music and the other disciplines thoroughly, go beyond the duties of architects and produce mathematicians. [...] Such men, however, are rarely met. We can point to Aristarchus of Samos, Philolaus and Archytas of Tarentum, Apollonius of Perga, Eratosthenes of Cyrene, Archimedes and Scopinas from Syracuse.[32]

Vitruvius is intent on building an image for his profession which situates it firmly within high culture. He makes it clear that architects are educated, fully-rounded individuals who can be trusted with the wider civic and political implications of their activities. Especially if he was an *apparitor*, the image of the architect he puts forth can be read as a paradigm of the gentleman technician, ready, willing and able to serve the state and climb the social ladder at the same time. That Vitruvius distinguishes architects from mathematicians is due not only to a consideration of how many things one can realistically be expected to do in a lifetime, but also to the emphasis he chooses to put on the architect's duties, his commitment to what is good and useful for the state. Geometry and arithmetic are justified on the basis of their *use*, not of their role in the advancement of knowledge.

Vitruvius' praise of mathematics is reprised and expanded at the beginning of book 9; once again, its value is derived from its actual benefits, from the things mathematics does. Athletes traditionally win fame and honour, yet, in Vitruvius' view, their achievements do not really signify anything for humankind. How much juster it would be, he comments, if similar honours were bestowed on learned men! He mentions Pythagoras, Democritus, Plato and Aristotle as examples, and recounts their many discoveries, 'which have been useful for the going forth of human life'. Probably on the basis of the *Meno*, Plato is reported as the author of the duplication of the square, which is introduced as a practical problem: 'Suppose there is a square area, or field with equal sides, and it is necessary to double it'. Vitruvius stresses that Plato provided a solution by means of lines, because the results produced through numbers, i.e. via simple multiplications, would not be correct. Next, he introduces a contrast between the mathematician and the craftsman:

> Then again, Pythagoras showed how to find a set-square without the constructions of a craftsman, and whereas the craftsmen who make a set-square with great labour are hardly able to get it close to the truth, the same thing corrected by procedures and methods is by him explained with instructions. For if three rulers are taken, of which one be 3 feet, another 4 feet and the third 5 feet, those rulers combined with each other will touch one another at their extremities making the shape of the triangle, and will form a corrected square. Moreover, if single squares with equal sides be described along the several rulers, when the side is of three, it will have 9 feet of area, the one of 4, 16, the one which will be 5, 25. [...] The same calculation, as it is useful in many things and measurements, so it applies to buildings in the construction of staircases, for the adjustment of the steps.[33]

Pythagoras and the artisan both want the same thing, but the former obtains a correct solution without 'great labour'. Vitruvius follows this up with the story of how Archimedes discovered a method to tell whether an allegedly golden crown made by a craftsman for king Hiero was made entirely of gold, or of an alloy of gold and baser metal. Finally, he praises Archytas and Eratosthenes for their solutions to the duplication of the cube, which, in his report, originated from a request by Apollo of an altar double the extant one. Although philosophy and poetry are also mentioned, the examples at the core of Vitruvius' argument are all mathematical, starting with the duplication of the square and concluding with that of the cube. Mathematics thus exhibits its utility for building, agriculture, the service of the king or devotion to the gods. Mathematics also allows Vitruvius implicitly to associate himself with famous figures from the past, and to contrast their virtue with the incapacity or moral defects of others.

Hero of Alexandria

Hero lived, probably in Alexandria, around AD 62 (the date of an eclipse he mentions). Of his works many have survived: *Automata*, *Pneumatica*, *Belopoeiika*, *Cheiroballista*, *Mechanica* (in an Arabic translation), *Dioptra*, *Metria*. There are also texts attributed to him whose authorship is debated: *Stereometria* and *Geometrica* may contain Heronian material but were put together at a later stage; the *Definitions* may be by Diophantus. Mathematics plays a number of roles in Hero's work: he deals with straightforward geometrical and arithmetical questions; his mechanics is heavily geometrized, and he has a number of pronouncements about mathematics and mechanics as forms of knowledge. I will discuss this latter aspect in chapter 6.

Let us start with the *Definitions*, which covers plane and solid figures, proportion, equality and similarity, infinity and divisibility and what it means to measure something. The introduction, addressed to a Dionysius, describes it as a preparation to Euclid's *Elements*, and indeed [Hero] explains and expands where Euclid had only concisely defined. One example is the definition of line, which in the *Elements* is 'length without breadth':

> Line is length without breadth and without depth or what first takes existence in magnitude or what has one dimension and is divisible as well; it originates when a point flows from up downwards according to the notion of continuum, and is surrounded and limited by points, itself being the limit of a surface. One can say that a line is what divides the sunlight from the shadow or the shadow from the lighted part and in a toga imagined as a continuum <it divides> the purple line from the wool or the wool from the purple. Already in customary language we have an idea of the line as having only length, but neither breadth nor depth. We say then: a wall is according to hypothesis 100 cubits, without considering the breadth or the thickness, or a road is 50 stades, only the length, without also concerning ourselves with its breadth, so that the calculation of that as well is for us linear; it is in fact also called linear measurement.[34]

The *Definitions* tends explicitly to relate basic concepts to external reality or to mathematical operations one may perform in the real world. Rather than axiomatico-deductive, its structure is demonstrative, in that the topics follow an increasing order of complexity, and taxonomic schemes are extensively used. While reporting as definitions things that in Euclid figure as postulates, axioms or even theorems, it does not cover the same ground as the *Elements*: it introduces many more geometrical objects (conics, to name but one) and occasionally provides alternative definitions altogether (for instance, of parallel lines). Moreover, some philosophical issues are addressed, such as the continuum principle mentioned above or the idea of divisibility:

> A part is a magnitude smaller than a greater magnitude, in the case when the greater is measured up exactly into equals. One says part here neither in the same sense that the earth is part of the

> universe, nor that the head <is a part> of a human being, but not even in the sense that, having drawn a perpendicular to the diameter of a circle from its extremities, we say that an angle taken outside the semicircle is a part of the perpendicular; for it is impossible for a right angle to be measured up exactly by this angle, which is called horn-shaped, since the horn-shaped <angle> is smaller than any rectilinear angle. We will thus rather take the part among magnitudes which are of the same kind and call the part among magnitudes accordingly, that is we say that the angle of a third of right angle is a part of the right angle. So we can leave aside the common sophistical saying, that if the part is what measures up exactly, then also what measures up exactly is part, but the solid is measured up exactly by lines one foot long, therefore the line one foot long is a part of the solid, which is absurd. A line one foot long measures not the solid, but the length of the solid and its breadth and its depth, which are of the same kind as the line itself.[35]

On the whole, although the question of the authorship may ultimately remain unsolved, the *Definitions* constitutes interesting evidence for the existence of accounts of basic mathematics other than Euclid.

As for Philo of Byzantium, mathematics is for Hero an important element in the development of military technology: through experience and adjustments, subtracting a little here, adding a little there, the engineers found 'harmonious' measurements which could be expressed in mathematical terms.[36] Indeed, Hero's machines are described as if they were geometrical objects, through lettered diagrams. To find the dimensions of the hole of a stone-thrower, one has to proceed as follows:

> Multiply by one hundred the weight in minas of the stone to be discharged; take the cube root of the product; and of whatever units you have found the root to be, add to what you have found the tenth part, and make the diameter of the hole that number of fingers. For instance let the stone be of eighty minas; one hundred times these produces 80,000; the cubic root 20 and the tenth of these 2 produces 22; of as many <fingers> will be the diameter of the hole. If the product does not have a cube root, it is necessary to take the nearest and add the tenth part.[37]

The same result can be obtained with the duplication of the cube, for which Hero provides a solution via a moving ruler.[38] We thus have an example of a problem solved in two different ways, neither of which seems to be privileged with respect to the other: a set of general instructions for calculation, followed by an arithmetical solution which takes approximation, and thus a certain degree of inaccuracy, into account; and a geometrical 'instrumental' solution whose results are substantiated but not necessarily handy for use. Along with the *Belopoeiika*, many of Hero's treatises (*Mechanica*, *Automata*, *Dioptra*, *Pneumatica*) describe machines for several purposes: lifting weights, enlarging or copying two- or three-dimensional figures, cutting thick pieces of wood, supporting a roof, putting up a puppet show, putting out a fire with a water-pump, producing music, measuring distances or the size of the moon. As with the catapults, Hero pays great attention to the materials and the construction specifications of his machines, and mathematizes them, reducing their components to geometrical objects. In some cases, their working itself is geometrically explained: for instance, a knowledge of centres of gravity and points of suspension, which Hero derives from Posidonius and Archimedes, can help to lift or support a weight. The hidden mechanism of a puppet-theatre must be constructed in respect of solid geometry; the functioning of a pump is explained by means of an indirect geometrical proof.[39] In all these cases, mathematics is put to use, and utility is not divorced from what we would consider simple entertainment. The *Automata* and *Pneumatica* describe self-moving statues of Dionysius, or 'bottomless' cups that keep spouting wine: they may have been built as part of a well-appointed banquet or a public celebration, where putting up an impressive show was one of the duties of the host or organizer. Entertainment was far from useless; it was an important way for the elite to display their wealth and power.

The *Dioptra* and especially the *Metrica* are the most mathematical of Hero's works. The dioptra is of course another machine, and the beginning of the book describes in detail how it is to be built, and duly proclaims its innumerable uses. After that, Hero puts it to work on a number of problems related to land-surveying, astronomy and engineering. To quote just a few examples, it can be used to measure the width of a river or the depth of a ditch, to help dig underground water conduits or build a well, to measure a piece of land or determine a boundary, or to divide pieces of land even when they are inaccessible because, for instance, they are thickly wooded. A mathematically constructed instrument, deploying geometrical principles, becomes the means through which the practitioner measures, counts, divides, and grasps a vast deal of the world that surrounds him. The facing page contains an example of dioptra-assisted land-division (see Diagram 5.3).

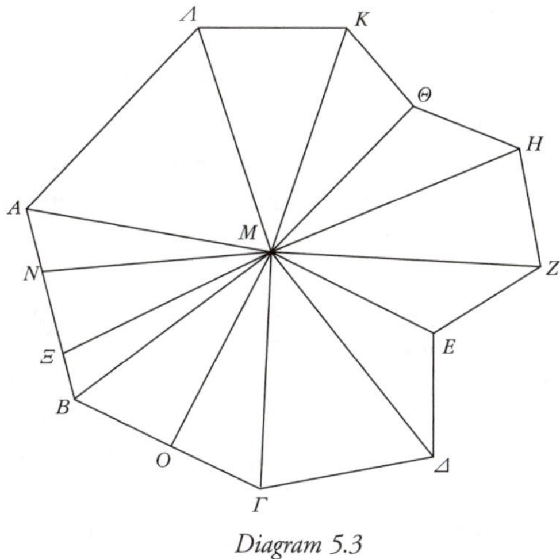

Diagram 5.3

To divide the given area through the given point in the given parts. Let the given point be for instance a water-spring, so that all the parts one takes use the same water. Let the given area be surrounded by straight lines *AB*, *BΓ*, *ΓΔ*, *ΔE*, *EZ*, *ZH*, *HΘ*, *ΘK*, *KΛ*, *LΛ*; if the lines surrounding the area are not straight, but some unordered line, let points in succession be taken on it, so that the lines between each successive <point> are straight. Let the given point be *M*, and let it be required to divide the area into seven equal parts through the point *M*. Let *MN* be drawn perpendicular to *AB* by means of the dioptra, so that if we imagine *MA*, *MB* joined, it will be possible to measure the triangle *AMB*. For the <rectangle> formed by *AB*, *MN* is double the triangle *ABM*. It is also possible to measure, as it has been written before, the whole area. If the triangle *ABM* is equal to a seventh part of the whole area, the triangle *ABM* will be one of the parts; if greater, it is necessary to subtract from it, having drawn *MΞ*, and one produces the triangle *AMΞ* equal to the seventh part of the whole area; if the triangle *ABM* is less than a seventh, it is necessary to subtract from the triangle *BΓM* the triangle *BMO*, which, together with the triangle *AMB*, will be a seventh part of the whole area [...] In this way we calculate the remaining triangles as well, and we divide the area into the given parts from the point *M*.[40]

The *Metrica* deals, as the title suggests, with measurement and division of plane and solid figures. Hero indicates from the start that his task is essential to geometry itself, which means, as the 'old account teaches us', the measurement and division of land. The universal utility of these operations propelled research at the hands of authorities such as Eudoxus and Archimedes, whose main discoveries (volume of the cylinder, area of the sphere) only confirm that geometry is quintessentially about measurement. His own work, Hero says, will be a combination of past results and personal contributions to the field.[41]

The structure of the *Metrica* will become clear if we look at one topic: the area of the triangle. The order of treatment is as follows: first the simplest case – the area of a right-angled triangle is half the area of a rectangle with the same sides. There follow the case of an isosceles triangle and of a scalene one, distinguished into the two sub-cases of height falling inside or outside the triangle. Hero then provides a general method, which allows one to calculate the area of *any* triangle, given its sides. This can be quoted as a good example of Hero's procedure throughout the *Metric.* First of all he demonstrates the method on a *specific* triangle, taking the reader through all the calculations, including the extraction of an approximated square root, which I have omitted here:

> There is a general method to find the area of any triangle whatever, given the three sides and without the height; for instance let the sides of the triangle be of 7, 8, 9 units. Add the 7 and the 8 and the 9; it makes 24. Of these take the half; it makes 12. Subtract the 7 units; the remainder is 5. Again subtract from the 12 the 8 <units>; the remainder is 4. And further the 9; the remainder is 3. Multiply the 12 by the 5; they make 60. These by the 4; they make 240; these by the 3; it makes 720; take the root of these and it will be the area of the triangle.

After the arithmetical part, Hero has a geometrical one, presented as a proof, and thus as a justification of what precedes it (see Diagram 5.4):

> The geometric proof of that is as follows: Given the sides, to find the area of a triangle. It is in fact possible to find the area of the triangle when one draws the height and obtains its magnitude, but it is required to obtain the area without the height. Let the given triangle be *ABΓ* and let each of *AB*, *BΓ*, *ΓA* be given; to find the area. Let a circle *ΔEZ* be inscribed in the triangle, whose centre is *H*, and let *AH*, *BH*, *ΓH*, *ΔH*, *EH*, *ZH* be conjoined. Therefore the <rectangle> formed by *BΓ EH* is double the triangle *BHΓ*, the

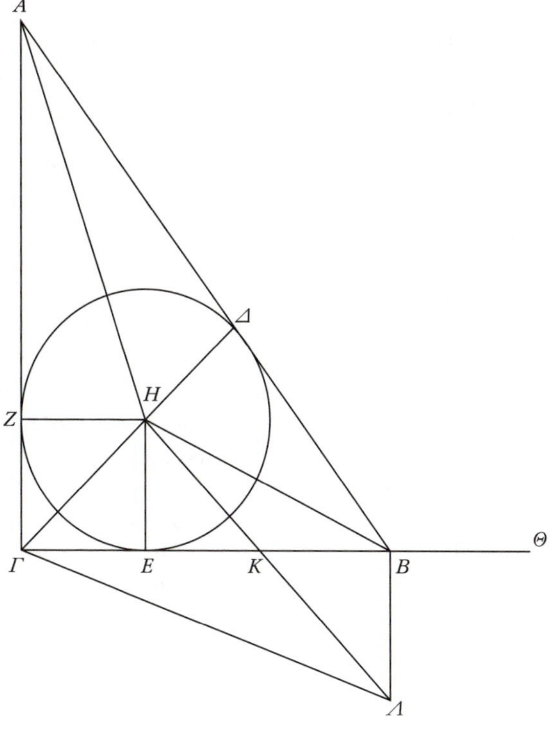

Diagram 5.4

<rectangle> formed by ΓA ZH is double the triangle $A\Gamma H$ and the <rectangle> formed by AB ΔH is double the triangle ABH. Therefore the <rectangle> formed by the perimeter of the triangle $AB\Gamma$ and by EH, that is the radius of the circle ΔEZ, is double the triangle $AB\Gamma$. Let ΓB be prolonged, and let $B\Theta$ be taken equal to $A\Delta$; therefore $\Gamma B\Theta$ is half the perimeter of the triangle $AB\Gamma$ since $A\Delta$ is equal to AZ, ΔB to BE, $Z\Gamma$ to ΓE. Therefore the <rectangle> formed by $\Gamma\Theta EH$ is equal to the triangle $AB\Gamma$. But the <rectangle> formed by $\Gamma\Theta$ EH is the side of the <square> on $\Gamma\Theta$ multiplied by the <square> on EH; therefore the area of the triangle $AB\Gamma$ multiplied by itself will be equal to the <square> on $\Theta\Gamma$ multiplied by the <square> on EH. Let $H\Lambda$ be drawn perpendicular to ΓH, BL to ΓB, and join $\Gamma\Lambda$. Since each of the <angles> $\Gamma H\Lambda$, $\Gamma B\Lambda$ is right, therefore $\Gamma HB\Lambda$ in the circle is a square; therefore the <angles> ΓHB, $\Gamma\Lambda B$ are equal to two right angles. Then also the <angles> ΓHB, $AH\Delta$ are equal to two right angles, since the <angles> around H are divided in half by AH, BH, ΓH and

> the <angles> ΓHB, $AH\Delta$ are equal to $AH\Gamma$, ΔHB and all the four right angles are equal; therefore the <angle> $AH\Delta$ is equal to $\Gamma\Lambda B$. And the right angle $A\Delta H$ is also equal to the right angle $\Gamma B\Lambda$; therefore the triangle $AH\Delta$ is similar to the triangle $\Gamma B\Lambda$. Therefore as $B\Gamma$ to $B\Lambda$, $A\Delta$ to ΔH, that is $B\Theta$ to EH, and conversely, as ΓB to $B\Theta$, $B\Lambda$ to EH, that is BK to KE since $B\Lambda$ is parallel to EH, and composing as $\Gamma\Theta$ to $B\Theta$, so BE to EK; thus also as the <square> on $\Gamma\Theta$ to the <rectangle> formed by $\Gamma\Theta$ ΘB, so the <rectangle> formed by $BE\Gamma$ to that formed by ΓEK, that is to the <square> on EH; in fact in the right-angled <triangle> EH is drawn from the right angle to the base perpendicularly; so that the <square> on $\Gamma\Theta$ multiplied by that on EH, the side of which is the area of the triangle $AB\Gamma$, will be equal to the <rectangle> formed by $\Gamma\Theta B$ multiplied by that formed by ΓEB. And each of $\Gamma\Theta$, ΘB, BE, ΓE is given: for $\Gamma\Theta$ is half of the perimeter of the triangle $AB\Gamma$; $B\Theta$ is the excess by which the half of the perimeter exceeds ΓB; BE is the excess by which the half of the perimeter exceeds $A\Gamma$; $E\Gamma$ is the excess by which the half of the perimeter exceeds AB, for this reason $E\Gamma$ then is equal to ΓZ, $B\Theta$ to AZ, because it is also equal to $A\Delta$. Therefore the area of the triangle $AB\Gamma$ is also given.

The geometrical proof is followed in its turn by a 'synthesis', once again a calculation carried out on a specific triangle, in this case one with sides 13, 14 and 15 units long.[42] We thus have both handy measuring procedures, in this case repeated twice, and the rationale behind them, the reason why those procedures work. These two parts are considered as a whole: the geometrical proof is the 'analysis' and the arithmetical calculation the 'synthesis'. Analysis and synthesis were distinct faces of the same coin. The fact that measuring happens in the real world, on objects which are not perfect geometrical entities, is never forgotten. Hero calculates the area of irregularly-shaped everyday objects by covering their surface with papyrus or linen; what is more, he measures the volume of irregular solid bodies by means of a method analogous to that devised by Archimedes. The reader is often reminded that the bodies in question are in fact a bathtub, a shell (in the sense of a decorative architectural element), a vault. In such a world no perfect accuracy is possible, and often one has to do with the best approximation, but the emphasis is less on the limitations than on the enormous versatility and power of mathematics when put to use for the necessities of life.

The other Romans

In parallel with the extensive Roman land-surveying activities, we have a sprawling body of works in Latin, known as the *Corpus Agrimensorum Romanorum*. The *Corpus* is a very mixed bag: it includes technical treatises; laws relative to land administration; abstracts from Euclid's *Elements* translated into Latin; lists of colonies. It is also a mixed bag of interpretative problems: its contents are difficult to date; its Latin is hard to decipher, and its manuscript versions are many and scattered around the world.

The earliest datable author in the *Corpus* is Julius Sextus Frontinus, a general, four times a consul, an augur and, around AD 97, a supervisor of the water supply for the city of Rome. He wrote three treatises: on land-surveying, on military stratagems, and on aqueducts. When Frontinus took up the job as water supervisor, he had no specific competence in the field and was ready to admit it – the treatise is the result of his desire to get acquainted with the task at hand. He found the water supply of Rome in a state of chaos; many pipes of many different sizes conveyed water from the public reservoirs to private establishments without authorization, a myriad unregulated rivulets escaping state control. The abuse took place with impunity because Frontinus' staff, the water-men, regularly profited from, and in fact encouraged, fraud by private citizens. In order to recover the situation, Frontinus employed mathematics extensively: he gave a streamlined description of the nine aqueducts Rome had at the time, each of them denoted by sets of measurements for length and capacity. Then he did the sums and, comparing input to output, saw that things did not add up: not all the water that went in at the source came out once inside the city. In fact, the extent itself of the abuse was detected because of his calculations.

It has been argued that the book on aqueducts served as a report of Frontinus' activity to the Senate.[43] If so, then the figures he abundantly provides would have been equivalent to rendering accounts: they demonstrated that he had the situation under control, and allowed the senators, if they so wished, to satisfy themselves that it was indeed so. The point is not so much whether they are an accurate reflection of the water supply in Rome *c.* AD 97 – indeed, as it has been observed, water output cannot be accurately measured on the basis of the diameter of the pipes, the way Frontinus does. The point is that mathematizing water supply played a major role in Frontinus' rhetoric of good administration. This impression is reinforced by the fact that he set out to standardize the pipes, choosing one type, the *quinaria*, as the official size. Up to then, different pipes had been used, making both repair and checking of irregularities extremely difficult. Frontinus decided that it was a good idea to set a standard, officially stamped, type of pipe, thus reducing the confusion of sizes to a common unit of measure. In his words, 'everything that is bounded by measure must be certain, unchanged and equal to itself'.[44] We find similar issues in his treatise on land-surveying, where he describes the measuring art as follows:

> It is impossible to express the truth of the places or of the size without calculable lines, because the wavy and uneven edge of any piece of land is enclosed by a boundary which, because of the great quantity of unequal angles, can be contracted or expanded, even when their number [of the angles] remains the same. Indeed pieces of land which are not finally demarcated have a fluctuating space and an uncertain determination of *iugera*. But, in order that for each border its type is established and the size of what is enclosed within is determined, we will divide the piece of land, to the extent allowed by the position of the place, with straight lines. [...] We also calculate the area enclosed within the lines using the method of the right angles. [...] Having assigned boundaries to its space, we restore the place's own truth. [...] For any smallest part of the land which is to be in the power of the measurer must be bound with the method of the right angles.[45]

The mathematical act of measurement allows an accurate division, and makes the business of apportioning and administering land easier. At the same time, mathematics has connotations of stability and certainty. In fact, geometrization is elevated by Frontinus to the status of a restoration of the essence itself of the land – the truth of the place consists in its being transformed into a geometrical object. The land-surveyor, like the water supply supervisor, is an agent of order over and against chaos and confusion at both a material and a moral level.

The *Corpus* also includes *Categories of Fields* by Hyginus Gromaticus (second century AD), a general treatise aimed at the professional which includes a history of land-surveying, an explanation of the surveyor's legal competences, as distinct from the magistrate's, and instructions on how to decipher inscribed boundary stones. Hyginus points out very often that surveyors make mistakes because of their lack of experience or knowledge. One of the hardest tasks is laying out the *decumanus* and *cardo* properly. Many people apparently were unable to find the true cardinal points; a good surveyor, however, should know some astronomy, and Hyginus, after citing Vergil, Lucan, and Archimedes, who 'wrote how much sand the world could contain if it was filled up', provides an orientation procedure that uses a gnomon and a simple geometrical construction.[46]

Like Vitruvius' architect, the land-surveyor in the *Regulation* must have an extensive field of competence, to include law, mathematics and astronomy.

It is the possession of this knowledge and of the related skills that distinguish the surveyor from other state officers, and the good surveyor from the bad one, all the more strongly in a sensitive field where mistakes could have political consequences.

Even keener on the necessity for the surveyor to have mathematical knowledge is Marcus Junius Nipsus (second century AD), author of *Measurement of a River*, *Replacing a Boundary* and *Measurement of an Area*. This latter work contains definitions of measures and angles, followed by a number of problems, mostly about triangles, including 'given an odd number, form a right-angled triangle', 'given an even number, form a right-angled triangle', and

> to measure the area of all triangles via one method, say, right-angled, acute-angled and obtuse-angled. We would find it out in this way. I unite into one the three numbers of any of the three triangles. That is, the right-angled one, whose numbers are given, the cathetus 6 feet, the base 8 feet, the hypotenuse 10 feet, I unite these three numbers into one, and they make 24. Of this I always take the half. It makes 12. This I set aside, and from this number, that is, from 12, I subtract the <other> individual numbers. I subtract 6: I put the rest under 12. Analogously I subtract the base, 8 feet, from 12: I put the rest under 6. I subtract the hypotenuse, 10 feet, from 12, the rest is two, I put it under four. Then I multiply 6 by 4. It makes 24. This I multiply by 2, and it makes 48. This I multiply by 12. It makes 576. Of this I take the root, and it makes 24. It will be the area. And the area of the other triangles will be calculated in the same way.[47]

The other mathematical problems in Nipsus' treatise are in the same vein: they read like a set of instructions, provide no proof and have as objects specific triangles, whose sides have a definite measure. A connection with Hero of Alexandria has been suggested, and is not unlikely; a similarity with the procedural style of other mathematicians from this period is evident.

A further version of mathematics for the surveyor is found in Balbus' book. He starts by telling his addressee Celsus of how he had been required to assist the emperor Trajan on a military expedition:

> After we took the first step on hostile territory, immediately, Celsus, the earthworks of our Caesar began to demand of me the calculation of measurements. When a pre-arranged length of marching had been completed, two parallel straight lines had to be produced at which a huge defensive structure of palisaded earthworks would

> rise up, for the protection of communications. By your invention, when a part of the earthworks was cut back to the line of sight, the use of the surveying instrument extended these lines. In regard to a survey of bridges, we were able to state the width of rivers from the bank close at hand, even if the enemy wished to harass us. Then it was that calculation, worthy of a god's respect, showed us how to know the heights of mountains that had to be captured. It was calculation that I began to revere more fervently, as if she were worshipped in all the temples, after experiencing and taking part in these great enterprises.[48]

Celsus was evidently himself a *mensor* – Balbus refers to 'our profession', and his tone is rather deferential, as if Celsus was the more senior or more expert of the two. The utility of surveying is displayed right from the start: Balbus comes on the scene as the busy servant of the state and of 'our most holy emperor', his expertise contributing to Rome's military supremacy. He takes the opportunity to promote the profession itself, because one of the inventions he uses is by Celsus, and it is their relationship, their shared knowledge, that enables him to deploy it. He also states the importance of calculation, indeed, his religious respect for it – in his view, it should become the object of universal worship, or, in other words, what surveyors do should gain everybody's recognition and respect.

Balbus' treatise was meant to be the first in a series, and it consists entirely of definitions; it presents a taxonomy of basic geometrical concepts, starting with a definition of measure and a list of units of measurement, and following with point and line, down to three-dimensional figures. While some of the definitions correspond to those in Euclid's *Elements* ('Point (*signum*) is that of which there is no part', 'Line is length without breadth, and the limits of a line are points'),[49] Balbus' account is geared to his chosen public, i.e. other surveyors. Thus we find:

> The kinds of lines are three, straight, circular and curved. A straight line is that which lies equally with respect to its straight points; circular, <that> whose path will be different from the arrangement of its points. A curved line is multishaped, like fields or ridges or rivers; the border of unsurveyed (*arcifiniorum*) lands is delimited by such lines, and similarly many things, which by nature are shaped by an irregular line.[50]

Balbus often explains a concept with examples (fields, boundaries, elements of the landscape) or vocabulary (*arcifinius*) drawn from the surveyor's experience. Also, he often refers to the Greek equivalents of the terms he is defining, transliterating them into Latin. In sum, his catalogue of definitions seems to have been compiled in order to enable experienced surveyors better to ground their knowledge. What was previously simply the line typical of an unsurveyed piece of land can now be subsumed into the geometrical domain as one of the three main lines, the 'curved' one. That some key terms are introduced in their Greek version may be meant to facilitate the understanding of Greek geometry in the original. In other words, Balbus' treatise is not about teaching the surveyor what to do in the field. His intended reader probably knew that already, and in any case, as it has been pointed out, one does not need to know the mathematical definition of right angle to be able to draw one. The treatise was rather about enhancing the surveyor's knowledge by enabling him to explain, order and classify concepts that he was already using. Another passage from the beginning of this text is quite revealing:

> It would seem shameful to me if, having been asked how many kinds of angle are there, I answered 'many': therefore I have examined the types, qualities, characters, modes and numbers of the things that are relevant to our profession, as much as my occupations allowed.[51]

Many surveyors practised their job in contexts where their authority could be disputed, or at least set against other sources of authority; knowing one's right angles may have boosted the claims of any exponent of 'our profession'. In sum, we could say that the surveyor envisaged by Balbus, and typified by himself and Celsus, is again a gentleman technician, who does his duty when he is required to, but can also appreciate returning to his *studium* and his *otium*, like any well-born, well-educated Roman of the time. His expertise is demonstrated not just by the fact that his devices work, but also by his being well schooled in the foundations of his art, including to some extent the Greek tradition of geometry.

We find similar issues in fields other than land-surveying: for instance, in the *Astronomy*, addressed by the author, (another) Hyginus (perhaps second century AD) to a Marcus Fabius. Hyginus praises Fabius' learning and discernment, and contrasts the education and judgement of the *periti* (experts) with those of the non-experts. The account he gives of astronomy is quite thorough, starting with basic definitions of sphere, pole and so on, and ending with a full star-catalogue, which takes up information from previous astronomers, including Eratosthenes, and expands the mythological stories behind many of the constellations' names. The star-catalogue comes complete with a calculation of how many stars there are in each constellation. For instance:

> The Snake has two stars at the top of the head, under the head four all in the same place; two towards the left hand of Ophiuchus [a neighbouring constellation], but that which is closer to its body is the brighter; and on the back of the Snake towards the junction of the body five, and in the first coil of the tail four, in the second coil towards the head it has six stars. So all together it is twenty-three stars.[52]

Astronomical expertise implies that one is able to recognize the stars, even those which are not very bright, assign them to the right constellation, and express their precise number. Apart from Hyginus, we have extensive literary evidence of the astronomical interests of many illustrious Romans: Cicero translated Aratus into Latin, and Germanicus wrote a commentary on the same text. Manilius chose Augustus as dedicatee of his *Astronomy*, which, as well as detailed arithmetical and geometrical procedures, contains an explicit parallel between the hierarchy of the stars and that of human society.[53]

Less lofty matters are treated by the jurist Lucius Volusius Maecianus in a short book he dedicated to Marcus Aurelius, with the aim to inform him on the divisions of the basic Roman currency unit – the *as*. After a brief introduction where he asserts the necessity for the emperor to acquaint himself with the topic because of its utility for inheritances 'and many other things', Maecianus goes into the details of each subdivision of the main money denomination, what part it is of the larger unit and what written sign denotes it (presumably in order to recognize it in accounts or on an abacus). Things are made more complicated by the fact that an *as*, for example, can be divided in two different ways: into equal parts, like half, thirds, and so on (up to twelfth parts, or *unciae*) or into unequal parts, like five-twelfths, seven-twelfths and so on (up to eleven-twelfths). Maecianus' premise is that the emperor is not completely at home in the complicated world of money, and that some dexterity is required fully to command division into parts, whose nature, as he says more than once, is infinite. The skills in question, one would think, were normally possessed by money lenders, secretaries and/or their slaves. Indeed, Maecianus says that he has got some of his information from 'reckoners' (*ratiocinatores*).[54]

The expectation that a powerful political figure ought to know his twelfth parts of an *as* may sound far-fetched. On the other hand, Quintilian (*c.* AD 35–90) insists on the necessity for the good orator to have rather extensive mathematical knowledge, both for its persuasive style of argumentation and for its contents, and checking accounts had been mentioned by Cato the Elder (234–149 BC) as one of the duties of landowners, to be done together and often by overseer and master.[55] Arithmetic was used in agricultural treatises to calculate the workforce necessary to run a certain establishment, or the time required to produce a certain yield. Here is a passage by Columella (mid-first century AD):

> This calculation shows us that one yoke of oxen can meet the requirements of one hundred and twenty-five *modii* of wheat and the same of legumes, so that the autumn sowing may total two hundred and fifty *modii*, and even after that seventy-five *modii* of three-months crops may still be sown. The proof of this is as follows: Seeds that are sown at the fourth ploughing require, for twenty-five *iugera*, one hundred and fifteen days' labour of the ploughmen; for such a plot of ground, however hard, is broken in fifty days, re-ploughed in twenty-five, ploughed a third time and then sown in forty days. [...] Forty-five days also are allowed for rainy weather and holidays, on which no ploughing is done; likewise thirty days after the sowing is finished, in which there is a period of rest. Thus the total amounts to eight months and ten days.[56]

Columella provides extensive information about land measurement too: he starts with a detailed description of units of measure, the *iugerum* and its many subdivisions, in a passage which reads rather like Maecianus' account of money. He then considers how to obtain the area of pieces of land of different shapes: square, rectangular, wedge-like, triangular, circular, semi-circular, curved but less than a semicircle, hexagonal, in increasing order of complexity. Finally, he explains how to calculate the number of trees a field of a given size can contain, given the distance between the plants. And yet, he says, land-surveying has been included in his account only out of friendship towards the addressee, Silvinus – Columella had already turned down a similar request made by another friend:

> I replied that this was the duty not of a farmer but of a surveyor, especially as even architects, who must necessarily be acquainted with the methods of measurement, do not deign to reckon the dimensions of buildings which they have themselves planned, but think that there is a function which befits their profession and another function which belongs to those who measure structures after they have been built and reckon up the cost of the finished work by applying a method of calculation. [...] But [...] I will comply with your wish, [Silvinus,] on condition that you harbour no doubt that this is really the business of geometers rather than of countrymen.[57]

Notice the dialectic between knowing how to do something and actually having to do that thing because that is your job – the architect knows how to measure the dimensions of buildings, but does not have to carry out that operation himself; the landowner does not have to survey his land, but it is good for him to know how surveying works. There is a definite idea that knowledge of these matters is valuable and entails a form of power, even when it is disassociated from actual application.

The ability to check accounts and land-surveys could be required of rich Roman citizens not just to keep their own house in order. Here is an official report of what Pliny the Younger set out to do on his arrival to Bythinia as imperial legate:

> [...] I did not reach Bythinia until 17 September. [...] I am now examining the finances of the town of Prusa, expenditure, revenues, and sums owing, and finding the inspection increasingly necessary the more I look into their accounts [...] I am writing this letter, sir, immediately after my arrival here.
>
> I entered my province, sir, on 17 September, and found there the spirit of obedience and loyalty which is your just tribute from mankind. Will you consider, sir, whether you think it necessary to send out a land surveyor? Substantial sums of money could, I think, be recovered from contractors of public works if we had dependable surveys made. I am convinced of this by the accounts of Prusa, which I am handling at the moment.[58]

Pliny's letters to Trajan are punctuated by requests for architects to sort out buildings which were falling apart, engineers for the same purpose, and, as in this case, *mensores*. All these mathematical experts appear to have been significant components of the administration of the provinces.

The Elder Pliny, Pliny the Younger's uncle, wrote a monumental *Natural History* addressed to the emperor, 'his' Vespasian. A self-declared collector of information, Pliny the Elder said that he had concentrated 2,000 works into the 36 books of his *History*. Indeed, the whole of the first book is a summary of contents, divided into topics covered and sources used, those latter in their turn distinguished between Roman and other. The summary is organized as an account, a quantified report of the extent of Pliny's knowledge: at the end of the summary of each book, we are given the total of facts, or famous rivers, or towns and peoples, or types of plants, that have been discussed. Yet, for other aspects quantification and calculation should not be pushed too far:

> Posidonius holds that mists and winds and clouds reach to a height of not less than forty stades from the earth [...] The majority of writers, however, have stated that the clouds rise to a height of

> ninety. These <figures> are really unascertained and impossible to disentangle, but it is proper to put them forward because they have been put forward already, although they are matters in which the method of geometrical gathering, which is never fallacious, is the only one that it is possible not to reject, if anybody likes to pursue these things further, not in order to establish a measure (for to want that is a sign of an almost insane whiling away one's time) but only a conjectural extimate. [...] And when they have dared to predict the distances of the sun from the earth they do the same with the sky [...] with the consequence that they have at their finger's ends the measure of the world itself. [...] This calculation is a most shameful business, because [the multiplication] produces <a number> which is even beyond reckoning.[59]

While inveighing against the heaven-measurers, Pliny praises Hipparchus for his catalogue of stars, and has absolutely no problem in reporting Eratosthenes' and Hipparchus' measurements of the Earth. Indeed, the *History*, whose introduction affects a casual attitude to knowledge and learning, is interspersed with passionate defences of their importance, and with lamentations on the ignorance of the time, when everybody is seeking profit rather than wisdom or learning.[60] Pliny also follows the practice, exemplified by many ancient historians, of using accurate figures to bolster some points. For example, in his section on metals he launches into a long discussion of the evils brought about by gold, especially its role in the corruption of old Roman customs, and adds:

> It follows that there was only 2,000 pounds [in weight] at most when Rome was taken [by the Gauls], in the year 364, although the census showed there were already 152,573 free citizens. From the same city 307 years later the gold that Caius Marius the younger had conveyed [...] amounted to 14,000 in weight [...] Sulla had likewise [...] carried in procession 15,000 in weight of gold.[61]

The numbers emphasize the parallel growth of Rome's power and of the menacing mass of metal. Pliny continues with cautionary tales of the excesses due to misuse of gold, in their turn punctuated by figures and accurate expenditures – he calculates just how out of proportion the exploits of people like Caligula or Nero were. He even comes to the conclusion that the very existence of numbers past a hundred thousand is a by-result of usury and of the introduction of coined money.[62] Similar mathematical moralizing we find in Seneca (*c.* 4 BC–AD 65). He denied that mathematics was a part of philosophy; in fact, its role consisted merely in measuring and counting the natural phenomena studied by philosophy. He also commented that the only study worth pursuing is that of wisdom, and that poetry, music, mathematics, astronomy are of only limited value:

> A geometer teaches me to measure my estate; but I should rather be taught to measure how much is enough for a man to own. He teaches me to do sums and put my fingers to the service of greed, but I should prefer him to teach me that those calculations have no importance [...] What good is there for me in knowing how to divide a piece of land into shares, if I know not how to share it with my brother? What good is there in adding together carefully the feet in a *iugerum* and including even something that has escaped the measuring rod, if I get upset by an arrogant neighbour who encroaches on my land? [...] O noble art! You can measure curved things, you reduce any given shape to a square, you enunciate the distances of the stars, there is nothing which falls outside your measure: if you are so good at your art, measure a man's soul, say how big or how small it is. You know what a straight line is; what good is that to you if you do not know what a straight life is?[63]

The reader can compare the pronouncements of some of the land-surveyors with these statements, which appear to deny that geometry or arithmetic can have any wider or higher significance. In conclusion, we can say that, contrary to the belief that there is no such thing as Roman mathematics, the Romans did a lot of counting and measuring, and they did a lot of thinking about counting and measuring. Some of them were experts whose knowledge included mathematics and whose service to the state depended on their mathematical knowledge; some of them used mathematics, sometimes critically, to articulate views about the relation of man to nature and the limits of human knowledge.

The other Greeks

Strabo, like Vitruvius a witness of the transition from Roman republic to empire, also opens his book by saying that the study of his subject, geography, requires extensive knowledge: philosophy, natural history, and especially astronomy and mathematics. He states that one need not know those last two disciplines thoroughly, but should be acquainted at least with some basics: for instance, 'what a straight line is, or a curve, or a circle, [or] the difference between a spherical and a plane surface'.[64] Mathematics also helps Strabo score points against other geography writers:

> Eratosthenes is so simple that, although he is a mathematician, he will not even hold fast to the solid opinion of Archimedes, who in his *Floating Bodies* says that the surface of every liquid body at rest and in equilibrium is spherical, the sphere having the same centre as the earth. All those who have studied mathematics at all accept in fact this opinion. [...] And as testimonies for [his] ignorant opinion [Eratosthenes] produces architects, even though the mathematicians have declared architecture a part of mathematics.[65]

Eratosthenes' crucial mistake was choosing the wrong supporter for his opinion: not Archimedes, evidently the chief authority on the topic, but architects. Appealing to the right source was by itself a sign of inclusion among those 'who have studied mathematics', because a shared tradition of results and canonical texts was part of what constituted them as a group. Strabo uses mathematics to cast doubt on other geographers elsewhere: at one point he sums up the distances they provide and demonstrates that the world thus obtained would be impossibly large, or shows that their data are incompatible with the relative positions and even climates of certain countries. Yet, he is keen to remark that criticisms based on mathematics are only fair when the theory to be criticized itself uses mathematical accuracy to gain credibility.[66]

Space in Strabo is often geometrized and enclosed within lines parallel or perpendicular to each other, to form a map. Countries are often assimilated to geometrical shapes. On several occasions, however, he makes it clear that geography is not a completely mathematical discipline. Knowledge can be gleaned from the experience of sailors, in preference or in alternative to geometrical reasonings. Or again, unless a country is well-defined by rivers or mountains, 'in lieu of a geometrical definition, a simple and roughly outlined definition is sufficient'.[67] On the other hand, Strabo launches in a full-length justification of the importance of mathematics in order even to define who the audience for his book is:

> The sailor on the open sea, or the man who travels through a level country, is guided by certain popular notions, and these notions impel not only the uneducated man but the man of affairs as well to act in the self-same way, because he is unfamiliar with the heavenly bodies and ignorant of the varying aspects of things with reference to them. For he sees the sun rise, pass the meridian, and set, but how it comes about he does not consider; for, indeed, such knowledge is not useful to him with reference to the task before him, any more than it is useful for him to know whether or

> not his body stands parallel to that of his neighbour. But perhaps he does consider these matters, and yet holds opinions opposed to the principles of mathematics – just as the natives of any given place do; for a man's place occasions such blunders. But the geographer does not write for the native of any particular place, nor yet does he write for the man of affairs of the kind who has paid no attention to the mathematical sciences properly so-called; nor, to be sure, does he write for the harvest-hand or the ditch-digger, but for the man who can be persuaded that the earth as a whole is such as the mathematicians represent it to be, and also all that relates to such an hypothesis. And the geographer urges upon his students that they first master those principles and then consider the subsequent problems; for, he declares, he will speak only of the results which follow from those principles; and hence his students will the more unerringly make the application of his teachings if they listen as mathematicians; but he refuses to teach geography to persons not thus qualified.[68]

In this passage, exposure to mathematical knowledge and the ability to be persuaded by its argumentations institute a hierarchy: harvest-hands, ditch-diggers, the uneducated at the bottom; the man of affairs perhaps in a sort of limbo – he only cares about what is useful and does not pay enough attention, but may be redeemable; finally, the students who can listen *as mathematicians* at the top, as Strabo's audience of choice. In my view, such pronouncements, especially since we do not find them actually applied in the *Geography*, whose position on mathematics is much more ambiguous, are an example of how claims about mathematics are also about something else. The remarks about not wanting to write for the natives of any given place or for the man of affairs signal Strabo's ambition that his account be universal rather than particular; disengaged from the mercantile outlook of, for instance, the tales of commercial travellers; made more authoritative by being made exclusive: one has to be qualified in order fully to understand it.[69] Mathematics was associated with all of these characteristics, so that invoking it, even without actually deploying it, helped Strabo articulate some of the features of his chosen way of doing geography.

Later than Strabo and thus a witness of the consolidation of Roman rule, Philo of Alexandria (late first century BC/early first century AD) was a Greek-speaking Jew, and most of his works revolve around interpreting the Bible, to which task he applied philosophy. The story of creation, for instance, is a Platonizing account of order emerging from chaos and, since 'order involves number', it is punctuated by numerical symbolism. Philo draws out the properties and wider significance of all the numbers involved: six days of activity for God, the heavens arranged on the fourth day, seven days to complete the creation. Thus, six is the first perfect number, being both the sum and the product of one, two and three; four is the base and source of ten (1+2+3+4=10) and of musical concords; seven is so rich of meanings that Philo himself does not know where to start.[70] As for the role of mathematics itself, Philo assigns geometry, along with grammar, as a maid to philosophy, since it needs philosophy to express definitions for its subject-matter. The study of geometry is deemed useful to learn about equality, symmetry, proportion and consequently justice;[71] indeed, equality and proportion are manifested everywhere in the world and are a sign of divine presence. In a sense, then, God is the best mathematician of all: not only does his creation exhibit symmetry, but he is the only one who can know the world, mathematically speaking, in a perfect way. Absolute accuracy in measuring or counting are only possible to God: *pace* the land-surveyors, He is the only one who can carry out a perfect division.[72]

Reading meanings in numbers, or numerology, was also a prominent feature in the *Arithmetical Theology*, now lost, by Nicomachus of Gerasa. He has variously been called a neo-Pythagorean or a neo-Platonist mathematician, and indeed he does mention Pythagoras and Plato often and favourably. Those categorizations, however, have had the unfortunate effect of pushing Nicomachus to the margins of modern histories of ancient mathematics, which tend to dismiss material associated with the neo-Pythagoreans as wacky mysticism. This is unfortunate, because Nicomachus was probably one of the most popular mathematicians of antiquity. The type of mathematics he did is widely represented in philosophical and literary works, he is often mentioned, and his success is attested by later commentaries and by an alleged translation into Latin, by Apuleius (*c.* AD 125–70), of his *Introduction to Arithmetic*.[73]

Apart from the *Introduction to Arithmetic*, we have an *Introduction to Harmonics* (an *Introduction to Geometry* and a *Life of Pythagoras* are lost). The *Arithmetic* discusses topics such as odd and even numbers, their various sub-species, including perfect numbers, division into factors, prime numbers, multiples and parts, ratios, triangular, square and polygonal numbers, and finally proportions and means. The argumentative style is discursive, with examples employing specific numbers, but no proof in the axiomatico-deductive style. Persuasion seems to be brought about by mere showing, and by enabling the reader to see for himself[74] that what is being said is verified in actual instances. Nicomachus also provides tables, for instance of multiples, for quick reference. An example will clarify his way of proceeding:

> We shall now investigate how we may have a method of discerning whether numbers are prime and incomposite, or secondary and

> composite [...] Suppose there be given us two odd numbers and some one sets the problem and directs us to determine whether they are prime and incomposite relatively to each other or secondary and composite, and if they are secondary and composite, what number is their common measure. We must compare the given numbers and subtract the smaller from the larger as many times as possible; then after this subtraction, subtract in turn from the other as many times as possible; for this changing about and subtraction from one and the other in turn will necessarily end either in unity or in some one and the same number, which will necessarily be odd. Now when the subtractions terminate in unity they show that the numbers are prime and incomposite relatively to each other; and when they end in some other number, odd in quantity and twice produced, then say that they are secondary and composite relatively to each other, and that their common measure is that very number which twice appears. For example, if the given numbers were 23 and 45, subtract 23 from 45, and 22 will be the remainder; subtracting this from 22 as many times as possible you will end with unity. Hence they are prime and incomposite to one another, and unity, which is the remainder, is their common measure. But if one should propose other numbers, 21 and 49, I subtract the smaller from the larger and 28 is the remainder. Then again I subtract the same 21 from this, for it can be done, and the remainder is 7. This I subtract in turn from 21 and 14 remains; from which I subtract 7 again, for it is possible, and 7 will remain. But it is not possible to subtract 7 from 7; hence the termination of the process with a repeated 7 has been brought about, and you may declare the original numbers 21 and 49 secondary and composite relatively to each other, and 7 their common measure in addition to the universal unit.[75]

If we compare the passage with its equivalent in Euclid's *Elements* (7.1), many differences will emerge: Euclid's proposition is enunciated in a way which requires proof, i.e. in the form of a theorem, and in fact it provides a proof, whereas Nicomachus just adds 'necessarily' a couple of times and simply states what is going to be the case. On the other hand, Euclid deals with two non-specified numbers, while Nicomachus appends two examples with specific numbers, whose function is both persuasive and pedagogic, because they enable the reader to go through the operation. You (my reader) will have started to notice that the presence of specific examples, exercises for the reader, as it were, with or without a corresponding general proof, is an usual feature in the authors of this period.

The first thing the *Introduction to Arithmetic* does is also what many other mathematical treatises did: justify its existence. Unlike Hero, who played the card of usefulness and social relevance of mathematics, Nicomachus presents it as an altogether nobler and more intimate enterprise, aimed at personal happiness, which, he concedes, is 'accomplished by philosophy alone and by nothing else'.[76] Philosophy, on the other hand, means love of wisdom, and wisdom looks for the truth in things, that is, it looks for what does not flow and change, but stays always the same: magnitudes and multitudes, or, formulated differently, size and quantity. In other words, real happiness can only be achieved through the study of mathematics, and awareness of this ultimate aim informs the whole book. As a premise to the demonstration that some numerical relations originate from the relation of equality, which is primary and seminal, Nicomachus states that he wants to present

> very clearly and indisputably [...] the fact that that which is fair and limited, and which subjects itself to knowledge, is naturally prior to the unlimited, incomprehensible, and ugly [...] it is reasonable that the rational part of the soul will be the agent which puts in order the irrational part, and passion and appetite [...] will be regulated by the reasoning faculty as though by a kind of equality and sameness. And from this equalizing process there will properly result for us the so-called ethical virtues, sobriety, courage, gentleness, self-control, fortitude, and the like.[77]

Acquiring wisdom amounts to recognizing that the universe and everything within it are well-ordered and good. The many interweaving relationships between numbers are a way, an excellent and straighforward way, for the human intellect to get to this essential truth.

Similarly enough, Ptolemy (mid- to late second century AD) opened what was to become his most famous work with the following praise of mathematics:

> [...] Aristotle divides theoretical philosophy [...] into three primary categories, physics, mathematics and theology. [...] the first two divisions of theoretical philosophy should rather be called guess-work than knowledge, theology because of its completely invisible and ungraspable nature, physics because of the unstable and unclear nature of matter; hence there is no hope that philosophers will ever be agreed about them; [...] only mathematics can provide sure and unshakeable knowledge to those who practise it, provided one approaches it rigorously. [...] With regard to virtuous conduct in practical actions and character, this science, above all things,

> could make men see clearly; from the constancy, order, symmetry and calm which are associated with the divine, it makes its followers lovers of this divine beauty.[78]

The divine beauty in question is that of the universe, whose regular motions Ptolemy takes upon himself to describe geometrically by means of uniform circular orbits. Understanding how the universe works holds the key to grasping its order and harmony, and thus constitutes a philosophical enterprise. As a deeply moral, indeed religious task, astronomy thus requires virtuous behaviour on the part of its practitioners. In concluding the *Syntaxis*, he reminds his addressee, Syrus, that his aim has been 'scientific usefulness', not 'ostentation'.[79] Part of this attitude translated into respect for the astronomers of the past (he quotes Archimedes, Apollonius, Eratosthenes and especially Hipparchus), and appreciation of what valid contributions they have made. Ptolemy even shows willingness to stand corrected by someone else's methods if they provide more accurate results than his.[80]

As well as accuracy, Ptolemy aimed at rigour and certainty, which he thought were attainable via a combination of geometrical and arithmetical proofs. Overall, the *Syntaxis*, parts of which are mathematically very complex, together with the *Planetary Hypotheses*, which gives further measurements of the distances of the planets, can be said to provide a total mathematization of the heavens. Ptolemy also turned his attention to the phenomena of vision and hearing, again mathematized in the *Optics* and *Harmonics*, and to the terrestrial world in the *Geography*, where each major city of the time is listed with its coordinates, and maps are provided which aim geometrically to represent the whole inhabited earth.

In terms of methodology, apart from deductive-style proofs, Ptolemy sometimes accompanied the general proof of a proposition with calculations on actual numbers, in order to show that the results matched.[81] He devoted large parts of the *Syntaxis* to the construction of tables (which circulated separately as *Handy tables*), so as to simplify calculations for time-keeping or chart-making. Like maps, tables were a kind of instrument – Ptolemy also explains how to build a quadrant, an equinoctial ring and an astrolabe.[82] Crucially, general proofs, arithmetical operations, tables, and observation, aided or not by instruments, were meant to work together. As an example, I report Ptolemy's instructions on how to find the position of the sun at any given time:

> we take the time from epoch to the given moment (reckoned with respect to the local time at Alexandria), and enter with it into the table of mean motion. We add up the degrees corresponding to the various arguments, add to this the elongation, 265 and 15 parts,

> subtract complete revolutions from the total, and count the result from Gemini 5 and 30 parts rearwards through the signs. The point we come to will be the mean position of the sun. Next we enter with the same number, that is the distance from apogee to the sun's mean position, into the table of anomaly, and take the corresponding amount in the third column. If the argument falls in the first column, that is if it is less than 180 parts, we subtract the [equation] from the mean position; but if the argument falls in the second column, i.e. is greater than 180 parts, we add it to the mean position. Thus we obtain the true or apparent [position of the] sun.[83]

Or again, the theorems which justify the way in which the values of a table of chords have been obtained, can in their turn be used to test and correct the values in the said table, if in the future they turn out to be incorrect because of mistakes in scribal transmission.[84]

Ptolemy could be seen as an ideal counter-point to Sextus Empiricus. Sextus was a sceptic: his surviving works are an outline of Pyrrhonian philosophy and a long attack on any form of knowledge possible, from ethics, rhetoric and grammar to the mathematical sciences. Traditionally, mathematics had been seen as producing certainty: who would doubt that two plus two is four? Well, Sextus would: as far as arithmetic was concerned, for instance, he aimed to undermine its foundations in order to pull down the whole edifice. In *Against the Arithmeticians*, he relentlessly criticized basic notions such as number, unit or addition and subtraction. He came to the conclusion that number is nothing.[85] A similar strategy was deployed against the geometers: although they consider themselves safe because they use hypotheses, Sextus says, the very use of hypotheses can be criticized in several ways. The definitions themselves of point ('a sign without dimensions') and line ('a flux of the point' or 'length without breadth') are ill-founded – how can a point, which is incorporeal and with no dimension, generate a line, which has dimension?[86] And even assuming for the sake of argument that basic definitions are valid, geometrical operations such as bisecting an angle are shown to be impossible.[87]

Who are the mathematicians targeted by Sextus? He mentions Eratosthenes, especially his definition of line as 'flowing' from a point, and, most frequently, Pythagoras and the Pythagoreans. With a few exceptions (e.g. the definition of point), the mathematical notions contained in Sextus look less like what we find in Euclid than in Nicomachus. Sextus also affords insights into the public aspect of mathematics, the roles it played in the community. His doubts that arithmetical or geometrical certainties are not as incontrovertible as the mathematicians would have them are not taken to their fullest consequences:

> Life judges everything on the basis of standards which are the measures of number. Surely if we abolish number, the cubit will be destroyed, which consists of two half-cubits and six palms and twenty-four fingers, and the bushel will be destroyed and the talent and the remaining standards; for all these, as composed of a multitude <of things> are at once forms of number. Hence all the other things too are held together by number, loans, testimonies, votes, contracts, times, periods. And in general it is infeasible to find anything about life that does not participate in number.[88]

A different strand of Graeco-Roman philosophy is represented by Alcinous (second century AD), who wrote an introduction to Platonism where mathematics figured as part of theoretical philosophy. He followed Plato's *Republic* in believing that mathematics, as well as being useful for practical purposes, sharpens the intellect, hones and elevates the soul and gives accuracy. While reiterating that the mathematical sciences are subordinate to philosophy, Alcinous also says that they are a 'prelude' to contemplation, compares them to gymnastics and even deems music, arithmetic, astronomy and geometry 'initiation rites' and 'preliminary purifications' of our spirit, before greater studies are begun.[89] A fuller idea of what mathematics Alcinous may have had in mind is gleaned from another, probably contemporary, author for whom Plato was also a major reference point: Theon of Smyrna.

In his *Account of Mathematics Useful to Reading Plato*, Theon has a detailed image of the mathematical disciplines (five in his case, he adds stereometry) as the stages of an initiation rite which culminates with philosophy. While following the *Republic* and the *Epinomis* quite closely, he quotes from numerous other authors, including the Pythagoreans, Eratosthenes and (twice, briefly) Archimedes. One of the images that recurs most often in his only partially preserved treatise is that of a correspondence between various levels of reality: man, the cosmos, the city, the physical elements. Consequently, some mathematical notions, such as the tetrad (a group of four, a four-some), are ubiquitous: the four elements are a tetrad, as are the geometric solids that correspond to them (here Theon follows Plato's *Timaeus*); the 'common things' (man, house, city quarter, city) are also four. Basic arithmetical terms are defined in a way similar to Nicomachus; methods are provided to find various types of numbers (for instance, square or perfect number) and to generate proportions or means; a general statement is often accompanied by a specific example by way of demonstration and both 'numerical' and 'geometrical' procedures are occasionally given to find the same result.[90] The text as we have it ends with a section on astronomy, including a detailed discussion of eccentrics and epicycles.

And, to conclude, Galen. He was one of the most famous doctors of his times, physician to the emperor Marcus Aurelius and author of a huge number of books. Galen believed that a good doctor should also be a philosopher,[91] and often paired medicine and philosophy when talking about supreme knowledge in general. Both dealt with the two inseparable aspects of a human being: body and soul; also, medicine was a particularly complex form of knowledge, which a philosophical mind would understand better than a non-philosophical one. Formulating a diagnosis or predicting the outcome of an illness implied reading from signs, making causal connections, distinguishing bad inferences from good ones; communicating one's response convincingly and often in competition with other doctors' responses demanded that one's medical skills be supported by rhetorical capacities, by the ability to demonstrate that what one was stating was right. The well-rounded education provided by philosophy thus served an important purpose in the life of a doctor. The problem with philosophy, however, was that nobody seemed to agree on anything, and different schools bickered endlessly; the medical world was equally divided into sects, whose exponents debated about causes of diseases, appropriate cures and just about anything else. Looking for some certainty and (like Ptolemy and like many philosophers of this period) for a criterion to distinguish truth from falsity, Galen came to admire the rigour of mathematical proofs, and the consensus they engendered among geometers, arithmeticians and astronomers. He was impressed by the fact that nobody for instance would doubt the results contained in Euclid's *Elements* or in his *Phenomena*.[92] In fact, mathematics, although hard to follow for the general public, seemed to achieve universal persuasion among not only its own practitioners, but also philosophers and rhetoricians.

On top of its compelling form of argumentation, and the positive consequences this had in the establishment of shared belief, mathematics deserved recognition because of its concrete workings in the world. Galen never lost sight of the fact that the people engaged in mathematical practices (calculators, geometers, architects, astronomers, musicians, gnomon-makers) produced something: predictions of eclipses, buildings, instruments like sundials and waterclocks. He brought out the full implications of this in the following passage:

> Imagine that a city is being built, and its prospective inhabitants wish to know, not roughly but with precision, on an everyday basis, how much time had passed, and how much is left before sunset. According to the method of analysis, this problem must be referred to the primary criterion, if one is to solve it in the manner that we learnt in our study of the theory of gnomons; then, one must go down the same path in the opposite direction

> in order to do the synthesis [...] When we have in this way found the path which is to be followed in all cases, and once we have realized that this kind of measurement of periods of time within the day must be carried out by means of geometric lines, we must then find the materials which will receive the imprint of such lines and of this gnomon. [...] If you have no desire to find out this method, my friend – what can one say? You have obviously failed to recognize your own conceit, and the fact that one who is ignorant of these problems will never discover anything in the whole course of a year, indeed, in the whole course of a life. For they were not discovered in the lifetime of a single man. Geometrical theory was there previously, and was first used to discover those theorems which are known as 'elements'; once they were discovered, the men who came later added to these theorems that most wonderful science to which I have attributed the name 'analytical', and gave themselves and anyone else who was interested a most thorough training in it. And they have yet to produce a more wonderful product of their ingenuity than those of the sundial and the water-clock.[93]

For Galen, mathematical truth is demonstrated both by its products and by its proofs, and its validity is guaranteed by the role it has in the community, by shared assent and collective persuasion. Assent in mathematical proofs is generated by the experience itself of going through the demonstration or of learning a certain method to solve geometrical problems – one can see that it works. Analogously with the embodied mathematics of sundials, water-clocks, predictions of eclipses or architectural calculations: one can see that they work, they too are proofs of the incontrovertible truths of mathematics, and a proof which is often out there in the street, under everybody's eyes.[94] Thus, measuring, counting, the most basic mathematical activities are identified as fundamental elements of humankind's shared knowledge of the world.

Notes

1 Pliny Sr., *Natural History* 2.3–4, Loeb translation with modifications.

2 Anonymous, *On Plato's Theaetetus* 29.42–31.28 (*P. Berol.* 9782, probably from Hermopolis Magna, dated to the second century AD), my translation.

3 *P. Oxy.* 1.29 (redated in Fowler (1999), 211 to the late first/early second century AD) has enunciation and diagram of *El.* 2.5, in a quickly written but competent hand on bad papyrus; *P. Fayum* 9, latter half of second century AD, containing *El.* 1.39 and 41, looks like a professional scribe's copy but is very different from the text of Euclid we have (see chapter 4); *P. Berol.* 17469, second century AD, has figures and enunciations of *El.* 1.8, 9 and 10, the figures drawn with a ruler, the text carefully written. All this in Fowler (1999), 209 ff.

4 *P. Vindob.* 19996, in Gerstinger and Vogel (1932), exercise 25, my translation. Cf. also Fowler (1999), 253 ff.

5 *P. Mich.* 3.145.3.6.5–8 (second century AD), tr. J.G. Winter, Ann Arbor 1936, with modifications.

6 *P. Mich.* 3.144 (early second century AD).

7 See Jones (1997a), (1997b) and (1998).

8 *P. Lond.* 130 (AD 81), translation in Neugebauer and van Hoesen (1959), 23–4.

9 See e.g. Cicero, *Against Verres* 2.2.69.170 ff., *Letters to Atticus* 5.21 and cf. Fallu (1973); *P. Amh.* 77 (AD 139), in *SelPap*; *P. Tebt.* 315 (second century AD), in *SelPap*; Fronto, *Letters to the Emperor* 5.

10 Cf. the charter of Tarentum, *CIL* 12.590 (bronze tablet, between 88 and 62 BC); Urso (Osuna) in Spain, 44 BC, *CIL* 12.594; and Málaga, AD 81–4, all in Lewis and Reinhold (1990), 1.414–5, 1.424, and 2.236, respectively.

11 *IG* 5.1 numbers 1432 and 1433, translation in Levick (1985), number 70, with modifications.

12 *CIL* 13.6247 (near Worms in Germany, not dated, Lupulius Lupercus); *CIL* 5.3384 (P. Caecilius Epaphroditus, Verona, not dated); *CIL* 14.472 (probably AD 144, the child was a house-born slave), respectively. Cf. Martial, *Epigrams* 10.62, who mentions a *calculator* with numerous pupils.

13 Purcell (1983).

14 E.g. the second-century Italian *alimenta* inscriptions and the tariff list from Coptus (AD 90) in Lewis and Reinhold (1990), 255–9 and 66, respectively, or the bilingual tax and duties inscription (in Greek and Aramaic) from Palmyra (AD 137) in Smallwood (1966), number 458. On the *alimenta* cf. Andreau (1999), 119 with more bibliography.

15 *CIL* 6.10048 (Rome, AD 146), translation in Lewis and Reinhold (1990), 146–7.

16 *The achievements of the divine Augustus*, translation in Chisholm and Ferguson (1981), number A1, with modifications. See also Lewis and Reinhold (1990), 170, 260, 264–5; Nicolet (1988); Andreau (1999), 49.

17 Cf. Buchner (1982) and *contra* Schütz (1990).

18 Domitian's interest in mathematics and its applications is attested by a treatise *On specific gravities* (only extant in Arabic translation) addressed to him. The author, Menelaus, also wrote a *Sphaerics* (again only extant in Arabic translation) cited by Ptolemy in his *Syntaxis*. In the *Gravities*, Menelaus mentions an earlier treatise on centres of gravity addressed to the emperor Germanicus, who in his turn wrote a (still extant) commentary on Aratus' *Phenomena*.

19 Pliny Sr., *Natural History* 36.72 f., Loeb translation with modifications.

20 Gibbs (1976), 17.

21 See Della Corte (1954). I thank Henry Hurst for his help on this point.

22 See Dilke (1971), 66 ff.; Panerai (1984), 116.

23 Dilke (1971), 73.

24 Fragments of similar maps, albeit for different purposes, have been found in Rome and Spain: see Chouquer and Favory (1992).

25 See also the examples mentioned in Hinrichs (1974); Smallwood (1966), numbers 434, 441, 446, 454, 465; Levick (1985), numbers 50, 51, 57; Sherk (1988), numbers 87, 91, 96; *O. Bodl.* 2.1847 in Fowler (1999), 231 ff.

26 *ILS* 5947a, from Lamia, in Smallwood (1966) number 447, my translation.

27 *CIL* 8.2728=*ILS* 5795 (AD 157, from Tazoult), my translation. The top of the stele is adorned with three figures of women, labelled 'Patience', 'Virtue' and 'Hope'. I take it that the beginning of the inscription (which is fragmentary) is a letter copied for information by Datus.

28 Frontinus, *On land-surveying* 18–19.
29 Cf. Panerai (1984).
30 Vitruvius, *On Architecture* 1 Preface 2. An identification with Mamurra, *praefectus fabrum* (officer of engineers) under Julius Caesar, has been proposed, see Purcell (1983), 156, who also argues for his being an *apparitor*.
31 Vitruvius, *On Architecture* 1.2.2; 1.6.6 ff.; 4.3.3; 5.5.1; 9.7.1 ff.; 10.11.1–2, respectively.
32 Vitruvius, *ibid.* 1.1.3–4, 17, Loeb translation with modifications.
33 Vitruvius, *ibid.* 9 Introduction, quotation at 6–8, Loeb translation with modifications.
34 Hero, *Definitions* 2, my translation.
35 Hero, *ibid.* 120, my translation.
36 Hero, *Belopoiika (Construction of Catapults)* 112.
37 Hero, *ibid.* 114, translation in Marsden (1971) with modifications.
38 The duplication of the cube at *Construction of Catapults* 115–18; the same solution in *Mechanics* 1.11.
39 Hero, *Mechanics* 1.24 ff.; *Automata* 1.6-7; *Pneumatica* 1.1, respectively. Cf. also the explanation of the five simple powers by means of properties of the circle, *Mechanics* 2.7.
40 Hero, *Dioptra* 26, my translation.
41 Hero, *Metrica* 1. Preface.
42 The whole proposition at Hero, *Metrica* 1.8, my translation. The same proposition also in *Dioptra* 30.
43 See DeLaine (1996).
44 Frontinus, *On the Aqueducts of the City of Rome* 34.
45 Frontinus, *On Sand-surveying* 15–16; text in Hinrichs (1992), my translation. See Cuomo (2000b).
46 Similar observations on the necessity of geometry and astronomy for town-planners in Strabo, *Geography* 1.1.13, 2.5.1.
47 M. Junius Nipsus, *Measurement of an Area* 300.11–301.5, my translation. For an argument that the Roman land-surveyors knew Hero's work, see Guillaumin (1992).
48 Balbus, *Explanation and Account of All Measures* 92.7 ff., translation in Sherk (1988), number 113, with modifications. Dilke (1971), 42, thinks that the emperor in question is Domitian.
49 Balbus, *Explanation and Account of all Measures* 97.15, 98.15, respectively.
50 Balbus, *ibid.* 99.3–10, my translation.
51 Balbus, *ibid.* 93.11–15, my translation. See also Athenaeus Mechanicus, *On Machines* 15.2–4, who comments that being able to explain clearly how a machine works earns more praise even than building it.
52 Hyginus, *On Astronomy* 3.87, my translation.
53 Manilius, *Astronomica* 5.734 ff.
54 L. Volusius Meacianus, *Distribution* 65.
55 Quintilian, *Handbook of Oratory* 1.10.3, 1.10.34–49; Cato, *On Agriculture* 2.2, 2.5, 5.4. Cf. also Cicero, *On the Orator* 1.128, 1.158.
56 Columella, *On Agriculture* 2.12.7–9. Similar passages in Cato, *On Agriculture* 10 ff. and Varro, *On Agriculture* 1.18.
57 Columella, *On Agriculture* 5.1.2–4, Loeb translation with modifications.
58 Pliny the Younger, *Letters* 10.17a–17b (*c.* AD 110). Pliny complains about his official duties ('I sit on the bench, sign petitions, produce accounts, and write innumerable – quite unliterary – letters.') at *ibid.* 1.10.9.
59 Pliny the Elder, *Natural History* 2.85–8, Loeb translation with modifications. On similar issues see also Lucian, *Icaromenippus* 6, with a probable reference to Archimedes' *Sand-Reckoner*.

60 Pliny the Elder, *Natural History* 2.95; 2.247; 14.4.
61 Pliny the Elder, *ibid.* 33.15-16, Loeb translation with modifications.
62 Pliny the Elder, *ibid.* 33.133.
63 Seneca, *Letters to Lucilius* 88.10–13, translation based on Loeb and on C.D.N. Costa, Aris and Phillips 1988.
64 Strabo, *Geography* 1.1.21.
65 Strabo, *ibid.* 1.3.11, Loeb translation with modifications.
66 Strabo, *ibid.* 2.1.23 ff. See also *ibid.* 2.4.3.
67 Strabo, *ibid.* 2.1.11; 2.1.30, respectively. A country which cannot be described in accurate geometrical terms is e.g. Italy, *ibid.* 5.1.2.
68 Strabo, *ibid.* 2.5.1.
69 See Clarke (1999).
70 Philo of Alexandria, *On the Creation* 13–14, 48, 89–128. Cf. also e.g. *On the Migration of Abraham* 198–205.
71 See especially Philo, *On Mating with the Preliminary Studies* e.g. 16, 75, 146.
72 Philo, *Who is the heir of divine things and on the division into equals and opposites* 141 ff.
73 The information is due to a late source, Cassiodorus, see *OCD* on Apuleius.
74 Or herself – Nicomachus' *Harmonics* is addressed to a woman, further indication of the wide appeal of his work.
75 Nicomachus, *Introduction to Arithmetic* 1.13.10–13, tr. M.L. D'Ooge, University of Michigan Press 1926.
76 Nicomachus, *ibid.* 1.2.3, translation as above. At *ibid.* 1.3.7, Nicomachus cites the passage in Plato's *Republic* 527 ff. where the useful side of mathematics is subordinated to its role in the elevation of the soul.
77 Nicomachus, *Introduction to Arithmetic* 1.23.4–5, translation as above.
78 Ptolemy, *Syntaxis* 1.1, tr. G.J. Toomer, Duckworth 1984, with modifications.
79 Ptolemy, *ibid.* 13.11.
80 On this latter point, see Ptolemy, *ibid.* 4.9.
81 E.g. Ptolemy, *ibid.* 3.3; instructions are given to find the position of the moon and of the five planets both geometrically and arithmetically in *For the Use of the Tables* 6–7, 9–10, respectively. A similar practice in Vettius Valens, *Anthology* e.g. 1.8, 1.9, 1.15 (second century AD).
82 Ptolemy, *ibid.* 5.1; 5.12; 8.3, respectively. Pappus, Theon and Proclus also describe the construction of astronomical instruments.
83 Ptolemy, *ibid.* 3.8, translation as above.
84 Ptolemy, *Syntaxis* 1.10.
85 Sextus Empiricus, *Against the Arithmeticians* 34; cf. also *Against the Physicists* 2.309.
86 Sextus Empiricus, *Against the Geometers* 22 ff., 29 ff.
87 Sextus Empiricus, *ibid.* 109 ff.
88 Sextus Empiricus, *Against the Logicians* 105–6, Loeb translation with modifications.
89 Alcinous, *Handbook of Platonism* 161–2.
90 Cf. e.g. Theon of Smyrna, *Account of Mathematics Useful to Reading Plato* 116.23–118.3.
91 The title of one of his works was *The Best Doctor is also a Philosopher* – at 53, he claims that, according to Hippocrates, astronomy and consequently mathematics are central to the study of medicine.
92 Galen, *On his own books* 11.39-41; *On the Sentences of Hippocrates and Plato* 112; 482–6; *Logical Institution* 12; 16; *On Prognosis* 72–4; 120
93 Galen, *The Affections and Errors of the Soul* 80–7, translation in Singer (1997), with modifications.
94 Cf. Galen, *On his own books* 11.41.

6

GRAECO-ROMAN MATHEMATICS: THE QUESTIONS

This chapter is about two well-known passages by Cicero:

> With the Greeks geometry was regarded with the utmost respect, and consequently none were held in greater honour than mathematicians, but we Romans have delimited the size of this art to the practical purposes of measuring and calculating.[1]

and Plutarch:

> But all this [the Roman engines at the siege of Syracuse, 212 BC] proved to be of no account for Archimedes and for Archimedes' machines. To these he had by no means devoted himself as work worthy of serious effort, but most of them were trifles of geometry at play, since earlier the ambitious king Hiero had persuaded Archimedes to turn some of his art from the things of the mind to those of the body [...]. For the art of making instruments, now so celebrated and admired, was first originated by Eudoxus and Archytas, who embellished geometry with subtlety, and gave problems which had not been solved by reason-like or geometrical proof the support of sense-like and instrumental arguments. [...] But Plato was incensed at this, and inveighed against them as corruptors and destroyers of the excellence of geometry, which had turned from incorporeal things of the mind to sensible things, and was making use, moreover, of bodies which required much vulgar handicraft. For this reason mechanics, having been banished, was distinguished from geometry, and being overlooked for a long time by philosophy, has become one of the military arts.[2]

Even though the two authors were not contemporaries, their statements are connected. First of all, they both set out divides, between practical and non-practical mathematics, and between Greece and Rome. Second, both Cicero and Plutarch have been immensely influential, and many modern views of ancient mathematics are still tinged by their opinions. Many historians have bought wholesale into the view that Greek mathematics 'typically' was pure and speculative, whereas the Romans, that 'characteristically' pragmatic people, only cared for concrete applications. The fact that one is talking about mathematics, an objective science, has clouded another important matter: the images of 'Greek' and 'Roman' knowledge, like all historical images of knowledge, were a construction, not a neutral reflection of reality. That they have been so successful does not indicate so much that they were true, as that they tapped into persistent beliefs about mathematics, knowledge and the world in general.

We will explore some of the issues arising from those two passages: the first section deals with Cicero, Plutarch, the myth of Archimedes and the Greek/Roman divide. The second section takes its start from the pure/applied divide and then moves on to look at the distinctions mathematicians themselves were making: what boundaries did they draw, how did *they* see themselves? In a sense, it is a contrast between a view from above and a view from below, of (in principle) the same thing: mathematical practices in the Graeco-Roman period.

The problem of Greek versus Roman mathematics

Like many upper-class Romans of his time, Cicero had travelled to Greece as a young man, attended the lectures of famous philosophers and rhetoricians and started to collect a library. He espoused the widely-held opinion that mathematics was one of the constituents of a liberal education, and, again like many Romans at the time, had an interest in astronomy (he translated Aratus' *Phenomena* and wrote a treatise on astrology). As a magistrate, and a land-owner engaged in business, he was numerate enough to understand accounts and detect financial frauds through them. In sum, Cicero was not a mathematician, but he was not completely ignorant of mathematics either; in line with his philosophical beliefs and everyday demands, he recognized mathematics as an important form of knowledge. I do not think that he had a specifically mathematical agenda – I think that what he said about mathematics was also about something else.[3]

Let us look at the context of the short passage we have quoted at the beginning. It is part of the *Tusculan Disputations*, written *c.* 45 BC at a moment of forced rest in Cicero's previously hectic public life. The work consists of five dialogues, inspired by the Platonic model, set in the peaceful atmosphere of a country villa and concerned with big themes such as death, pain and virtue. Each dialogue is preceded by an introduction; the passage on mathematics comes from the first of those, where Cicero recalls his own major role in making Greek philosophy accessible to the Romans.[4] Admitting that the masters of the world had something to learn from another people was not unproblematic, and Cicero qualifies the Romans' debt to the Greeks in several ways: Romans are better at war and household management, they are doing really well at oratory and, although they have nothing as ancient as Homer and Hesiod, their literature is now flourishing. In fact, if some areas of Roman culture are underdeveloped it is for lack of appreciation, not of talent – Cicero laments that not enough honours are paid to poetry, painting, music and (our passage) geometry.

At a simple level, then, Cicero is matter-of-factly comparing two cultures: the balance is rather even, because, if the Greeks have started earlier, the Romans have learnt faster, and sometimes improved upon their teachers. The statement about Greek and Roman mathematics need not mean more than what it says; Cicero's picture of the situation is in fact pretty accurate, in that there is absolutely no evidence to contradict it on the Roman side. I would be hard put to adduce a Latin equivalent of Euclid, Archimedes or Apollonius. And yet, I do not think that is the main point – it is not the question that interests me here. Cicero's sentence should not be taken as a simple description of a state of things any more than any other general statement about Greeks and Romans – in fact, no more than what Cicero says in the same passage about philosophy and music, only to contradict it later. After declaring that music is virtually an unknown art among the Romans, Cicero claims that, thanks to Pythagorean influences, the early Romans practised singing on a large scale. Philosophy is first said to have been successfully introduced to the Romans from the Greeks, yet later Cicero laments that it is far from being praised and appreciated for its proper value.[5] As historical testimony, Cicero's picture, while superficially not entirely false, fails to take into account the fact that practical activities were not limited to the Romans: a lot of mathematics done in Greek was about measuring and counting. He ignores the views measurers and calculators, Roman or otherwise, had of themselves and their activities – as we shall see in the next section, they did not agree with his. In other words, the point of Cicero's picture is not that it tells the story as it was, because it does not, or not to the extent that historians would have it do. Its huge significance lies rather in what it can tell us about the perception, and uses of, mathematics at Cicero's time. A passage by Plutarch will help me set the scene for discussion.

In the *Life of Cicero* (paired with the *Life of Demosthenes*), Plutarch describes Cicero's early years and his educational trip to Greece:

> In Rhodes [Cicero's] teacher in oratory was Apollonius [...] It is said that Apollonius, who did not understand Latin, asked Cicero

> to declaim in Greek. Cicero was very glad to do so [...] So he made his declamation and, when it was over [...] finally Apollonius said: 'Certainly, Cicero, I congratulate you and I am amazed at you. It is Greece and her fate that I am sorry for. The only glories that were left to us were our culture and our eloquence. Now I see that these too are going to be taken over in your person by Rome'.[6]

The equation between acquisition of Greek culture and acquisition of power was arguably as valid for Cicero as it was for Plutarch, albeit in different ways, given the lapse of time between the two. Mastering the Greek language and literature, including philosophy and scientific treatises, was an important way for second- and first-century BC Romans to affirm their entitlement to master the Greek world itself. This is stated explicitly in a letter to Cicero's brother Quintus, who in 60 BC had been confirmed as proconsul of the main Greek province for the third year running. Cicero reminds Quintus that rulers, according to Plato's doctrine, should be philosophers – there should be no hegemony without education.[7] And in the *Tusculan Disputations* themselves, he calls for a 'birthday' of philosophy in Latin, for his fellow Romans to 'wrest' philosophy away from the failing grasp of Greece. An important element in the justification of the Romans' rule over the Greeks was that they had assimilated and espoused Greek values, at a time when the Greeks themselves were in decline. But there is more. Greek culture was not used just so that the Romans could construct their entitlement to rule over the Greeks and *a fortiori* over the barbarians. There was a battle to be fought at home too: at the same time that Rome was expanding into the Mediterranean and negotiating its encounters and co-existence with the Greek-speaking world, enormous changes were taking place within Rome and Italy itself.[8] Acquiring an empire had created a lot of new wealth; almost uninterrupted civil wars throughout the first century BC had caused a quick turnover in the small group of those who were rich and powerful enough to be either targets or emissaries of political persecution. As a result, the senatorial and equestrian orders saw a number of sudden departures and new arrivals. Cicero himself was a 'new man', the first generation of his family to enter the Senate. For him (or indeed for his brother) a good education, being at the centre of a cultural circle, possessing a well-stocked library, knowing Greek philosophy to the point actually to introduce it to other, less sophisticated Romans, were all fundamental ways of signalling status and political clout. Take the two brothers as represented in the letter to Quintus: they are philosophers, who have learnt the Greek lesson, and that is precisely why they are now entitled to rule over Greece, *and* why they are better qualified to do so than other Romans.

A Greek education enabled Romans to justify their power not just to the Greeks but also to other Romans. Cicero's divide was not just about Greek and Roman, and not just about pure and applied mathematics, but also about class and internal social conflict. After all, how does he describe *publicani*, tax officers who must have done a lot of calculating? They are powerful and have become indispensable, but they are also universally hated and a headache for a provincial governor who has to reconcile the local populations to their presence. Again, Cicero's works contain many stories about accounts that have been cleverly doctored – another example of mathematical ability that can turn to the service of corruption. On occasions, secretaries (*scribae*) are described as 'entrusted with the public accounts and the reputations of our magistrates', and presented as dangerous upstarts, examples of social mobility gone wrong.[9] Again, how does Cicero depict land-surveyors? We have some insight into that question thanks to three speeches given in 63 BC, first to the Senate, then to the popular assembly, about the necessity to oppose an agrarian law put forth by the tribune of the plebs Publius Servilius Rullus. In the latter's proposal, a commission of ten people was to be set up and equipped with technical staff, with the task of administering all public land: selling it, distributing it, founding new colonies if they thought it necessary. Cicero did not like the proposal for several reasons: he claimed that it did not make financial sense, and would in fact impoverish the state; his arch-enemy Julius Caesar was rumoured to be behind it; and above all it would relinquish enormous power into the hands of people whose social credentials were not immaculate.

The second speech significantly opened with a long apology of the social status of Cicero himself. He was, as we have said, a *homo novus*, but he emphasized that he had earned his status and reputation through good service, thus he would continue his good service to the people by warning them against the dangers of the agrarian law. He insisted that the excessive power trusted to the agrarian officers would subvert the whole structure of the state:

> [Rullus] gives the decemvirs [the agrarian officers] an authority which is nominally that of the praetors but is in reality that of a king. [...] he provides them with apparitors, clerks, secretaries, heralds, and architects [...] he draws money for their expenses from the treasury and supplies them with more from the allies; two hundred surveyors from the equestrian order, and twenty attendants for each are appointed as the servants and henchmen of their power. [...] Just observe what immense power is conferred upon them; you will recognize that it is [...] the intolerable insolence of kings. [...] They are allowed to establish fresh colonies, to restore old

> ones, to fill all Italy with their own; they have absolute authority to visit all the provinces, to confiscate the lands of free people, to sell kingdoms [...] [they can] delegate their power to a quaestor, send a surveyor and ratify whatever the surveyor has reported to one man by whom he has been sent.[10]

Surveyors – also characterized by Cicero as 'a picked band of young surveyors' and Rullus' 'handsome surveyors'[11] – are here a hefty crowd of henchmen, ready to divide out and sell the whole world to the highest bidder. That Rullus' surveyors should belong to the equestrian order is evidence of the growing importance of that group, and of its association with *publicani* and *apparitores* – they were threatening to push against boundaries that Cicero and people like him may have wanted to keep in place.[12] 'Surveyor' is elsewhere in Cicero almost a term of abuse, to the point of making me think that it need not necessarily imply that the targets of criticism *were* actual practitioners. In his speeches to the Senate against Mark Antony, Cicero calls Antony's brother Lucius, among other things, 'a heap of crime and iniquity' and, sarcastically, 'that most just surveyor' (*decempedator*, literally 'wielder of a ten-foot rod', pun involuntary?). Other associates of Antony include Nucula and Lento, 'the parcellers of Italy', and Decidius Saxa, a tribune of the plebs from Celtiberia and a 'cunning and expert land-surveyor', who started in the army, measuring out the camp, and now is hoping to measure out the city of Rome herself.[13] In all cases, surveyors are socially mobile individuals: children or grand-children of equestrians, army veterans, natives of semi-barbaric provinces come good through political intrigue.

On the other hand, Archimedes, the quintessential Greek mathematician, is for Cicero an example of virtuous life. He plays a big role in the *Tusculan Disputations*: he is likened to Plato's demiurge, because he built a sophisticated astronomical sphere in imitation of the heavens, and appears again in the fifth dialogue, where Cicero's contention is that only the practice of virtue can make one really happy. For instance, Dionysius of Syracuse, a man who had wealth and power but no virtue, was deeply unhappy.

> With the life of such a man, and I can imagine nothing more horrible, wretched and abominable, I shall not indeed compare the life of Plato or Archytas, men of learning and true sages: I shall call up from the dust on which he drew his figures an obscure, insignificant person belonging to the same city [...] Archimedes. When I was quaestor I tracked out his grave, which was unknown to the Syracusans (as they totally denied its existence), and found it enclosed all round and covered with brambles and thickets; for

> I remembered certain [...] lines inscribed, as I had heard, upon his tomb, which stated that a sphere along with a cylinder had been set up on top of his grave. Accordingly, after taking a good look all round [...] I noticed a small column rising a little above the bushes, on which there was the figure of a sphere and a cylinder. And so I at once said to the Syracusans (I had their leading men with me) that I believed it was the very thing of which I was in search. Slaves were sent in with sickles who cleared the ground of obstacles, and when a passage to the place was opened we approached the pedestal fronting us; the epigram was traceable with about half the lines legible, as the latter portion was worn away. So you see, one of the most famous cities of Greece, once indeed a great school of learning as well, would have been ignorant of the tomb of its one most ingenious citizen, had not a man of Arpinum pointed it out.[14]

The contrast between Greeks and Romans is inscribed within wider contrasts: the virtuous and content life of the wise man and the evil and miserable existence of the non-philosophized one; the people who ought to exert power because they have wisdom and those who in this bad, bad world end up sitting on a throne while wallowing in ignorance. What side Cicero was on needs no explaining – remember that he was writing while on exile from a political arena where he had not received his due deserts and had been unjustly persecuted. His pilgrimage to the tomb of Archimedes is both the cathartic act of expiation of a Roman towards the Greek genius unjustly extinguished by an earlier Roman,[15] and an identification with the deceased, a sort of looking through the thickets for a representation of himself, the wise man neglected by his own country fellows.

The *Tusculan Disputations* is not the only work by Cicero where we find Archimedes, and Cicero is not the only author who writes about him. The Syracusan mathematician had become a symbol of indomitable intellect and amazing ingenuity since at least Polybius, who describes his machines at length, and Livy. The details varied, but the story of him being killed more or less by mistake, while distracted by a mathematical demonstration, is common to various versions. Plutarch, to whom we now turn, was writing in an already long tradition, and some of the features of his account fit with earlier reports. Yet, in no other source do we find the remarks I have quoted at the beginning about what kind of mathematics Archimedes preferred, or what he really thought about his machines. The natural question would then be, is Plutarch's account accurate – is that what Archimedes really thought?

Let us say, first of all, that Plutarch probably had no way of knowing Archimedes' inner thoughts, any more than we do, unless he had access to some work (now lost) where Archimedes voiced them. It is unlikely that he did, and it would have been I think even more unlikely not to mention this Archimedean source, had he had it. Thus, Plutarch's report may or may not be true, but there is no sure way of finding out. Historians have of course drawn on Archimedes' extant works and on the role mechanics plays in them to side one way or the other, but the evidence is inconclusive. So, as in the case of Cicero, I prefer to shift the question from, is Plutarch's report true, to, what does it tell us about the way mathematics was perceived and used at his time?

In general, mathematics for Plutarch brought pleasure to those who practised it, but was also, in Plato's mould, a serious business, and there was a right way and a wrong way of going about it. For instance, while he did engage in numerological speculations, on several occasions he poked fun at people who attributed meanings to numbers inappropriately.[16] One of the *Table-Talks*, which purport to be reports of conversations held over dinner, has as subject 'What Plato meant by saying that God is always doing geometry'. One of the dinner guests likens the mathematical sciences to 'smooth and undistorted mirrors' where the objects of intellectual knowledge, basically Plato's Forms, can be seen reflected, and in this connection repeats a story we have heard before:

> Plato himself reproached Eudoxus and Archytas and Menaechmus for setting out to remove the problem of doubling the cube into the realm of instruments and mechanical constructions, as if they were trying to find two mean proportionals not by the use of reason but in whatever way would work. In this way, he thought, the good of geometry was dissipated and destroyed, since it slipped back towards sensible things instead of soaring upward and laying hold of the eternal and immaterial images in the presence of which god is always god.[17]

This passage is followed by a description of how the Platonic god eliminates arithmetical proportion, which corresponds to a desire for political and economical equality, and sustains geometric proportion, a different kind of equality where everybody is rewarded according to their (opportunely assessed) worth. The political meaning of the right sort of mathematics is clearly expressed, and linked to the reproach that hits the 'practical' mathematicians Eudoxus, Archytas and Menaechmus. Plutarch's discourse is not just about mathematics, of the cosmically and religiously proper sort – mathematics is used to say something about the pursuit of philosophy in the context of a particular social and political order. If one proceeds with the text, one finds good geometry as the guarantor of cosmic order itself:

> Matter is always struggling to break out into unboundedness, and seeking to avoid being subjected to geometry; but reason seizes upon it and encloses it in lines and marshals it in the patterns and distinctions which are the source and origin of all that comes to be.[18]

Recent studies of Greek literary culture between the first and the second century AD have argued that the flourishing of interest in grammar, lexicography and rhetoric, with its emphasis on a revival of atticizing Greek, was 'building defences around educated Greek'.[19] It can be argued along similar lines that Plutarch was building defences around what he saw as the 'right' mathematics and science, and deploying heroical examples from the past (Archimedes but also Plato) to that purpose. It will be useful to quote some more from the *Life of Marcellus*:

> And yet Archimedes possessed so great a mind and so deep a soul and such a wealth of theories that, although he had gained from those [from the machines] name and reputation not human, but of some superhuman being, he did not want to leave behind something written about them, but considering the business of mechanics and all the arts that as a whole touch upon utility low-born and fit for vulgar craftsmen, he directed his ambition only to all the things in which the beautiful and the extraordinary are not mixed with the necessary.[20]

The siege of Syracuse marked an important step in the Roman subjugation of the Greek world. As a result of the victory, Archimedes was killed, the city sacked, and its art treasures transported to Rome. Marcellus, however, is depicted as an unwilling conqueror: he weeps over Syracuse about to be pillaged, he tries to save Archimedes' life, he is, as Plutarch points out, the living example that even Romans could harbour gentleness and humanity. Marcellus is in a sense converted by Archimedes' martyrdom, and acknowledges the value of Greek culture first by mourning his death and then by showing good taste in his choice of Syracusan pieces of art as triumphal spoils. 'Before this time', Plutarch comments, 'Rome neither had nor knew about such elegant and exquisite productions, nor was there any love there for such graceful and subtle art'.[21] In fact, if Archimedes is made to embody the conflict between disinterested knowledge and dire pragmatism, after his death Marcellus takes up a similar role, and becomes the promoter of exquisite taste and love of beauty against the criticism of his compatriots:

> They blamed Marcellus [...] because, when the population [of Rome] was accustomed only to war or agriculture, and was inexperienced in luxury and recreation [...] he made them leisurely and loquacious about arts and artists, so that they spent a great part of the day in discussions. Marcellus however spoke of this with pride even to the Greeks, declaring that he had taught the ignorant Romans to admire and honour the wonderful and beautiful productions of Greece.[22]

To conclude: both in the case of Cicero and Plutarch, the divides they constructed and deployed were much more complicated than simple Greek/Roman or pure/applied dichotomies. Those divides had a political significance, not just in a cross-national, but also in a cross-social-strata sense. For instance, they upheld leisure, detachment, 'playing' with knowledge – all luxuries that not everybody could afford. Plutarch's emphasis on cosmic order is unequivocal, as are his general attitude to technical knowledge, and his remarks about the low and vulgar artisans. In short, mathematics gave upper-class Romans and Greeks a way of articulating their positions about the state and the individual, power and knowledge. But what did the mathematicians themselves, whether Greek or Roman, think they were doing?

The problem of pure versus applied mathematics

This section is a survey, in chronological order; I will draw some conclusions at the end.

Let us start with Geminus (first century AD). We know from the extent of the citations in the scientific and philosophical literature well into the fifth century AD, that his work was very influential, although the *Introduction to the Phenomena* is the only thing that has survived. The *Introduction* deals with astronomy in a rather discursive style, thus following a well-established *genre* of which Aratus, with his poem, was one of the most popular exponents. Geminus, however, emphasizes that mathematicians and physicists are different from poets and also from philosophers, and that astronomy and meteorology, which Aratus had put together, are in fact quite distinct. Objects of knowledge (the heavenly phenomena) and purposes of knowing (weather forecast, astrological predictions) may be shared, but primacy is claimed for mathematical and physical explanations. For instance, the grammarian Crates is criticized because, basing himself on Homer, he had affirmed that the Ocean was situated between the tropics. What is more, he had had the effrontery to claim that this view was supported by mathematics. Instead, Geminus thinks that Crates' theory is contrary both to physical and to mathematical accounts, and remarks that it had not been adopted by any of the ancient mathematicians. Thus, he presents himself as a more reliable interpreter of the past mathematical tradition than Crates; in fact, Geminus even dares occasionally to criticize the ancients. His rebuttal can be read as a delimitation of the relative spheres of competence of on the one hand poets, grammarians and (in a later passage) philosophers, and on the other hand of mathematicians like himself.[23] Geminus is just as cutting in his description of what he calls the meteorology of the inexperts: their calendars are based on repeated experience, not on well-defined principles; they lack *techne* and their conclusions are not necessary.[24] On the contrary, Geminus' mathematical version of astronomy enjoys greater certainty and reliability; its predictions of an eclipse or of the rising of a star can be depended upon. Moreover, it has great ethical significance, as established in the introduction, where the regularity and uniformity of the heavens' motions is compared to the irregularities and continuous changes of human existence.[25]

Geminus exemplifies features which we will find, in various modes and combinations, in the rest of this section: he draws a line between mathematics and other disciplines (further lines are sometimes drawn within the mathematical domain between different modes of practice); he adds authority to his statements by appeal to a tradition of which he is a representative; he emphasizes the ethical, social or political significance of his form of knowledge.

Vitruvius engages in similar boundary-drawing. As well as proclaiming what the good architect is, Vitruvius clarifies what he is not: for a start, he knows a good deal of mathematics, but is not a mathematician; he knows about building and making things, but is not a craftsman. Those two boundaries are far from well-defined, since the passages where he takes his distance from mathematicians and craftsmen could be juxtaposed to other parts of the text, where we find a slightly different story. At the beginning of the sixth book, Vitruvius tells the anecdote, widely circulated in this period, of the philosopher (various names in various versions) who was shipwrecked and cast on a strange shore, where he saw geometrical diagrams written in the sand. He exulted at the sight, because they were the signs of human civilization. The point of the tale is that mathematics is a possession for ever, 'immune from the stormy injustice of fortune, the changes of politics and the ravages of war'.[26] In fact, learning a craft also fits that description and Vitruvius reveals that he was trained in the architectural art (although he underlines that his education was both literary *and* technical). The system of apprenticeship common in the crafts has, he says, the advantage that future practitioners are vetted not only on the basis of their skill, but also of their reliability and loyalty. The good craftsman, skillful, loyal and immune from the storms of fortune, is then contrasted with the bad architect, who lacks both experience and training and gives a bad name to the profession as a whole.[27]

In other words, the relation between architects and mathematicians on the one hand, and architects and craftsmen on the other hand, is defined and re-defined instrumentally. When Vitruvius targets excessive specialization or unreliability, he emphasizes differences. When his aim is to mark out good architecture against bad, or assert the quasi-philosophical value of architectural knowledge, he flashes his mathematical credentials and reminisces about his technical roots. These issues are explicit in another passage, where Vitruvius seems to arrange the practice of the arts, and consequently their practitioners, in a sort of spectrum, ranked on the basis of fortune and social standing, from the poor and obscure to the wealthy and officially recognized. Crucially, one's position in this spectrum is not determined by mere skill or industry – out of a list of painters and sculptors, some became famous, while others remained obscure. Yet, those latter lacked neither industry nor devotion to the art nor skill: but their reputation was hindered, either by scanty possessions, or poor fortune, or the victory of rivals in competitions.[28]

The real value of craftmanship is not appreciated by the general public, because they can only judge what they see, and what they see is unfortunately not 'manifest and transparent' – the non-experts are not able to recognize real expertise. If the validity of the arts basically cannot speak for itself, social graces and adulation are what influences people's judgement. In other words, what should be a matter of greater art or skill, thus ultimately of greater (technical) knowledge, becomes a matter of wealth or social status. One needs to fight back, and this is why Vitruvius published his book: to display 'the excellence of our knowledge' and make a stand for technicians whose well-rounded education, honesty, significance for the community and roots in a long-standing tradition ought to counteract the respectable family, good connections, and vacuous charms of the bad architects.

Another mathematician in a pugnacious mood, Hero of Alexandria, opened his *Belopoeiika* as follows:

> The largest and most essential part of philosophical study is the one about tranquillity (*ataraxia*), about which many researches have been made and still are being made by those who pursue learning; and I think research about tranquillity will never reach an end through reasonings. But mechanics has surpassed teaching through reasonings on this score and taught all human beings how to live a tranquil life by means of one of its branches, and the smallest – I mean, of course, the one concerning the so-called construction of artillery. By means of it, when in a state of peace, they will never be troubled by reason of resurgences of adversaries and enemies, nor, when war is upon them, will they ever be

> troubled by reason of the philosophy which it provides through its engines.[29]

Everybody yearns for a tranquil, undisturbed life, and pursues happiness. A person, however, cannot enjoy true *ataraxia* without political and social security, and tranquillity must be reached both at an individual and at a communal level. The existence of this common ground, shared by everybody, makes Hero's comparison between forms of knowledge possible; if they aim at the same thing, they can be assessed one against the other on the basis of their results. In other words, he creates a context for competition between philosophy and mechanics where none need exist, because he is keen to show the wider relevance of his form of knowledge for society. He employs a string of oppositions: philosophy uses reasonings, which are mere words, mechanics uses machines, which you can see and touch, and so can the enemy; the largest part of philosophy is pitted against the smallest branch of mechanics. The rest of his account, as we have seen in chapter 5, employs mathematics to confer certainty on the workings of catapults, and is organized as a progressive history, where the achievements of engineers accumulate over time producing better and stronger machines. On the whole, mechanics is depicted as a science which can bring about peace and happiness for the individual and for the state, produces tangible results, justifies them on reliable mathematical bases and is practised by people who belong to a glorious tradition.

Reliance on previous authors and claims to general utility are also found in the *Pneumatics*, which deals with the properties and effects of vacuum and of moving air (*pneuma*, as distinct from still or elemental air). In the introduction, Hero again compares philosophers and mechanicians. Pneumatics, he says, has been studied by both, 'the former proving its power discoursively, the latter from the action of sensible bodies'.[30] The contrast is the same as in the *Belopoeiika*, empty words on one side, concrete and tangible effects on the other. Hero repeats it when he later opposes the fabricated arguments of those who assert things on the nature of vacuum but do not offer any sensible proof, and those who can sustain what they say by appeal to sensible phenomena.[31] Yet another piece of evidence from Hero is the introduction to the third book of the *Metrica*, about division of geometrical figures. After reminding the reader, as he had already done at the beginning of the first book, of the Egyptian origins of geometry, he continues:

> To assign an equal region to those who are equal and a greater <region> to those who deserve it according to proportion is considered completely useful and necessary. The whole earth is in

> fact divided by nature herself according to merit; in fact on her live a great people who have been assigned a great region, some on the other hand have a small region, being small in relation to the former ones; no less the individual cities are also divided according to merit; to those who govern and to the others who are able to rule, a greater part according to proportion, to those on the other hand who are not able to do that, small places and villages are left, and to those with smaller personalities farms and things like that; but those things somehow take on an extremely gross and useless proportion; if on the other hand you want to divide the regions according to the given ratio, so that, as it were, not even a grain of millet of the proportion exceeds or falls short of the given ratio, one needs geometry alone; in it <are> equal fitting, justice to the proportion, the uncontested proof about these things, which none of the other arts or sciences promises.[32]

The themes of reliability and usefulness of mathematics recur; its significance is stated with respect to the order itself of things in the world. Hero establishes a correlation between nature, the characteristics of peoples, the size of the regions they live in and their political role. He then inserts mathematics in this scenario, by identifying division of areas as a primary concern. As in the introduction to the *Belopoeiika*, he uses contrasting terms: great peoples and great regions, small peoples and small regions, those who rule and those 'with smaller personalities' who do not even have the potential to rule,[33] cities on the one side and villages and farms on the other. Reading between the lines, one could see this as a disillusioned depiction of a world dominated by inequality; in particular, dominated by a great people who had acquired enormous quantities of land. Like many non-Romans faced with the inevitability of Roman rule, Hero invokes nature and merit as justifications of the *status quo*. Yet, his images of mathematics and mathematicians as the guarantors of tranquillity and justice over uncertainty and disproportion can be read as a powerful statement that guarantees were needed, and tranquillity and justice strongly demanded.

Whereas Hero tends to compare mathematicians and mechanicians with 'outsiders', i.e. philosophers, in astrological texts different types of boundaries are drawn. As we have seen, both material and textual sources point to the existence of different levels and different traditions of astrological practice. A first distinction between astrologers is introduced by the degree of accuracy of their calculations and predictions – our authors tend to equate accuracy with the possession of mathematical skills.[34] A particularly well-articulated statement of these issues is in Ptolemy. He distinguishes astronomy from astrology thus:

> Of the means of prediction through astronomy, o Syrus, two are the most important and valid. One, which is the first both in order and in power, is that whereby we apprehend the aspects of the movements of sun, moon and stars in relation to each other and to the earth [...]; the second is that in which by means of the natural character of these aspects themselves we investigate the changes they bring about in that which they surround. The first of these, which has its own theory, valuable in itself although it does not attain the result given by its combination with the second, has been expounded to you as best we could in its own treatise in a demonstrative way. We shall now give an account of the second and less self-sufficient <means of prediction> in a properly philosophical way, so that one whose aim is the truth might never compare its perceptions with the sureness of the first, unvarying one.[35]

In order to demonstrate that astrology is feasible at all, he points out that celestial phenomena do affect what happens on earth: the sun and moon obviously, but also constellations or planets, whose appearance in the sky was made to correspond, ever since Hesiod's time, to changes in the weather. Astrology for Ptolemy is the reading of signs in the heavens for their consequences in the terrestrial region. Yet, due to the imperfect nature of the physical elements down here, which introduce unpredictable changes and disruptions, astrology can only be conjectural. Still, in principle, it should get more credence than other, generally accepted, types of sign-reading, such as farmers and herdsmen observing the heavens for agricultural purposes, or sailors detecting the onset of a storm. Astrologers are more reliable than all those people, because of the amount of knowledge they bring into their reading of signs. Ptolemy calls this knowledge 'theory', because it is based on a correct understanding of the times, places and periodic motions of the planets. A hierarchy of semiotic knowledge is thus constituted, with 'dumb animals' at the bottom, then farmers, herdsmen and sailors whose job entails regular observations, and finally people who have been thoroughly trained in mathematical knowledge of the heavenly phenomena. They can refer to the same things as the 'ignorant men' above, but with more accuracy and certainty. The skilled astrologer is still not infallible, but he is less likely to err in his predictions. A second distinction is drawn within the field of astrology itself. Bad practitioners, Ptolemy says, bring the profession into disrepute, either because they are not properly trained ('and they are many, as one would expect in a great and manyfold science'[36]) or because they give the name of astrology to divinatory practices that have got nothing to do with it.

In sum, Ptolemy upholds the value of astrology, which had come under many attacks,[37] by emphasizing its links with mathematical knowledge (of the kind expounded in the *Syntaxis* and condensed in the *Handy Tables*) and its value for the pursuit of a happy and tranquil life.[38] At the same time, he restricts the field of those who can practice it to thoroughly-trained people who are not only interested in their own gain.

Slightly later than Ptolemy, Galen, though not a mathematician in a strict sense, argues, in ways analogous to what we have encountered so far, that mathematics gives certainty, agreement and concord, while philosophy does not, and that the consensus thus generated coagulates a community around canonical texts (especially Euclid's works) and accepted procedures. Galen's sympathy partly stemmed from the fact that mathematics, like medicine, was a *techne*. In his *Exhortation to the Arts*, medicine, together with a whole host of mathematical disciplines, from geometry and arithmetic to architecture and gnomon-building, is part of the band of Hermes, the 'lord of the word'. The god's followers are well-behaved, keep their place and are honoured not on the basis of political reputation, noble family and wealth, but of a good life and of excellence in their profession. They are pitted against the followers of the goddess Fortune, whose mob includes the great tyrants of the past (Cyrus, Dionysius) as well as demagogues, betrayers of friends, murderers and robbers.

The ethical and political significance of mathematical professions is again clearly expressed by some of the authors in the *Corpus Agrimensorum*. We have already drawn attention to Frontinus and Balbus. Another example is Siculus Flaccus with his *Categories of Fields*. The treatise seems addressed to other surveyors, in order to make them aware of the various elements involved in working on the terrain, not only the administrative differences between types of land, but also the huge number of methods people in different regions have of marking boundaries. Flaccus frequently insists that the surveyor must pay attention to, and, whenever possible, respect, the customs of the place; he suggests that his expertise consists not only of technical abilities, but also of diplomatic skills and of experience in the ways of the world, gained from having travelled widely in the service of the Empire.[39] The knowledge of the surveyor encapsulates Roman history itself: sometimes the names given to a type of land bring the imprint of past events. For instance, Flaccus derives the etymology of *territorium* from the 'terror' felt by the populations when they abandoned their land, after being defeated.[40] Some other times the reader is alerted to the difficulty of disentangling successive layers of land management: for instance, the Gracchan and Sullan surveyors in the second and early first century BC respectively set boundary markers which sometimes were still in place in the second century AD and needed to be recognized. Or again, the vicissitudes of history, including the civil wars of the first century BC, could cause confusion in the central archive, supposed to contain maps of all the territories of the empire. The quick succession of events between Julius Caesar and the ascent of Octavian Augustus had left its trace on the records of land ownership, and only through calculation of the total areas involved was Flaccus able to set the record straight.[41] The business of surveying taps into history in an even more seminal way: the reason why fields are classified into various categories, about which the surveyor has to be knowledgeable, is that some peoples have fought the Romans; others, having experienced Roman military might, have settled for peace; others still have traditionally been allies of the Romans, to the point of fighting alongside them.[42] The land these different peoples occupy will have different statuses, in fact acquire a different nature in the surveyor's eyes.

> Thus, the natures of the fields are many and diverse, and they are unequal either because of the circumstances of war, or because of the interests of the Roman people or, as they say, because of injustice.

Flaccus also observes that 'wars are the reason why fields were divided'.[43] What Hero had put down to nature Flaccus unflinchingly attributes to war, personal interest and injustice; what the two have in common, though, is the recognition that, one way or the other, mathematicians, land surveyors, the exponents of *their* knowledge, had a fundamental role to play in society.

Let us start to draw some conclusions: we seem to have two opposing fronts. On the one hand, Seneca, Pliny the Elder, Cicero, Plutarch are questioning or downright denying the significance of certain mathematical practices. On the other hand, the authors we have grouped in the second section are saying that certain mathematical practices *have* great moral and political significance. In some cases, they even argue specifically that their version of mathematical knowledge is *better* than other forms of knowledge at attaining moral and political goals, that it gives concrete, reliable results in fields of common interest. For instance, philosophy, according to some *the* knowledge proper of political rulers, is made the butt of criticism not only for its supposed inability to do anything for anybody in the real world, but also because its exponents kept squabbling with each other.[44] In sum, the point of contention was not just, which mathematics is better – it was a debate about knowledge and its role in society where mathematics was only a part of the picture, and it was a debate about the role within society of the people whose knowledge identified them as philosophers, geometers, or architects. Some of the more vocal mathematicians (Vitruvius, the land-surveyors) are known to have belonged to strata of Roman society which were socially on the up, or aspired to be so. Yet, this is not a straightforward, simplistic, conflict of upper-class snobs vs. middle-class social climbers. The positions in the social spectrum of our protagonists are infinitely more complicated than that, as are their individual variations on the theme 'knowledge is power – but what knowledge'? An interesting case, for instance, is Frontinus, a *homo novus* like Cicero, and probably an example of upward mobility through the army: as a member of the elite, a senator and a consul, surely he was not expected to write a treatise on land-surveying. Let us focus, however, on his preoccupation with wanting to match the expertise of his subordinates:

> There is nothing as dishonourable for a decent man as to conduct an office entrusted to him on the basis of the prescriptions of his assistants, which it is necessary to do, every time that the ignorance of the person in charge has recourse to the experience of those who, even though they are parts necessary to the task, should still be like some sort of hand and instrument of the agent.[45]

That must have been in fact a real, if rarely recognized, problem: the knowledge that the elite looked down upon (and we are talking land-surveying, accounting, machine-building, the nitty-gritty of administration in general) was also the knowledge needed by the elite to run the state and thus maintain their elite status. On a different level, it was the knowledge needed to manage an estate or a workshop; management was usually delegated to slaves or freedmen, but they could not be trusted completely. Cicero, Columella, and others remind the reader and themselves more than once of the fine line they had to tread between leaving accounts, lists, decisions, money administration in the hands of their expert subordinates, and still retaining some control over them. It is the paradox embodied by Columella not wanting to talk about land measurement because it is below his dignity but still talking about it because one needs after all to know one's own land; by Pliny the Younger who would like to write literature and is forced to check accounts, but at the same time checking the accounts is what signifies his power with respect to, for instance, the population of Bythinia; it is the paradox of Cicero's *scribae* in whose hands and within whose public accounts the reputation of a magistrate lay. It is behind, I think, Galen's words:

> [People] prize horses trained for war, and dogs for hunting, more than any other kind; and they usually have their household staff trained in some skill (*techne*), often at considerable expense. None the less they neglect their own education. But is it not disgraceful that the slave should be worth as much as 10,000 drachmas, while

> the master is not worth one? One drachma? No one would take such a fellow even as a gift.[46]

Vitruvius, Hero, Balbus, Flaccus realized what they were worth and the wider implications of their knowledge. Consequently, they claimed recognition for it and for themselves. In the process, they often took care to come across as a group – the land-surveyors had a well-defined juridical identity – and/or as a tradition, stretching back sometimes to the remote past. For instance, Vitruvius saw mechanics and architecture as having crucially contributed to human civilization itself.[47] Interestingly, Archimedes, appropriated by some authors as the hero of good, philosophical, abstract mathematics is to some extent reclaimed by the mathematicians and mechanicians. Vitruvius and Hyginus mention his achievement and its importance for their work; Hero inscribes him squarely within the tradition of measuring and dividing, which is how *he* sees geometry. In a sense, they were all deploying a man from the past to talk about their present – they were all writing their own histories of mathematics for contemporary use.

Notes

1 Cicero, *Tusculan Disputations* 1.2.5, Loeb translation with modifications.
2 Plutarch, *Life of Marcellus* 305.4-6, Loeb translation with modifications. The life parallel to Marcellus' is Pelopidas'.
3 I adapt one of the main arguments in Edwards (1993).
4 An achievement recognized by e.g. Beard and Crawford (1999).
5 Cicero, *Tusculan Disputations* 4.3–5 and 5.5–6, respectively.
6 Plutarch, *Life of Cicero* 863a, tr. R. Warner, Penguin 1972. Cf. Swain (1990b).
7 Cicero, *Letters to his Brother Quintus* 1.27 ff. Cf. Ferrary (1988).
8 See e.g. Hopkins (1978), Wallace-Hadrill (1998).
9 See e.g. Cicero, *Against Piso* 61; *Letters to his Brother Quintus* 1.32 ff.; *Against Verres* 3.184. Cf. also Badian (1972); Andreau (1999), and, for *scribae* as *apparitores*, Purcell (1983).
10 Cicero, *On the Agrarian Law* 2.32–34, Loeb translation with modifications.
11 Cicero, *ibid.* 2.45, 53.
12 See Nicolet (1970), Badian (1972), Alföldi (1975).
13 Cicero, *Philippics* 11.12, 13.37, 14.10 (delivered in 43 BC).
14 Cicero, *Tusculan Disputations* 5.64–6. The passage with Archimedes as the demiurge at *ibid.* 1.63. It is worth noting that the Syracusans' neglect of Archimedes contradicts Cicero's initial statement that the Greeks held geometry in great honour. We find the same issues (Archimedes' globe, Archimedes and Dionysius) in *On the Commonwealth* 1.21–2 and 1.29, respectively.
15 See Gigon (1973).
16 Plutarch, *Table-Talk* 738d–739a.
17 Plutarch, *ibid.* 718e–f, Loeb translation with modifications.
18 Plutarch, *ibid.* 719e.
19 Cf. Swain (1997).
20 Plutarch, *Life of Marcellus* 307.3–4, Loeb translation with modifications.

21 Plutarch, *ibid.* 310.1.
22 Plutarch, *ibid.* 310.6–7, Loeb translation with modifications.
23 Geminus, *Introduction to the Phenomena* 16.22–23, 17.32, 7.18, 7.22.
24 Geminus, *ibid.* 17.1–25.
25 Geminus, *ibid.* 1.19–21.
26 Vitruvius, *On Architecture* 6 Preface 2. Cf. Cicero, *On the Commonwealth* 1.29; Galen, *An Exhortation to Study the Arts* 8.
27 Vitruvius, *On Architecture* 6 Preface 6–7.
28 Vitruvius, *ibid.* 3 Preface 3, Loeb translation with modifications.
29 Hero, *Construction of Catapults* 71–73.11, translation in Marsden (1971) with modifications. See also Athenaeus Mechanicus, *On Machines* 4-5, where the many words of the philosophers are contrasted with the really useful things produced by the mechanicians.
30 Hero, *Pneumatics* 1 Preface.
31 Hero, *ibid.* 1 Preface.
32 Hero, *Metrica* 3 Preface, my translation.
33 The term I have translated with 'having small personalities' is *mikropsuchos* and is not used very much outside Aristotle's work. Guillaumin (1997) argues for Platonist, Aristotelian and Stoic echoes in this passage.
34 See Vettius Valens, *Anthology* e.g. 4.11, 5.6, 9.9.
35 Ptolemy, *Tetrabiblos* 1.1, Loeb translation with modifications.
36 Ptolemy, *ibid.* 1.2.6, Loeb translation with modifications.
37 Cf. Cicero, *On Divination*; Sextus Empiricus, *Against the Astrologers.*
38 This latter especially at Ptolemy, *Tetrabiblos* 1.10 ff.
39 Siculus Flaccus, *Categories of Fields* e.g. 59, 71, 135, 145. As for other instances of diplomatic skills, Hyginus (thought to be not the same as Hyginus 'Gromaticus', also in the *Corpus*) recounts the episode of a surveyor assigned by Trajan to the distribution of land to veterans, who, as well as inscribing the size of the allotments on bronze tablets, noted down their borderline and position, so as to avoid fights and arguments among them: Hyginus, *Categories of Fields* 84.
40 Siculus Flaccus, *Categories of Fields* 26.
41 Siculus Flaccus, *ibid.* 272–6.
42 Siculus Flaccus, *ibid.* 5–7.
43 Siculus Flaccus, *ibid.* 33 and 210 respectively, my translation.
44 This according not just to mathematicians, but also to people like Galen and Lucian, in e.g. *Philosophies for Sale*, who in fact would consider themselves philosophers.
45 Frontinus, *On Aqueducts of the City of Rome* 2.
46 Galen, *An Exhortation to Study the Arts* 6, translation as above.
47 Vitruvius, *On Architecture* 10.1.4

7

LATE ANCIENT MATHEMATICS: THE EVIDENCE

> We cannot say that the number six is perfect because God perfected his work in six days, but that God perfected his work in six days because the number six is perfect. In fact, even if his works did not exist, that number would still be perfect.[1]

The period I will be examining in the present chapter (third to sixth century AD) begins with an Empire still in place and ends after its dissolution in the West. By the fourth century, Christianity had become the religion of the emperor (with the brief exception of Julian), and the Church emerged as a more and more powerful institution, with buildings, books, schools, canonical authors of its own. The interaction between new and old religion, between the various Christian sects, and between church and state are some of the key issues in the history of this period. One of the most dramatic ways in which a mathematician could be affected by religious changes is exemplified by the death of Hypatia of Alexandria, who taught philosophy and mathematics. She was lynched by a Christian mob in AD 415, for reasons that probably had to do with her being an educated, politically visible, pagan woman.[2] Other, more subtle, traces of Christianity in the field of mathematics will be explored in the next chapter. Here we devote individual sections to Diophantus, Pappus, and Eutocius, and general ones to 'Philosophers' (mainly Iamblichus and Proclus) and 'Rest of the world'. To begin with, as usual, I will review the material evidence, which will include (new to this chapter) legal sources.

Material evidence

As for earlier periods, the great majority of our papyrological sources come from Egypt. A particularly interesting cluster of third-century AD papyri, at present scattered around at least five different collections, has been identified as belonging to the archive of a rich landowner called Appianus, whose estates were in the Arsinoite nome. Many of the documents are accounts of a rather complex type. In fact, it has been said that 'the system of accounting on the Appianus estate [...] is the most sophisticated presently known from the Graeco-Roman world'.[3] Each administrator on each sub-division of the estate drew up his own little accounts, for day-to-day running of the farm, payment of the workforce and so on. The information relative to the production of crops, the sale of produce, the use of animals, the general expenditure on the staff (including the administrators themselves), was then summarized from these smaller accounts into larger monthly ones, in their turn pasted together (literally, as pieces of papyrus roll) into one big yearly account for each particular sub-division of the estate. Entries were arranged by sector, with cash expenses and gains extrapolated from all the different sectors.

It has been argued that the function of these accounts was not only to verify the honesty of the administrators, or to keep a record of things, but also to plan the management of the estate. Appianus and his own staff of secretaries could at little more than a glance get a clear idea of where profits and losses were being made, what parts were not productive, and even compare different sub-divisions of the estates for productivity. Accounts of this kind gave the owner the opportunity to take better economic decisions because the information was purposefully selected and arranged. It has also been observed that the accounting systems across different sub-divisions of the estate or across different estates look very similar, although variations may have been deliberately introduced by at least one administrator, Heroninos.[4] This suggests that the training undergone by the various administrators must have also been similar.

In sum, it seems that third-century Egyptian administrators were highly numerate, that their skills could be an instrument of social advancement (Heroninos came up through the administrative ranks, his son Heronas followed in his footsteps), and that they may have undergone a streamlined system of accountancy training. We have quite abundant evidence – papyri, wax tablets, wooden tablets bound together to make a booklet, potsherds – that financial mathematics was extensively taught at an elementary level. In terms of quantity (and it could of course be a mere accident of transmission) we have in fact more documents of this kind from late antiquity than from any other post-pharaonic period, and they are in Latin and Coptic as well as in Greek. Most of them contain division, multiplication and currency-conversion tables, very often together with literary material ranging from paraphrases of Homer to psalms. Sometimes the tables contain verifications that the result, usually provided without explanation, is indeed correct. In some cases we are given the name of the author of the document: thus, a wooden schoolbook probably from the sixth century AD is inscribed with the words 'Papnouthion <son> of Iboïs'.[5] One booklet contains both addition tables and a fragment of what looks like a land-survey;[6] a sixth-century wooden board preserves division tables on one side and nine problems (written in three different hands) on the other. The problems, specific rather than general and set in question-and-answer form, with the solutions written between the lines just above the questions, deal with everyday situations such as calculating interest on loans, or payments on leases of land.[7]

> A man once owning one island, a man came ... I say that I want to sow the island. I offer you in respect of half the island ... artabae of lachana per arura; in respect of the rest, for the third part 4 [or 1] artabae of lachana per arura; and in respect of the remaining sixth part 3 artabae of lachana. He came at the time of the harvest ... artabae of lachana. Find the number of arurae he [150] sowed.[8]

The fourth century saw a number of administrative reforms, perhaps initiated by the emperor Diocletian and definitely followed by his successors; they included a new way of assessing tax liability.[9] Traditionally, the amount of tax, in cash or kind, was assessed on the basis of the size of the land, its quality and its crops. There is evidence from Egypt that a tax on the person (a poll tax) was also levied. The system we find at work in the fourth century combined these two types of taxation into a unified system, whereby the total tax liability of a town or rural community could be expressed. Land was measured in *iugera*, or other local standards which in some cases were fixed by law; whereas people were counted by heads, *capita*. There is evidence that, as well as the two separate units, *iugera* and *capita*, there was a general unit, the *iugera vel capita*, which implied abstracting from what was actually being measured (property or people) for the purpose of calculating tax. In other words, the late ancient taxation system may have involved a sort of mathematization of resources, but evidence is too scant to make any definite claims. We can assume that, whatever the details of the system, if indeed there was a unified, in some sense 'abstract', numerical, unit of measure for both people and property, formulas for conversion between them must have been devised, taught, learnt and applied by skilled people. Calculations and account-keeping must have been required, and demand for them may have effectively increased from the end of the third century onwards.

The so-called archive of Aurelius Isidorus from Karanis (late third–early fourth century AD) offers more evidence for the use of mathematics in everyday administrative practices (mostly accounts), as well as providing insights into the fiscal system: for instance, it documents measurements of estates apparently for tax purposes. In three of these cases, the surveys were carried out by the same *geometrai*, Aurelius Aphrodisius and Aurelius Paulinus, in the presence of state officials.[10] This suggests that fiscal reforms may have boosted the need for the services of surveyors, as well as of calculators.

We have quite a lot of other evidence about land-surveying in late antiquity: for instance, an entry in the so-called Price Edict (AD 301), with which Diocletian tried to establish a maximum for prices and salaries, refers to wages for a teacher of *geometria*;[11] a letter by Pope Gregory I about a boundary dispute (AD 597) recommends recourse to a surveyor, as does another letter by Cassiodorus (which will be discussed in the section on the rest of the world later in this chapter).[12] *Mensores* are mentioned as part of governmental staff by the *Notitia dignitatum*, a sort of catalogue of the late ancient administrative machine.[13] Most of our information, however, can be gleaned from the law codexes. Under the emperors Theodosius II (AD 401–450) and Justinian (AD 483–565), jurists were employed to collect decrees and rescripts from the time past, organizing them under topical headings for quick reference. Thus, the *Codex* and *Digesta* of Justinian comprise laws that date from the third century, with names like Ulpian, Paul and Modestinus, to the sixth, with recent or current emperors and provincial governors.

Using legal evidence can be tricky for a variety of reasons: the range of applicability of the law is not sure, both because some laws seem to be limited to a particular province (and we do not know whether similar laws existed for other provinces), and because we cannot easily chart the relation between what a law suggests and actual circumstances. For instance, a law establishing harsh punishment for thieves may indicate that there was an increase in theft, that the current emperor wanted a zero tolerance policy for propaganda reasons, or that he particularly hated thieves. Whether laws were actually applied is difficult to tell: law enforcement must often have been impossible. That said, legal sources contain extremely interesting information about mathematics and mathematicians. In general, I assume that laws promoting architects, for instance, signify that the social status of architects was, or had potential to be, on the up. Thus, as far as land-surveying is concerned, a law to the effect that 'The chief of the land surveyors after completing two years [of service] is assigned the lowest office of *agens in rebus*' confirms the impression that surveyors were an important part of the administrative body, within which they had a regulated career path.[14] We know that, as in the past, land-surveyors could act as main arbitrators in boundary disputes.[15] On the other hand, an entry on 'If a land-surveyor declares the wrong size' established various sanctions against land-surveyors who did not do their job properly. The jurists (Ulpian and Paul) mentioned architects and accountants (*tabularii*) as parallel cases where specialized expertise was required but had to be legally controlled.[16]

An important group of laws concerns the fiscal status of various 'professionals', including people engaged in mathematical practices. Citizens were supposed to contribute to the upkeep of the Empire in cash, kind or in taking responsibility for the maintenance of roads, billeting of soldiers and so on. Certain categories of people, however, were exempt from some or all fiscal obligations, on various grounds that are not always specified but may have depended on other services they paid to the general welfare of the Empire, or on some special status they may have enjoyed. Thus, fiscal privileges were granted to, on the one hand, soldiers and shippers of grain cargoes, on the other hand, to the Christian clergy (in AD 313, following Constantine's conversion). The choice of worthy categories was a matter for continuous negotiation, through which we can not only assess the official view of some activities, but also realize that definitions could be problematic and involved issues of self-image and of the ethical import of one's knowledge. Take the case of philosophers: we have scattered evidence that people did sometimes allege their being a philosopher as a reason not to pay tax.[17] On the other hand, the laws often scoff at philosophers, who, far from being exempt from paying, should scorn money and thus be more than happy to contribute to the welfare of their fellow-citizens.[18]

Various categories of mathematicians figure in the extant tax legislation: apart from occasional occurrences of *geometres* where the term does not seem to denote a land-surveyor, and must thus denote a 'real' geometer, we have architects/mechanicians, astrologers/astronomers, and teachers of arithmetic, geometry, astronomy and music. Those latter were affected by blanket laws which aimed to promote education and thus established fiscal immunity for them (teachers of liberal arts), for rhetoricians, occasionally grammarians, and often also for doctors.[19] Every city with the exception of Rome was apparently given a fixed quota of teachers to whom immunity could be granted, and hopeful individuals may have had to undergo a public assessment.[20] As we have said, however, distinguishing between worthy and not-so-worthy was a complicated process, well exemplified by the following passage:

> The governor of a province regularly settles the law on salaries, but only for the teachers of the liberal studies. [...] Rhetors will be included, grammarians, geometers. The claim of doctors is the same as that of teachers, perhaps even better, since they take care of men's health, teachers of their pursuits. [...] But one must not include people who make incantations or imprecations or [...] exorcisms. For these are not branches of medicine [...] But are philosophers also to be included among teachers? I should not think so, not because the subject is not hallowed, but because they

> ought above all to claim to spurn mercenary activity. [...] Although also masters of an elementary school are not teachers, nonetheless, the custom has arisen that cases involving them should be heard, also those involving archivists and shorthand writers and accountants or ledgerkeepers. But the governor of a province must not hear outside the regular system cases of workers or craftsmen in fields other than those involving writing or short-hand.[21]

This law is an exercise in boundary-drawing: first to include liberal studies, then to include doctors, but also to exclude dubious healing practices. Philosophers are excluded on moral grounds, but primary teachers and other borderline categories of literate and numerate people such as *librarii, notarii* and *calculatores* are accommodated. At the same time, workers or artisans whose tasks have nothing to do with written signs are kept out. Thus, the criteria of inclusion seem to be literacy/numeracy, usefulness or recognition of an actual state of things which would be difficult to change; the criteria of exclusion may reflect social and moral concerns, some sort of notion of what constitutes 'liberal' or 'illiberal' pursuits. Yet, such a distinction could be blurred: while the law quoted above is inclusive of 'applied' mathematicians, another law excludes teachers of civil law and geometers from privileges accorded to philosophers, doctors, rhetoricians and grammarians, and a regulation issued by Diocletian and Maximian states that a certain exemption law includes teachers of liberal studies, but excludes calculators.[22] On the other hand, we know that accountants, called *numerarii* or *tabularii*, were very visible: practically every department of the administration, whether civilian or military, had a couple of them on its staff.[23] Another entry in the codexes underlines that public accounting was a key job, not to be left into servile hands:

> We prescribe by general law that, if [...] *tabularii* are needed, free men be appointed and that apart from them nobody who is subject to servitude be given access to this office. And if some master has allowed his slave or tenant to deal with public acts – for we want to punish collusion, not ignorance – the person himself, to the degree to which it would benefit public utility, should pay the penalty for the computations which were dealt with by his slave or tenant, while the slave, having damaged the *fiscus*, should be taken to the suitable flogging.[24]

The case of astronomy/astrology also presents several layers. While we know that the practice of astrology continued well into Christian times and beyond, late ancient documents abound in scathing comments on, and harsh punishments for, astrologers.[25] But they also introduce distinctions between a good and a bad use of the knowledge of the skies: time-keeping and calendar-making are allowed, if not encouraged.[26] Constantine toned down his own regulation, specifying that it was a bad thing to use magic arts 'against the well-being of people' but that there was nothing wrong in forecasting the weather for agricultural purposes. On the whole, the difference was made by the *use* one made of one's knowledge – the import of such knowledge being somewhat immaterial, what mattered were the ethic or civic qualities of the *mathematicus*, whether he wanted to do evil or to benefit 'the divine duties and men's work'.[27] Another explicit distinction was made by Diocletian and Maximian: 'It is to the public advantage to learn and to practice the art of geometry. But the damnable astrological art is prohibited under any circumstances'.[28] The praise of geometry, which here could mean land-surveying, suggests once again that this form of knowledge had a rather high profile. A similar view of the relation between knowledge and its uses in society underlies a group of laws which establish full fiscal immunity for architects, doctors, painters, carpenters and other thirty-three technical categories, with the exhortation that, in the spare time from their activities, they teach their profession to other people, in particular to their children.[29] One of these laws explicitly maintains that 'one needs as many architects as possible' and exhorts the prefect of the African provinces to encourage towards that career any youths in their twenties who have had a taste of the liberal studies; another describes the collective duties of 'mechanicos et geometras et architectos' as administering boundaries, measuring (it is not specified what) and looking after aqueducts.

In sum, late antiquity saw a continuation or even, arguably, an intensification, due to the government's fiscal policies and to its emphasis on efficient administration, of mathematical practices such as accountancy, land-surveying, and architecture. These forms of knowledge had a legal identity and a public profile, which was ambiguously articulated, but employed a pretty constant language of utility and public benefit.

Diophantus

We have two manuscript traditions, Greek and Arabic, for Diophantus' main work, the *Arithmetic*. Of the original thirteen, six books have been preserved in Greek, seven in Arabic, but, while the first three books of both traditions are the same, the remaining ones do not match. Scholars seem to be of the opinion that the Arabic tradition has kept the original sequence, and that the books numbered four to six in the Greek are extracts from the original books eight to thirteen.[30] A rather recent article has attributed the pseudo-Heronian *Definitions* to Diophantus, thus moving him to an earlier third-century or even a mid- to late first-century AD date, in this latter case to have him a contemporary of Hero. The earlier third-century date would identify the Dionysius to whom both *Arithmetic* and *Definitions* are addressed as the leader of the Christian school at Alexandria.[31]

The debate surrounding the real text and the real date of Diophantus mirrors the uncertainty about his place within Greek mathematics. He has anachronistically been hailed, since at least early modern times, as the founder of algebra, and consequently 'ahead of his times', whereas little study has been devoted to situating him and his work within an ancient context (whether third or first century AD). In fact, the arithmetic we find in Diophantus looks rather unlike what we have encountered so far, if only for its complexity and the dexterity of his solutions. An example will be useful at this point – I have tried to reproduce the symbols used in the Greek manuscript tradition, some of which (crucially, the minus sign and the sign for an unknown quantity) are introduced by Diophantus in the preface to book 1. Numbers are underlined when they are constant and not linked with the unknown quantity:

> To divide the prescribed number into two numbers twice, so that one <number> from the first division has to one from the second division a given ratio, while the remaining <number> from the second division has to the remaining from the first division a given ratio. Let it be prescribed then to divide the 100 into two numbers twice, so that the greater from the 1st division is 2ice the lesser from the 2nd division, while the greater from the 2nd division is 3 times the lesser from the 1st division. Let the lesser from the 2nd division be posited, x, therefore the greater from the 1st division will be 2x; the lesser from the 1st division therefore will be 100 – 2x; and since the greater from the 2nd division is triple of this, it will be 300 – 6x. Besides, the sum of the 2nd division makes 100; but the sum makes 300 – 5x; these are equal to 100, and it makes the x <equal to> 40. To the initial matter. The greater from the 1st division is posited 2x, it will be 80; while the lesser from the same division 100 – 2x will be 20; while the greater from the 2nd division 300 – 6x will be 60; while the lesser from the 2nd division x will be 40. And the proof is evident.[32]

A general problem is enunciated, which is then solved on specific numbers – that is, rather than a general method to find all possible solutions to the problem, one specific solution is provided. The move from general to particular takes place as the enunciation of the problem is followed by a clause which puts forth specific quantities and determines the givens (in the case above, the given ratios). It is a move somewhat parallel to the setting-out in Euclidean-style geometry, where a general enunciation is applied to a specific diagram. After a number of manipulations that result in the unknown quantity being made equal to a known quantity, a clause indicates a return of the argumentation 'to the initial matter', or, in other cases, a moving on to the 'positions', that is, a return to the enunciation which is finally filled in with solutions. In the proposition above, we also have a concluding reference to an 'evident' proof. What is Diophantus referring to? Perhaps he means that the specific solution ought to persuade the reader that the problem in general is solvable (and indeed, a problem may be considered solved if *one* solution is found), or also that it is left to the reader to work out a more general procedure.

> [An example of problem with four unknown quantities] To find four numbers so that the sum of the four squares, if any of them is added or subtracted, produces a square. Since the square on the hypotenuse of all right-angled triangles, if one adds or subtracts the two <squares> on the <sides> around the right angle, produces a square, let one search first of all four right-angled triangles which have equal hypotenuses; this is the same as dividing a square into two squares, and we have learnt to divide the given □ into two □s in infinite ways. Now then let us produce two right-angled triangles from the least numbers, such as 3, 4, 5; 5, 12, 13. And multiply either of the <triangles> posited by the hypotenuse of the other, and the 1st triangle will be 39, 52, 65; the <other> 25, 60, 65.[33] And they are rectangles which have the hypotenuses equal. Also, the 65 naturally is divided into squares in two ways, into the 16 and the 49, but on the other hand also into the 64 and the 1. This happens because the number 65 is the product of the 13 and the 5, either of which is divided into two squares. Now of the posited <numbers>, the 49 and the 16, I take the roots; they are 7 and 4, and I form the right-angled triangle from two numbers, the 7 and the 4, and it is 33, 56, 65. Similarly also the roots 8 and 1 of the 64 and the 1, and I again form from those a right-angled triangle where the sides <are> 16, 63, 65. And four right-angled triangles are produced which have the hypotenuses equal; returning to the problem at the beginning, I posit the sum of the four, 65x, any of these four being $4x^2$ times of the area, that is the 1st $4056x^2$, the 2nd $3000x^2$, the 3rd $3696x^2$, and the 4th $2016x^2$. And the four are

> $12768x^2$ equal to 65x and it makes x equal to 65 of the 12768 part [x = 65/12768]. To the positions. The 1st will be 17136600, the 2nd 12675000 of the same part, the 3rd 15615600 of the same part, the 4th 8517600 of the same part; the part <will be> 163021824. [x1=17136600/163021824; x2=12675000/163021824, and so on][34]

This proposition is another example of Diophantus' complex manipulations: the reader is invited to go through it, play around with the numbers, perhaps try alternative numbers and see if they fit the bill and can lead to a solution. The *Arithmetic* have no evident axiomatico-deductive structure: we are presented with some operative notions at the beginning, and then plunged into a sequence of problems which invite repetition or individual experimentation. Persuasion is achieved through rehearsing, or even reproducing, the mathematics in question. All solutions are specific, rather than general; in fact, the knowledge embodied by the treatise and that ideally should pass from the author to the reader is not so much a wealth of results, of truths that can be memorized, as a set of problem-solving skills. There is a rather strong sense then that the *Arithmetic* was an advanced training ground – Diophantus himself says as much in the introduction to the first book:

> Knowing that you are anxious, my most esteemed Dionysius, to learn how to solve problems in numbers, I have tried, beginning from the foundations on which the subject is built, to set forth the nature and power in numbers. Perhaps the subject will appear to you rather difficult, as it is not yet familiar, and the minds of beginners are apt to be discouraged by mistakes; but it will be easy for you to grasp, with your enthusiasm and my teaching; for keenness backed by teaching is a swift road to knowledge. [...] Now let us tread the path to the propositions themselves, which contain a great mass of material compressed into the several species. As they are both numerous and very complex to express, they are only slowly grasped by those into whose hands they are put, and include things hard to remember; for this reason I have tried to divide them up according to their subject-matter, and especially to place, as is fitting, the elementary propositions at the beginning in order that passage may be made from the simpler to the more complex. For thus the way will be made easy for beginners and what they learn will be fixed in their memory.[35]

And indeed Diophantus occasionally intervenes in the text with instructions,[36] or comments 'this is easy',[37] or divides problems into categories, so that the reader has a pointer to the solution: if two problems belong to the same category, they can be solved along similar lines; plus, if one learns to recognize a certain category of problems, one will also know how to go about solving them.[38] Or again, the reader is told what to do in the case of equations with terms of the same species but different coefficients, and how one should try and reduce everything to the equation of one, or at most two, terms. Some of the manipulations most frequently used in the text are clearly set out at the beginning of the first book, which also contains basic notions, for instance of the different types of numbers or of the minus sign. While defining this latter, Diophantus explains that in multiplication a minus by a minus makes a plus and a minus by a plus makes a minus. The focus throughout is on acquiring problem-solving arithmetical knowledge, with the help of practice, repetition, a certain arrangement of the material, and several tricks of the trade. It is a fascinating combination: amazing complexity and great abstraction on the one hand, as represented above all by the introduction of what we can call negative quantities. On the other hand, practical concerns, as represented for instance by proposition 5.30, which is cast as an epigram:

> A person has mixed eight-drachmae and five-drachmae cups, and has been ordered to make a profit for people who sail together. For the price of all of them, he has paid a square number which, added to a given number, gives you yet again a square whose root is the total number of cups. Distinguish, O youth, and say how many were the eight-drachmae and how many the five-drachmae cups.[39]

The apparent 'anachronicity' of Diophantus' work may be striking if we set it against Euclid's or, for that matter, Nicomachus' arithmetic, but it diminishes if we look at other, less obvious, texts for terms of comparison. For instance, mathematical problems in the form of little everyday anecdotes or in verse are known from other late ancient sources: for instance, the booklet mentioned on p. 214, or book 2 of Pappus' *Mathematical Collection* (see the next section). There are also, of course, parallels with Egyptian mathematics and with Hero's *Metrica*, especially for its recourse to specific solutions rather than general procedures. That Diophantus was familiar with the format of deductive mathematics as well emerges from the much shorter treatise *On Polygonal Numbers*, whose argumentative structure is Euclidean in style. His work might then be characterized as a hybrid – half-way between Greek and Egyptian arithmetic, between theory and practice, between general and particular, or even between the Christian bishop to whom it may have been addressed, and the pagan martyr Hypatia who wrote a commentary on it. A hybrid, perhaps – but then no more so than many other ancient mathematical texts, late or otherwise.

Pappus

Pappus of Alexandria's surviving works are a *Mathematical Collection* in eight books, seven of which are extant; commentaries on Ptolemy's *Syntaxis*, books 4 and 5, and on Euclid's *Elements*, book 10 (this latter in an Arabic translation); and a *Geography* (in an Armenian paraphrase). The *Collection* may have been put together at a later stage, since the various books differ in character, are not all addressed to the same person and cover many different subjects: arithmetic (book 2); means, including the two mean proportionals (book 3); the five regular solids (books 3 and 5); mathematical paradoxes (book 4); linear curves (book 4); isoperimetrism (book 5); astronomy (book 6); advanced geometry (book 7); mechanics (book 8). Much of the material comes from, or relates to, earlier texts. The relation between Pappus and his sources is very complex: he can endorse them, criticize them, or rework them in several ways. We will take a closer look at his use of the past in chapter 8 – here for reasons of space we will limit ourselves to a description of some of the contents of the *Collection*.

Pappus preserves pieces of ancient mathematics that would be little known otherwise: linear curves and 'analytical' geometry are just two examples. Take the first: Archimedes devoted a book to the spiral, and Eutocius talks about the cissoid and the cochloid/conchoid in the context of doubling the cube, but Pappus provides additional information about the properties and further uses of these curves, for instance, in the trisection of the angle. Moreover, he describes the quadratrix (see Diagram 7.1):

> A certain line was applied to the squaring of the circle by Deinostratus and Nicomedes and some other more recent ones; it takes the name from its characteristic property; for it is called quadratrix by these <people> and it has the following origin. Let a square $AB\Gamma\Delta$ be put forth and around the centre A let a circumference $BE\Delta$ be drawn, and then let AB be moved in such a way that $B\Gamma$, remaining always parallel to $A\Delta$, follows the point B which is carried along BA, and at the same time AB being moved uniformly traverses the angle $BA\Delta$, that is the point B <traverses> $BE\Delta$, and $B\Gamma$ passes through the straight line BA, that is the point B is brought along BA. It clearly happens that the straight line $A\Delta$ coincides with both AB and $B\Gamma$. This movement having been originated,

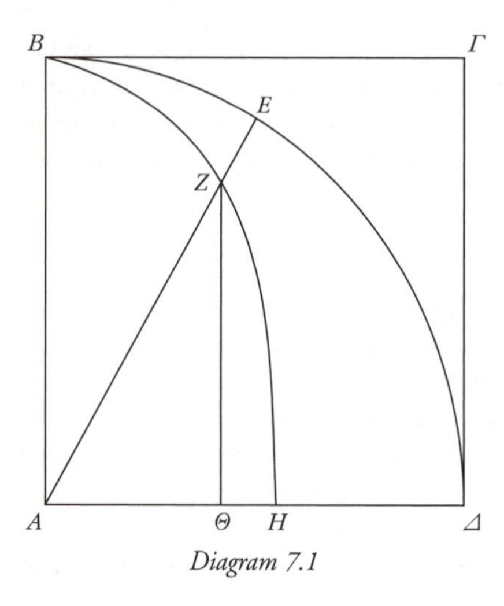

Diagram 7.1

> the straight lines $B\Gamma$ BA cut each other in transit along a certain point which is always being carried along with them, by which point a certain line is drawn in the space between the straight lines $BA\Delta$ and the circumference $BE\Delta$ concave with respect to those, <a certain line> which is BZH, which seems to be useful to find the square equal to a given circle. Its original characteristic property is the following. If any <straight line> whatever is drawn to the circumference, the circumference will be to $E\Delta$ as the straight line BA to $Z\Theta$; this in fact is evident from the origin of the line.

Pappus goes on to report that the description of the origin of the quadratrix was criticized by Sporus on the basis both of circularity and of indeterminacy. According to the first objection, the curve cannot be constructed unless the ratio of the velocities of AB and $B\Gamma$ is already given, or unless the ratio of a radius to a quadrant is already given (Sporus seems to assume that the two are equivalent). But the condition that the ratio of a radius to a quadrant be given cannot be satisfied without circularity, because it amounts to squaring the circle, and the quadratrix is itself being devised in order to square the circle. Sporus' second objection revolves around the fact that the exact position of the points of intersection which constitute the curve cannot be determined:

> The extremity of the curve which is used for the squaring of the circle, that is the point by which the line $A\Delta$ is cut [*sc.* the point H], is not found. [...] Even if the lines ΓB and BA, having been drawn, come back to the same point, they will coincide with the

> line $A\Delta$ and will produce no section by intersecting each other. The intersection in fact ends before the coincidence along the line $A\Delta$, the same intersection which instead should become the limit of the curve, in which the two lines meet the line $A\Delta$. Unless someone says that the curve is imagined prolonged until it meets $A\Delta$ – this does not follow from the principles set at the beginning, but rather would assume the point *H*, having taken as a premise the ratio of the line to the circumference.[40]

Pappus subsequently tries to get round the problem of indeterminacy with an alternative proposition. The sections concerning the quadratrix are part of a larger account, book 4, that includes mathematical paradoxes and also the following classification:

> We say that there are three kinds of problems in geometry, and some of them are called planar, some solid, some linear. Those which can be solved by means of straight line and circumference are justly called planar, because the lines by means of which these problems are discovered have their origin in a plane. Those problems whose solution is found applying one or more of the conic sections are called solid, because for their construction it is necessary to employ surfaces of solid figures (I mean it is necessary to employ conic surfaces). A third kind of problems remains, the so-called linear; other lines in fact, apart from the ones we mentioned, are applied to the construction [of a problem]; their origin is more variegated and rather constrained and they are generated from more disorderly surfaces and from interwoven motions. Those are the lines found in the so-called loci with respect to a surface, and others more variegated than these [...]. They have many wondrous properties. [...] lines of this kind are spirals, quadratrices, cochloids, cissoids. It seems to the geometers that it is no small mistake when [the construction of] a planar problem is discovered by someone by means of conics or of linear [curves], and in general when it is solved by means of a kind not its own.[41]

Although curves like the quadratrix were problematic objects, not easily definable, they were at the same time a fundamental tool for problem-solving. As in Diophantus, teaching or learning about problem-solving implies some kind of ordering of the material – for instance, the definition of common procedures or of the tools employed, which subsumes functional, yet variegated and unorganized, curves into a systematized and comprehensive whole. Moreover, the classification reported above is prescriptive, because, right at the end, it formulates a rule about problems and the procedures appropriate to their solution. It is thus an attempt at ordering in a strong sense. Operations of this kind required Pappus to have a whole wealth of material at his disposal, so that he could take stock of the different solutions that had *already* been given to problems of each kind and put them into categories which, he then claimed, had a value for *future* problem-solving.

The *Collection* abounds in 'meta-mathematical' passages where Pappus steps outside first-order discourse and classifies, defines or systematizes. For instance, he opens book 7 with a definition of analysis and synthesis, and in the course of the text also provides short definitions of problems, theorems, porisms, loci, *neuseis*. It is a way for him not only to clarify things for the reader, but also to set reference points. Book 7, sometime called the 'treasure of analysis', complements a group of canonical texts including Euclid's *Data* and *Porisms*, Apollonius' *Conics, Plane loci, Section of a Ratio* and *Section of an Area* and Eratosthenes' *On Means*. Book 6 is a similar companion text, this time to a set of astronomical works, ranging from Autolycus' *Moving Sphere* to Theodosius' *Sphaerics*. In both cases, the mathematical results from the past are annotated in various ways. The wealth of materials from the past in book 7 is subject to various orderings: there is a sense of an historical order, because the canon or treasure of analysis at hand is presented as the outcome of a process that has taken place over time, through the contributions of several people. Or again, Pappus orders each of the texts in book 7 as exemplified here by his description of Apollonius' *Section of an Area*:

> The first book of the section of an area has 7 modes, 24 cases, and 7 diorisms, of which 4 are maxima, three minima. And there is a maximum in the second case of the first mode, as well as in the first case of the 2nd mode and in the 2nd of the 4th and in the third of the 6th mode. The one in the third case of the third mode is a minimum, as is that in the 4th of the 4th mode, and in the first in the sixth mode. The second book of the section of an area contains 13 modes, 60 cases, and for diorisms those of the first, because it reduces to it. The first book has 48 theorems, the second 76.[42]

A list is compiled of how many cases, modes and theorems a book contains, and sometimes it is specified what the principal propositions are. All this amounts to a systematization of the canonical texts; knowledge of them has to be not only comprehensive, but has to follow the right order, use the right reference points. Of the several things at work in book 7, one is definitely the desire to enable, or indeed enpower, the reader to solve problems in advanced geometry; it is implied that this derives not just from what you know but from how you know it, from the contents *and* from a certain approach to them. I have mentioned above that the mathematical results from the past are annotated by Pappus in various ways. I report an example, again from book 7, relative to Apollonius' *Section of a Ratio* (see Diagram 7.2):

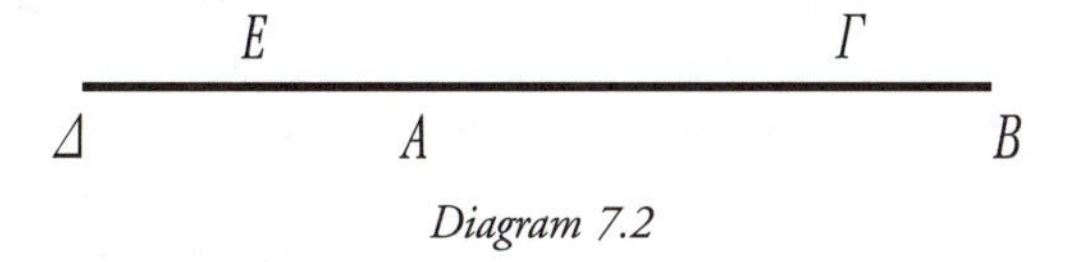

Diagram 7.2

> [Problem for the second of the section of a ratio, useful for the summation of the 14th mode]. Given two straight lines *AB*, *BΓ*, and producing line *AΔ*, to find a point *Δ* that makes the ratio *BΔ* to *ΔA* the same as that of *ΓΔ* to the excess by which *ABΓ* together exceeds the line that is equal in square to four times the rectangle formed by *ABΓ*. The combination cannot be made in any other way, unless *ΔE*, *AΓ* together are equal to the excess *EA* and all *ΔA* to all *AB* and also that *EA*, *AB*, *BΓ* have to one another the ratio of a square number to a square number, and that *ΓB* is twice *ΔE*. Let it be done, and let the excess be *AE*, for we have found this in the previous. Then as *BΔ* is to *ΔA*, so *ΓΔ* is to *AE*. And alternating and separating area to area, therefore the rectangle formed by *BΓ*, *EA* is equal to that formed by *ΓΔE*. But that formed by *BΓ*, *EA* is given. Therefore that formed by *ΓΔE* is also given. And it lies along *ΓE*, which is given, exceeding by a square. Therefore *Δ* is given. It is synthesized as follows [...][43]

Unfortunately, in this as in the majority of cases, we do not have the original text against which to assess the extent of Pappus' intervention. Let us note, however, his preoccupation that all the elements of a problem should be fully determined, and his attention to different cases and modes, with related different conditions of possibility and impossibility of a solution to the problem. Taken together with Pappus' emphasis on counting and itemizing the contents of a book, and on definition and classification, it would seem that he aimed to map the territory of mathematical knowledge as extensively as possible. That Pappus' interests were wide-ranging is evidenced not just by the sheer number of sources that seem to have been available to him, but also by the variety of topics he chose to discuss. Book 2, for instance, which as we have it is incomplete, deals with rules for the multiplication of multiples of ten. It consists almost exclusively of specific examples rather than general propositions, and ends with an epigram whose letters are transformed into the equivalent numbers (remember that he was using the Milesian notation) and then multiplied by each other. The following is an example:

> Let a multitude of numbers be the multitude *A*, of which <numbers> each is less than a hundred and divisible by ten, and another multitude of numbers, the multitude *B*, of which <numbers> each is less than a thousand and divisible by a hundred, and let it be required to tell the product of the *A*s and *B*s without multiplying them. Let then the <numbers> *H*, the units 1, 2, 3 and 4, be basic numbers of the <numbers> *A* and let the <numbers> *Θ*, the units 2, 3, 4 and 5, be basic numbers of the <numbers> *B* and assuming that the product of the basic numbers, of 2, 3, 4, 2, 3, 4, 5 that is, *E*, is of 2880 units, let the multitude of <numbers> *A* added to twice the multitude of <numbers> *B* be first divided by four, with the result *Z*, for it divides them <exactly>. And Apollonius proves that the product of all the *A*s and *B*s is as many myriads as there are units in *E* with the same name as the number *Z*, that is 2880 triple myriads. In fact one myriad with the same name as *Z*, that is triple, multiplied by *E*, that is 2880, makes the number of the products of the <numbers> *A* and *B*. Indeed, the product of the numbers of the *A*s and the *B*s is as many myriads as there are units in *E*, with the same name as the number *Z*. But let the multitude of <numbers> *A*, added to double the multitude of <numbers> *B*, divided by four, first have the remainder of one. And Apollonius gathers that the product of the numbers *A* and *B* is as many myriads with the same name as *Z*, multiplied by ten times *E*; but if the aforesaid multitude divided by four has the remainder of two, the product of the numbers of the *A*s and *B*s is as many myriads with the same name as *Z*, multiplied by a hundred times the number *E*; if the remainder is three, the product of those numbers is equal to as many myriads with the same name as *Z* multiplied by a thousand times the number *E*.[44]

Apart from the references to Apollonius (from a work now lost), the only justification of the validity of Pappus' multiplication method is that, once it is carried out over and over again on several sets of specific numbers, one can see that the method works. Thus, the arithmetic of book 2 does not seem to belong to an Euclidean or a Nicomachean tradition, if indeed those labels correspond to anything definite, and it is tempting to see a parallel of its many exercises and of its final piece of poetic mathematics in the work of Diophantus. The 'applied', 'concrete' mathematics of counting and calculating is also a possible scenario. Another part of Pappus' work with a probably close relation to practice is book 8, devoted to mechanics, a congerie of disciplines (more than a single discipline) which is praised extensively in the introduction. Pappus declares that mechanics has been highly esteemed by philosophers and by anybody interested in learning; emphasizes that Archimedes, far from looking down on it as some have said, owed his enormous reputation to his mechanical feats, and concludes by saying:

> Geometry is in no way injured, but is capable of giving content to many arts by being associated with them, indeed it, being as it were mother of the arts, is not injured by dealing with construction of instruments and architecture; indeed it is not injured by being associated with land-division, gnomonics, mechanics and scenography, on the contrary, it seems to favour these arts and also to be appropriately honoured and adorned by them.[45]

These statements about the complementarity and mutual benefit of mechanics and geometry are representative of Pappus' approach throughout book 8. Although he does not come across as an actual machine-builder (unlike, say, Hero), he is very favourable to the use of instruments in geometry. In fact, his own version of the duplication of the cube employs a moving ruler, and he claims that such procedures are often much more convenient than methods involving conics. He occasionally discusses mathematical problems useful for architecture – an example from book 8 follows (see Diagram 7.3).

In mechanics the so-called instrumental problems are separated from the domain of geometry, for instance [...] that proposed by the architects about the cylinder damaged along both bases. For they ask, given part of the surface of a right cylinder, of which no part of the circumference of the bases is preserved whole, to find the thickness of the cylinder, that is the diameter of the circle from which the cylinder has originated. It is found with the following procedure. Let two points *A B* be taken on the given surface and with these centres with one opening <of the compass> let first the point *Γ* be marked on the surface, and again with these centres *A* B and the opening greater than the first let *Δ* be marked, and with another opening *E*, and with another *Z*, and with another *H*. The

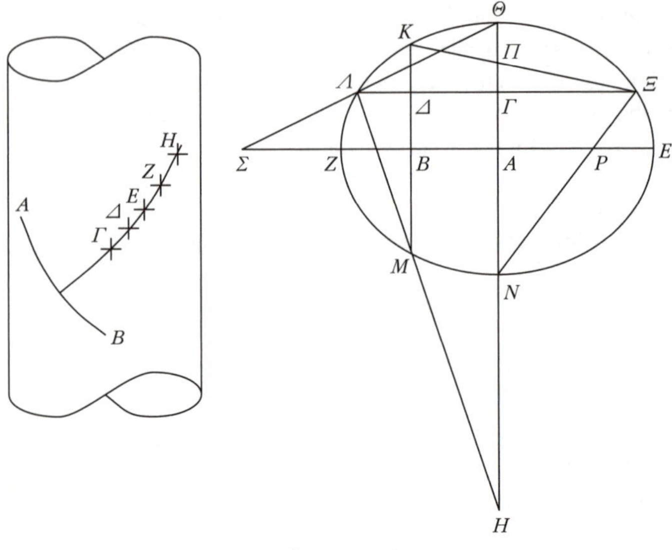

Diagram 7.3

5 points *Γ Δ E Z H* will then be in one plane because the <line> joining each of them as vertex of an isosceles triangle to the half-point of the straight line joining *A B* as common basis of the triangles is perpendicular to *AB* and the 5 lines have originated in one plane, and it is clear that <so have> the points *Γ Δ E Z H*. We set these in a plane as follows; from the three straight lines joining the triangle *Γ Δ E* in the plane let *ΘKL* be put together, and from the three joining *Δ E Z* <let> *KLM* <be put together>, from the three joining the points *E Z H* let the triangle *LMN* be put together; therefore the three triangles *ΘKL KLM LMN* will be taken instead of the triangles *ΓΔE ΔEZ EZH*. If then we draw an ellipse around the points *Θ K L M N*, its lesser axis will be the diameter of the circle which produced the cylinder.[16]

In sum, Pappus' work is testimony to the variety of sources, interests, even audiences of late ancient mathematics. Book 3 was sent to a woman, Pandrosion, who taught mathematics, and it mentions the philosopher Hierius, who was interested in geometry. It also criticizes a contemporary for an incorrect solution to the duplication of the cube. Book 5 again criticizes certain philosophical approaches to mathematics – the tone of the argumentation indicates that the addressee, Megethion, may have been a non-specialist.[47] Books 7 and 8, written for Hermodorus, are instead more advanced. Books 2 and 8 suggest links with mathematical practices such as calculation and architecture; the commentary on Ptolemy includes a rather detailed description of how to make an astronomical instrument. In sum, while Pappus' works looked back to the past, they were also firmly rooted in the present.

Eutocius

Eutocius of Ascalona (sixth century AD) wrote commentaries on some of Archimedes' works and on Apollonius' *Conics*, and probably taught philosophy. He may have been connected to the group of architects who built the cathedral of Hagia Sophia in Constantinople, because he mentions one of them, Anthemius of Tralles.[48] The commentary on Archimedes' *Sphere and Cylinder* is addressed to an Ammonius, who is described as an expert of philosophy in general and mathematics in particular, while Eutocius represents himself as young and relatively unexperienced. He also states that this, the first commentary he writes on Archimedes, is the first commentary on Archimedes to be written at all. This may be due to objective difficulty: Eutocius says at the beginning of the commentary on *SC* 1 that Archimedes' mathematical style is rather obscure and requires a lot of explanations and filling in of demonstrative steps. The *Measurement of the Circle*, on the other hand, to which he devotes himself next, is much clearer because it deals with one issue; also, squaring the circle, Eutocius says, has been of great interest to the philosophers of old and, according to a biographer of Archimedes, is necessary for the needs of life. Its few propositions, unlike those of *SC*, do not skip any steps and can be understood by the average reader.

Eutocius' declared intention is carried out in the body of his work: he tends not to comment on whole propositions, but to pick up a detail that needs clarifying. A representative example is given below.

> [On *SC* 1.13: some phrases are quoted and then explained (see Diagram 7.4)] *Let us imagine then a <polygon> circumscribed and inscribed in the circle* B, *and circumscribed to the circle* A *and similar to that circumscribed to the circle* B. How a polygon similar to another inscribed is inscribed in a given circle, is clear, and is stated also by Pappus in the commentary to the Elements; on the other hand, <on the topic of how> to circumscribe to a given circle a

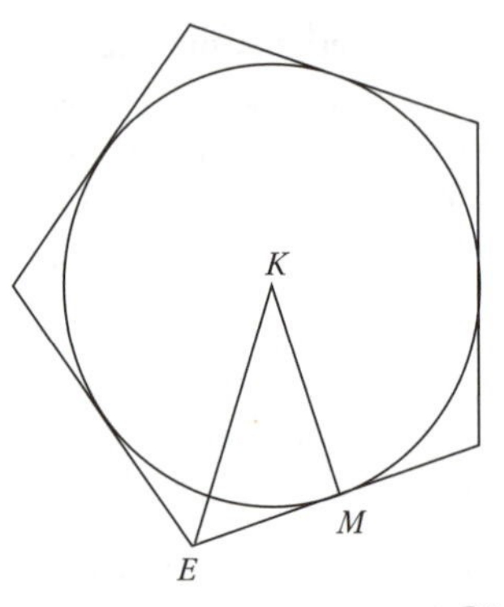

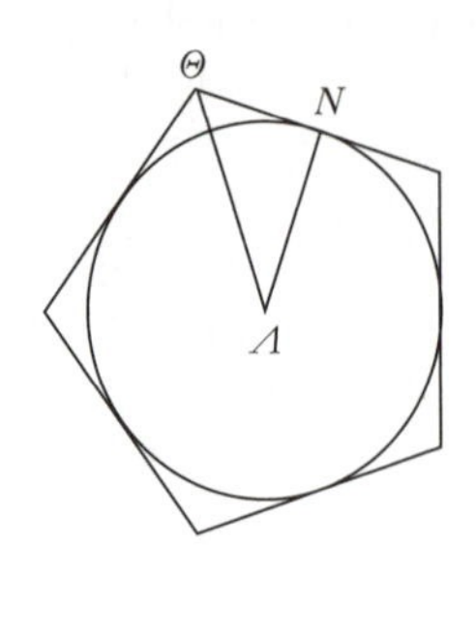

Diagram 7.4

polygon similar to that circumscribed to another circle we do not have anything analogous that has been said; which is to be said now. Let us inscribe then in the circle *A* a <polygon> similar to that inscribed in the circle *B*, and to the same <circle> *A* <a polygon> similar to that inscribed, as in the theorem 3; and it will be similar also to that circumscribed to *B*. *And since the rectilinear figures circumscribed to the circles* A, B *are similar, they have the same ratio as the power of their radii.* This has been proved for the inscribed <figures> in the Elements, but not for the circumscribed ones; it will be proved thus. Let us imagine the inscribed and circumscribed rectilinear figures separately and from the centres of the circles let *KE*, *KM*, *ΛΘ*, *ΛN* be drawn; it is evident then that *KE*, *ΛΘ* are radii of the circles around the circumscribed polygons and that to each other they are in power as the circumscribed polygons. And since the <angles> *KEM*, *ΛΘN* are half of the angles in the polygons, while the polygons are similar, it is clear that they are also equal. But also the <angles> *M*, *N* are right; therefore the triangles *KEM*, *ΛΘN* have equal angles, and it will be as *KE* to *ΛΘ*, so *KM* to *ΛN*; so also the <squares> on them. But as the <square> on *KE* to that on *ΘΛ*, so the circumscribed <polygons> to each other; and therefore as the <square> on *KM* to that on *ΛN*, so the circumscribed <polygons> to each other.[49]

Eutocius will also explore the validity of a statement by considering every possible variation in the diagram, or assess the appropriateness of the terminology, or add demonstrations that confirm that Archimedes' propositions remain valid in any case. He will supply a synthesis when Archimedes only has an analysis; or insert extra demonstrative steps when Archimedes has left them implicit; or, in the numerous cases where, in an exhaustion procedure, Archimedes considers circumscribed and inscribed figures at the same time, he will consider them separately. On occasions, the clarification of a detail turns into a long excursus: at *SC* 2.1, Archimedes had assumed that two mean proportionals, necessary for a certain construction, could be found, but had not explained how. Eutocius supplies twelve different solutions ascribed to earlier authors ranging from Plato (a likely misattribution) to Pappus. Some of the solutions are quite similar to each other (a fact underlined by Eutocius himself), so the rationale behind his providing a full anthology may have been, at least in part, to show that he had access to a wealth of ancient sources.

In the commentary to the 'much clearer' *MC*, Eutocius concentrates on the numerical proposition 3, supplying all the calculations of which Archimedes only provides the end result, and often indicating (they are all approximated values) how much they fall short or in excess of the more exact figure. In his work on Archimedes' *Equilibria of Planes*, he adds a proposition whereby the geometrical objects in question are given a number, i.e. measured; he can then take a certain element in the diagram as the unit and work out a numerical relationship between that unit and the other elements in the proposition.[50]

In the commentary on Apollonius, who himself had a marked interest in generalizations and in the study of sub-cases, Eutocius again takes the reader through variations on the elements of a diagram, in fact produces sub-cases from a manipulation of the diagram, and confirms the validity of the proposition in each case. He occasionally quantifies the total number of possible cases.[51] Or again, he invites the reader to reflect on the overarching demonstrative structure – for instance, he remarks on how the first ten propositions of Apollonius' *Conics* work as a sequence.[52]

As well as the contents, Eutocius is keen to explain what the intent of a mathematical treatise was, and that some interpreters misunderstood the aims of earlier authors, and thus did not read their works correctly:

> The numbers mentioned by [Archimedes] have been sufficiently explained; one must know on the other hand that Apollonios of Perga in the Okytokion proved the same by means of different numbers and with a greater approximation. If this seems to be more accurate, on the other hand it is not useful for Archimedes' purpose; for we have said that his purpose was in that particular book to find the approximation for the uses in life. Thus neither Sporus of Nicaea will be found right blaming Archimedes for not

> having found with accuracy to what straight line is equal the circumference of the circle [...] [and] saying that his teacher Philo from Gadares had led to more accurate numbers than those found by Archimedes [...]; all these people, one after the other, seem to ignore his purpose.[53]

Eutocius establishes a very close connection between himself and Archimedes: on top of explaining the details of his proofs, and on top of deciding which of his texts were obscure and which were not, he claims to understand Archimedes' real intentions about the uses of his work. If indeed this was the first commentary on Archimedes to be written, Eutocius emerges as the only authority; and definitely so, by contrast with others who have allegedly misunderstood the Syracusan mathematician.

There is yet another aspect to Eutocius' approach: we know (he tells us) that the very same works of Apollonius which he was clarifying were being edited by him. He had several versions of the *Conics*, and compared and collated them to obtain one text. Not only was Eutocius' intervention on the past a primary feature of his mathematics, the past itself as he had it (and probably as we have it now) was to some extent the result of Eutocius' intervention.

The philosophers

Iamblichus of Chalcis (mid- to late second century AD) was one of the most influential philosophers of his time. He is said to have inspired the policies of the emperor Julian; a letter mistakenly attributed to the same emperor declares Iamblichus 'the saviour, so to speak, of the whole Hellenic world'.[54] Several of his books survive, including a paraphrase (usually called a commentary) of Nicomachus' *Introduction to Arithmetic*,[55] a *Life of Pythagoras*, an *Exhortation to Philosophy*, a treatise on *General Mathematical Knowledge* (the four were originally part of a vast work *On Pythagoreanism*), and an account of the *Egyptian Mysteries*. We also have fragments of his commentaries to several works by Plato and Aristotle, and an *Arithmetic theology* has been attributed to him, but is probably spurious and loosely based on Nicomachus' lost work by the same title.

Iamblichus' most important reference points seem to have been 'the divine' Plato, to some extent Aristotle, whose psychology he was influenced by,[56] and above all Pythagoras. His philosophy, generally classified as neo-Platonism or neo-Pythagoreanism, is far too complex to be summarized here: let us just say that Iamblichus espoused a general Platonist cosmology, as described in the *Timaeus*. He believed that there is an order in the world, and that there is a correspondence between human souls, the divinity and the cosmos. Specifically, souls have fallen into the bodies but aspire to go back to where they belong (a sort of heaven) and in order to do so they must live philosophically, rid themselves of material things and perform theurgic rites, through which they can experience a sort of ecstasy. This did not exclude participation in political life: in fact, as in Plato's *Republic*, the wise man's love for other human beings leads him to counsel and guide them.

The role of mathematics in all this was enormous. Not only did Iamblichus take up some Platonic views on the question (mathematics as *dianoia* or as gymnastics that trains the soul to turn from material to abstract things): he developed them to the point of positing the One and the Many (*plethos*, literally 'multitude', 'quantity') as the very principles of the universe and of any knowledge we may have of it. As in the *Timaeus*, the demiurge produced the world mathematically, therefore time, space, the motions of the heavenly bodies, the generation and corruption of plants and animals, the musical concords, all are regulated according to mathematical proportions. The soul itself is a number for Iamblichus, but not in an arithmetical sense, because it embraces geometry, music and sphaerics (here equivalent to astronomy) as well; the theurgic rites which are indispensable for the soul's ascent to the divinity seem to have included mathematical rituals. Mathematics was thus in a sense even more fundamental than philosophy itself: this is neatly encapsulated by various statements of Iamblichus' to the effect that Plato took second place to Pythagoras or to the Pythagoreans, and his doctrines to their pervasive notion that everything is number.

If mathematics was so important, it was also important to state clearly what mathematics one was talking about, and what it consisted of. This is where the books on Nicomachus and on *Common Mathematical Knowledge*, plus some lost works on mathematics in ethics and politics, come in. As far as strictly mathematical content goes, Iamblichus concentrates on definitions of odd, even, odd-even, even-odd, polygonal, cubic, perfect and other types of numbers, their properties and classification. Arithmetical notions are often visualized via geometrical figures such as a line, an angle or a gnomon, or more pictorial images. He describes a square number as a race-course, formed of successive numbers from the unit to the root of the square, which is a 'turning-point'.[57] We do not have much about geometry *per se*, and Iamblichus' definition of point and line as flowing from a point fit in with his Nicomachean model. Again like Nicomachus', Iamblichus' treatises do not have an axiomatico-deductive structure, and he uses specific examples rather than general proofs. In fact, he shows some disrespect for traditional mathematical authorities:

> Here then it is a blatant mistake by Euclid not to distinguish even-odd from odd-even, nor to recognize that one of them on the one

> hand is opposed to the even in an even way, the other on the other hand is the mixture of both [...] For he says thus: an even-odd number is that measured by an even number odd-times. Odd-even is the same; for it is measured by an odd number even-times, as for instance 6; for if on the other hand we say twice three, it is even-odd, if instead three times two, it is odd-even; it is totally simple. But in the third of the Arithmetic books he confuses the three in one, being subject evidently to the appearance of the name. For he says: if an even number has the half odd, then it is odd even-times and even odd-times, evidently saying the same as those before. Next he adds: if an even <number> has neither the half even nor is produced by multiplication from the unit, then it is at the same time even even-times and odd even-times and even odd-times. And Euclid thus; for us on the other hand let rather what is commonly formed and shaped from both [...] be called the third type.[58]

This passage is not the only occasion on which Iamblichus criticizes Euclid, and is remarkable for several reasons. First, it is testimony to the fact that mathematics, like other cultural practices, was multi-stranded, multi-layered and conflictual. Even well-respected authors like Euclid could be attacked (and no mincing of words) for alleged incompetence. Second, it is evidence of the fact that there was something to get competitive about: mathematics was so important within Iamblichus' conception of the world that it was crucial to get it right. The points of contention (odd-even, even-odd) may appear banal to us, but for Iamblichus they must have had such a weight as to render Euclid's mistakes extremely significant. Third, by the time Iamblichus was writing, Euclid represented a (prestigious) mathematical past. Iamblichus was keen to create continuities between his philosophy and that of Pythagoras and Plato. In mathematical terms, that meant often attributing one discovery or another to the Pythagoreans, or enlisting to the Pythagorean field people who did not belong to it, such as Eudoxus, or composing micro-histories of particular fields, such as means theory, where Pythagorean import was presented as crucial. That the promotion of one's view of mathematics should take place through a rewriting or appropriation of the mathematical past is one of the prominent features of this period, and we will discuss it in chapter 8.

The reasons behind Iamblichus' mathematical choices, such as that of Nicomachean arithmetic (basically assimilated to Pythagorean arithmetic) over a Euclidean one, are explained as follows:

> Pythagorean mathematics is not like the mathematics pursued by the many. For the latter is largely technical and does not have a

> single goal, or aim at the beautiful and the good, but Pythagorean mathematics is preeminently theoretical; it leads its theorems towards one end, adapting all its assertions to the beautiful and the good, and using them to conduce to being.[59]

If one believes that mathematics has to do with the larger order of the universe, that it is much more than mere crunching theorems or attaining formulas for areas or volumes, Iamblichus' choice makes perfect sense. Neo-Pythagorean or neo-Platonist mathematics has often been dismissed as rather incoherent and unoriginal, but passages like the one quoted above show that in fact these philosophies valued mathematics very highly, because of the possibility through it to acquire knowledge of the universe and ultimately of your own self. That said, it has to be kept in mind that neo-Pythagoreans and neo-Platonists were not sharply differentiated, and that they did not present a unified front. In fact, their positions on mathematics, and particularly on what version of mathematics should be adopted, were rather varied. Proclus, for instance, although (or so the story goes) it was revealed to him in a dream that he was the reincarnation of Nicomachus, chose Euclid as the authority to comment upon. He also studied Ptolemy, on whose *Harmonics* the earlier neo-Platonist Porphyry had written a commentary. A fellow-student of Proclus, Domninus of Larissa, wrote a mathematical treatise which has been seen as a return to Euclid against current Nicomachean trends.[60]

Let us turn to Proclus. As well as the *Commentary on the First Book of Euclid's Elements*, his rather prolific production includes *The Elements of Theology*, *Physics*, *A Sketch of the Astronomical Positions*, and a number of other commentaries, mainly to Plato's works. Proclus believed that mathematics was intrinsic to the fabric of things, and that its language enabled one to articulate the knowledge of reality: the relation between whole and parts could be expressed in terms of divisibility; the bonds between disparate parts of the same universe could be seen as proportions;[61] the generation of beings from the divine Monad, the One which is the principle of everything, was a sort of multiplication, and so on. Mathematical objects themselves are intermediate between the intelligibles and the sensibles and consequently reflect the human condition – in fact, the soul itself is intermediate between indivisible principles and divided corporeal ones, between eternal existence and temporal activity.[62] Mathematics is then the knowledge most appropriate to human nature. Proclus also talks of a 'general' or 'whole' mathematics, which should deal with the whole spectrum of its applications, from lower forms such as mechanics, optics and catoptrics to the knowledge of divine beings.[63]

The declared aim of Proclus' commentary to the *Elements* is to show that Euclid is at bottom a Platonist. It opens with a long discussion of the nature and use of mathematics, and of the necessity for the good geometer to make the right distinctions and categorizations. When dealing with Euclid's basic definitions and first principles especially, Proclus' mode of operation is not different from his philosophical style of commentary: he expands the meaning of the text in a number of possible directions, comparing and contrasting the views of previous interpreters, eventually to show (in the majority of the cases) that every utterance of the author works from various angles, from the cosmological to the ontological. This is an indication that all levels of existence, the microcosm and the macrocosm, are connected and part of the same reality. Thus, explaining the definition of point amounts to showing what that definition means ontologically, cosmologically, epistemologically – it amounts to showing that it *means* something in each one of those domains, that it is much more than just a geometrical notion, and that by understanding it in full we can increase our knowledge in a number of unsuspected directions. The fact that everything in Proclus means a lot of things at the same time has elegantly been termed semantic superabundance.[64] An example of this, about the distinction between equilateral, scalene and isosceles triangle, is the following:

> From these classifications you can understand that the species of triangle are seven in all, neither more nor less. [...] You can also understand from the differences found in their sides the analogy they bear to the orders of being. The equilateral triangle, always controlled by equality and simplicity, is akin to the divine souls, for equality is the measure of unequal things, as the divine is the measure of all secondary things. The isosceles is akin to the higher powers that direct material nature, the greater part of which is regulated by measure, whereas the lowest members are neighbours to inequality and to the indeterminateness of matter; for two sides of the isosceles are equal, and only the base is unequal to the others. The scalene is akin to the divided forms of life that are lame in every limb and come limping to birth filled with matter.[65]

The relation between mathematical objects and their wider meanings is indicated by terms like 'analogy', 'symbol', 'akin' or 'likeness'. Proclus' style of commentary is slightly different when he is working on the propositions themselves of the first book of the *Elements* (rather than the preliminary material): once again he informs the reader of other interpretations and assesses them, but he also glosses the text by explaining anything that in Euclid is left unsaid. For instance, what principles or previous propositions the proof appeals to; what other propositions the statement at hand is used for; in the case of indirect proofs, what principles the proof contradicts. He sometimes underlines the relevance of a problem or theorem for other fields, such as land-surveying or astronomy,[66] or the rationale of Euclid's vocabulary, structure and phrasing. An example is given in the following section.

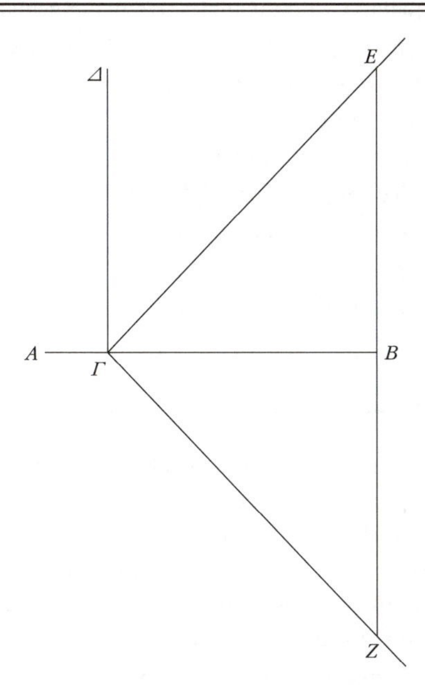

[Commenting on *Elements* 1.14, 'If with any straight line, and at a point on it, two straight lines adjacent to one another and not lying on the same side make the adjacent angles equal to two right angles, the two straight lines will be in a straight line with one another', (see Diagram 7.5).] That it is possible for two adjacent lines drawn at the same point on a straight line and lying on the same side of it to make angles on the straight line equal to two right angles we can demonstrate thus, after Porphyry. Let *AB* be a straight line. Take any chance point on it, say *Γ*, and let *ΓΔ* be drawn at right angles to *AB*, and let angle *ΔΓB* be bisected by *ΓE*. Let a perpendicular *EB* be dropped from *E*, let it be extended, and let *BZ* be equal to *EB* and *ΓZ* be joined. Then since *EB* is equal to *BZ*, *BΓ* is common, and these sides contain equal angles (for they are right angles), base *EΓ* is equal to base *ΓZ*, and all corresponding parts are equal. Angle *EΓB* is therefore equal to angle *ZΓB*. But angle *EΓB* is half of a right angle, for a right angle was bisected by

> $E\Gamma$; hence $Z\Gamma B$ is half of a right angle. Angle $\Delta\Gamma Z$ is therefore one and one-half of a right angle. But $\Delta\Gamma E$ is half of a right angle; therefore on line $\Gamma\Delta$ and at point Γ on it there are two adjacent straight lines ΓE and ΓZ lying on the same side of it and making with it angles equal to two right angles, ΓE making an angle equal to half of a right angle, and ΓF an angle equal to one and one-half of a right angle. Thus to prevent our drawing the impossible conclusion that ΓE and ΓZ, which make angles with $\Delta\Gamma$ equal to two right angles, lie on a straight line with one another, our geometer had added the phrase 'not lying on the same side'. Hence the lines that make with a line angles equal to two right angles must lie on opposite sides of the line, though starting at the same point, one extending to this and the other to that side of the straight line.[67]

Furthermore, Proclus is interested in classifying and arranging his material, for which purpose he deploys a whole battery of labels and definitions: he distinguishes between theorem and problem, returning to the issue over and over again; he divides a mathematical proposition into parts; he has names for various kinds of converse propositions, for interventions on the text in the form of sub-cases or objections; he defines locus theorems and among them distinguishes plane from solid ones.[68] As in Iamblichus, establishing the proper order is fundamental, it is what distinguishes the good geometer from, for instance, some past authors who, in his view, had muddled categories or misunderstood the place of some principles or statements. Alternative proofs are sometimes rejected because they violate the order of the *Elements* as Proclus had them, in that they present things that are not simple enough for that part of the account, or not appropriate for the introductory tenor of the book. For him, the structured progression of the proofs, the architecture of demonstration, is a likeness of the progress of the soul towards knowledge and self-elevation: the order of geometry has to be kept because it corresponds to an epistemic and cosmological order.[69]

In fact, Proclus' *Elements of Theology* are arranged according to an Euclidean model, with enunciation followed by a demonstration, often conducted via *reductio ad absurdum* or complete with clauses like 'therefore' (*ara*) and 'it is clear that' (*phaneron oti*). The propositions or theorems are sometimes deductively connected, in that some proofs directly appeal to ones which have been previously established.

We should not take mathematics, however, to be the supreme form of knowledge for Proclus, because he knew that its application to reality was problematic. He took issue with the lack of accuracy of some physical arguments; in his discussion of astronomy, he listed the problems tackled by Ptolemy and others, critically assessed their solutions and rejected some of their tenets, including the precession of equinoxes, eccentrics, and epicycles (those latter had the additional fault of not being mentioned by Plato).[70]

In sum, both our representative philosophers, Iamblichus and Proclus, assigned mathematics a tremendous role in their cosmology, epistemology and ethics. They turned to previous mathematical authors in order to comment and reflect on them, and in order to inscribe them in their own framework, or reject their contribution. While some of their positions or choices may be quite different from what we find in authors like Pappus, the way they approached the past, their emphasis on classification, right order, teaching, suggest once again that sharp boundaries between philosophy, mathematics and between different mathematical or philosophical traditions have to be problematized.

The rest of the world

Apart from our legal sources, we have evidence for 'applied' mathematics in some literary sources. Yet another author called Hyginus wrote a short and now fragmentary manual on the division of military camps. He provided various size specifications, depending on the terrain on which the camp was to be built and on the number of people to be accommodated. Those latter were measured in military units such as the *cohors*. He also specified what parts of the army should be located in what part of the camp, and mentioned the intervention of army surveyors, who lay out the camp with the help of a *groma*.[71] The work may have been addressed to the emperor of the day; Hyginus declares himself a trainee in the subject, but that does not prevent him from suggesting a few innovations in the way the right specifications for a certain number of legions are to be calculated.[72]

A fourth-century source on land-surveying is Agennius Urbicus, who again distinguishes his form of knowledge from other ones, by remarking that, while the Stoics assert that the world is one, if one really wants to know what the world is like, and how big, one needs geometrical knowledge.[73] He describes the methodology of his form of knowledge as follows:

> Thus, of all the honourable arts, which are carried out either naturally or proceed in imitation of nature, geometry takes the skill of reasoning as its field. It is hard at the beginning and difficult of access, delightful in its order, full of beauty, unsurpassable in its

> effect. For with its clear processes of reasoning it illuminates the field of rational thinking, so that it may be understood that geometry belongs to the arts or that the arts are from geometry.[74]

More than contributing to land-surveying lore, Urbicus is enhancing the status of *geometria*, which is presented as beautiful but also effective, an inquiry into reason but also one of the arts. These themes are well known to us from the land-surveyors of earlier periods, as is his insistence on the ethical significance of land-surveying:

> In making a judgement the land surveyor must behave like a good and just man, must not be moved by any ambition or meanness, must preserve the reputation both by his art and by his conduct. [...] [F]or some err because of inexperience, some because of impudence: indeed this whole business of judging requires an extraordinary man and an extraordinary practitioner.[75]

Apart from his continuation of traditional themes, Urbicus wrote commentaries on Frontinus' land-surveying works, operating on the text in such a way that it is almost impossible to distinguish commentary from original. From some remarks of his, it would also seem that he was involved in teaching. In sum, like many other late ancient mathematicians, his practice is a combination of old and new: he appropriates traditional values and traditional texts in order to show the contemporary relevance of his form of knowledge. Further appreciation of land-surveying comes from Cassiodorus. In a letter (*c.* AD 507–11) written on behalf of King Theodoric, he invites two *spectabiles viri* between whom a boundary dispute had arisen to entrust their case to the capable hands of a land-surveyor, who will solve it 'by means of geometrical forms and land-surveying knowledge' rather than with weapons. Perhaps in an attempt to drive home its authoritativeness, Cassiodorus appends a micro-history of land-surveying: how it originated with the Chaldeans, how it was taken up by the Egyptians and eventually by Augustus, who carried out an extensive programme of land-division; he mentions the 'metrical author' Hero as the person who 'made it into a written doctrine'. Land-surveying enjoys a great reputation indeed when compared to other branches of knowledge: arithmetic, Cassiodorus says, is taught to empty schoolrooms; geometry 'insofar as it discusses heavenly things' is only known to scholars; astronomy and music are learnt just for their own sake, but the *agrimensor* 'shows what he says, and proves what he has learnt'.[76] The emphasis on actual performance is paired with a reconstruction of the discipline's past, of its tradition.

And indeed, all the mathematicians we find in this period look back to previous work. In his two books, *On the Section of a Cylinder* and *On the Section of a Cone*, both addressed to a Cyrus, Serenus of Antinoopolis looks at Euclid and above all Apollonius as main reference points. His declared aim is to rectify some erroneous opinions the public has about Apollonius and to research further the topics that Apollonius has already dealt with. There is some element of ordering of the earlier text (he mentions propositions from the *Conics* by number) and an emphasis on generality: for instance, he wants to examine the sections of an oblique, not just a right, cylinder. Analogously, he contrasts the Euclidean definition of cone, which only describes a right cone (Euclid is however not named explicitly) with Apollonius', which includes oblique cones as well.[77] Again, Theon of Alexandria, as well as 'editing' Euclid's *Elements* (see the section on the real Euclid in chapter 4 p. 125), wrote two commentaries on Ptolemy's *Handy Tables* and one on the *Syntaxis*, in collaboration with his daughter Hypatia. In the commentary on the *Syntaxis*, he remarked that his readers were not very competent in mathematics, and this has often been taken to apply to late ancient readers in general. Yet, apart from the fact that those are rather stock remarks, typical of introductions,[78] we know that late ancient readers of Ptolemy, for instance, had a whole variety of interests and mathematical abilities. They included 'proper' mathematicians like Pappus who too wrote a commentary on the *Syntaxis*; Platonist philosophers such as Porphyry, who commented on the *Harmonics* and Proclus, who however had various problems with Ptolemaic astronomy; and of course astrologers like Paul of Alexandria and Hephaestio of Thebes, who mention Ptolemy as an example of accuracy in calculations.[79] On the other hand, the very fact that Theon wrote two different commentaries to the same work (the *Handy Tables*) should alert us to his desire to cater for a diverse public. The 'big' commentary in five books, now lost, must have been for a more advanced level of readership, while the simpler version, as he himself says, was aimed at those of his students who had difficulty following multiplications and divisions, and knew little geometry. The main interest of this not-terribly-mathematical public was probably in the use of Ptolemy's tables for astrological forecast.[80]

Yet another likely late reader of Ptolemy was Synesius of Cyrene, bishop of Ptolemais. He studied with Hypatia in Alexandria, and corresponded with her afterwards. The astronomical training he must have received is evident from a letter to Paeonius, an officer he had met at the imperial court in Constantinople. The letter, usually entitled *On the Gift*, was accompanied by an astrolabe, made by skilled craftsmen to Synesius' own specifications. The object, he said, included all that he had learnt from Hypatia. After praising Paeonius for combining political power and philosophy, Synesius extolled astronomy as the stepping stone to theology, its demonstrations riddles, the fictional situations being historical, mythological or just everyday. Many are problems of distribution into different parts, couched in terms of dividing golden apples among goddesses, or inheritances among ordinary mortals; there are also calculations of the length of a journey, the duration of one's life (in the form of a horoscope or an epitaph) or of the time it takes for a spout to fill up a fountain tank. A mythological example:

> Heracles the mighty was questioning Augeas, seeking to learn the number of his herds, and Augeas replied: 'About the streams of Alpheius, my friend, are the half of them; the eighth part pasture around the hill of Cronos, the twelfth part far away by the precinct of Taraxippus; the twentieth part feed in holy Elis, and I left the thirtieth part in Arcadia; but here you see the remaining fifty herds'.[84]

Another example, this time from real life:

> O woman, how hast thou forgotten Poverty? But she pressed hard on thee, goading thee ever by force to labour. Thou didst use to spin a mina's weight of wool in a day, but thy eldest daughter spun a mina and one-third of thread, while thy younger daughter contributed a half-mina's weight. Now thou providest them all with supper, weighing out one mina only of wool.[85]

Intriguingly, we find 'advanced' examples of poetic mathematics in association with illustrious mathematicians like Archimedes (author of the *Cattle Problem*) and Diophantus:

> This tomb holds Diophantus. Ah, how great a marvel! the tomb tells scientifically the measure of his life. God granted him to be a boy for the sixth part of his life, and adding a twelfth part to this, he clothed his cheeks with down; He lit him the light of wedlock after a seventh part, and five years after his marriage He granted him a son. Alas! late-born wretched child; after attaining the measure of half his father's life, chill Fate took him. After consoling his grief by this science of numbers for four years he ended his life.[86]

The rest of the manuscript, as we have said, contains various materials, including riddles which are not mathematical – something similar to what we find, on a smaller scale, in the so-called educational texts, with division tables on one side and paraphrases of Homer on the other. One has to conclude that these mathematical games could be read and solved (or not, as the case may be – some of them have not been solved to this day) by the general educated public, and used both for diversion and advanced instruction, in a way analogous to gnomic poetry.[87] Indeed, the Diophantus puzzle, like gnomic poetry, has a moral message, about the caducity of life and the tragedy of death and loss – a further example of how use of the mathematical tradition could work in many different ways.

Notes

1 Augustine, *The Genesis Interpreted Literally* 4.14, my translation.
2 Hypatia was not a unique example: book 3 of Pappus' *Mathematical Collection* is addressed to a woman, Pandrosion, whose pupils are also mentioned, and several examples of women philosophers are mentioned in Eunapius, *Lives of the Philosophers*.
3 Rathbone (1991), 331.
4 *Ibid.* (1991), 348.
5 Boyaval (1973); in general see Fowler (1999).
6 Brashear (1985).
7 Crawford (1953).
8 Translation in Crawford (1953), with modifications; the text is very difficult to reconstruct. The 150 between the lines is the solution.
9 There is debate as to the effective origin of the fiscal reform, see Carrié (1994).
10 *P. Cairo Isidor.* 3, 4 and 5 (all AD 299).
11 The meaning of *geometria* here (land-surveying or geometry?) is problematic, see Cuomo (2000a).
12 Both letters in Dilke (1971).
13 *Notitia dignitatum Orientis* 7.66, 11.12.
14 *Codex of Justinian*=*CJ* 12.27.1 (AD 405). The duties of the *agentes in rebus* included making reports on the provinces, and serving as couriers between the court and the provinces. They gained a reputation as secret police (see *OCD s.v.*).
15 *Codex of Theodosius*=*CT* 2.26 (from AD 330 to 392); *CJ* 12.27.1 (AD 405); *Digesta* 10.1 (including Gaius and Julian from the second century AD, Paul, Ulpian, Modestinus and Papinianus from the third).
16 *Digesta* 11.6.7 (Ulpian).
17 Pliny Jr., *Letters* 10.58; Dio Cassius, *History* 78.7.3; Aelius Aristides, *Sacred Discourse* 73; Philostratus, *Lives of the Sophists* 490 and 622–3; *P. Lips.* 47 (*c.* AD 372).
18 *Digesta* 27.1.6 (Modestinus, quoting Antoninus Pius, Paul, Ulpian, Commodus, Severus and Antoninus), 50.13.1 (Ulpian); *CJ* 10.42.6 (between AD 286 and 305), 10.52.8 (AD 369).
19 E.g. *CT* 13.3 (from AD 321 to 428); *Digesta* 50.5.10.2 (Paul); *CJ* 10.52 (from second century AD to AD 414).
20 *Digesta* 27.1.6.
21 *Digesta* 50.13.1, translation University of Pennsylvania Press 1985, with modifications.
22 Ulpian, *De excusatione* 149 and *CJ* 10.52.4 (between AD 286 and 305), respectively.
23 *CJ* 10.12.49.2; 10.12.49.4; 10.72.13; 12.7.1; 12.28.3; 12.29.3.1; 12.49.2; 12.49.4; 12.49.6; 12.49.10; 12.49.12; 12.59.3; 12.60.6. See also the diagram at the end of Seeck's edition of the *Notitia Dignitatum*.
24 *CT* 8.2.5 (AD 401), my translation.
25 *CT* 9.16 (from AD 319 to 409); *CJ* 9.18 (between second century AD and 389); 1.4.10 (AD 409). Cf. also MacMullen (1966); Straub (1970); Grodzynski (1974).

26 For portable dials from around the fourth century see de Solla Price (1969), Evans (1999). As already explained in chapter 5, the construction of sundials required some mathematical knowledge in order to produce plane projections of time lines. For calendars, see Salzman (1990); Maas (1992), ch.4.
27 *CJ* 9.18.4 (AD 321).
28 *CJ* 9.18.2 (between AD 286 and 305), my translation.
29 *CT* 13.4 (from AD 334 to 374); *CJ* 10.64 (AD 337 and 344); see also *Digesta* 50.6.7 (Tarruntenus Paternus, late second century AD).
30 See the entry on Diophantus in the *OCD* (by G.J. Toomer).
31 The later dating is based on Michael Psellus (eleventh century), according to whom Anatolius, bishop of Laodicea from AD 270, had dedicated a treatise on Egyptian arithmetic to Diophantus, see entry on Diophantus in *DSB* (by K. Vogel). For the earlier dating, see Knorr (1993). I am not sure as to the attribution of the *Definitions*, and take Diophantus' date to be the traditional mid- to late third century AD.
32 Diophantus, *Arithmetica* 1.12, my translation.
33 Take two numbers, a and b. Let us posit $x=a^2 + b^2$; $y=a^2 - b^2$; $z=2ab$. It will be $x^2=y^2 + z^2$. One says that the triangle with sides x, y and z is formed by the numbers a and b.
34 Diophantus, *ibid.* 3.19, my translation.
35 Diophantus, *ibid.* Preface, translation in Thomas (1967), with modifications.
36 Diophantus, *Arithmetica* 3.10, 3.15.
37 Diophantus, *ibid.* 5.1.
38 E.g. Diophantus, *ibid.* 2.11, 2.13, 3.12, 3.13, all examples of 'double equation'.
39 Diophantus, *ibid.* 5.30, my translation.
40 I.e. that the point H is given depends on the ratio of line and circumference being given. Pappus of Alexandria, *Mathematical Collection* 4.250.33–252.25, 254.10–22, my translation. Heath (1921), I 229–30 thought that 'both Sporus's objections are valid'; van der Waerden (1954), 192, disagreed.
41 Pappus, *Mathematical Collection* 4.270.3–31, my translation.
42 Pappus, *ibid.* 7.642.6–18, tr. A. Jones, Springer 1986, with modifications.
43 Pappus, *ibid.* 7.700.14–702.1, translation as above.
44 Pappus, *ibid.* 2.6.6–8.11, my translation.
45 Pappus, *ibid.* 8.1026.21–1028.3, my translation.
46 Pappus, *ibid.* 8.1072.31–1076.11, my translation.
47 Cf. Cuomo (2000a).
48 Eutocius, *Commentary on Apollonius' Conics* Introduction.
49 Eutocius, *Commentary on Archimedes' Sphere and Cylinder* 13, my translation. See Decorps-Foulquier (1998).
50 Eutocius, *Commentary on Archimedes' Equilibria of planes* 2.2 (176–8 Mugler).
51 Eutocius, *Commentary on Apollonius' Conics* e.g. 1.262.28–264.2, 264.22, 264.25–26, 266.24.
52 Eutocius, *ibid.* 1.214.6–216.12, cf. also 1.285.1–288.6.
53 Eutocius, *Commentary on Archimedes' Measurement of the Circle* 3 (162–3 Mugler), my translation.
54 [Julian], *Letters* 78.419a.
55 At least three more commentaries on Nicomachus' *Arithmetic* have survived from late antiquity: one by Asclepius of Tralles, one by Johannes Philoponus, and an anonymous one.
56 Shaw (1995).
57 Iamblichus, *On Nicomachus' Arithmetical Introduction* 75.22 ff. A detailed description in Heath (1921), I 113 ff.
58 Iamblichus, *Ibid.* 23.18–24.17. Cf. also 20.10–14; 25.24; 30.28; 74.23 ff.

59 Iamblichus, *On Common Mathematical Knowledge* 91.3–11, translation in Mueller (1987a).

60 Proclus' commentary to the first book of Euclid's *Elements*, which is all we have, was perhaps part of a commentary on the whole of the *Elements*, cf. Proclus, *Commentary on the First Book of Euclid's Elements* 398.18-19, where he may be referring to a commentary on the second book; 432, where he seems to leave the task to others. For Domninus of Larissa, see Tannery (1884) and (1885).

61 Proclus, *Commentary on Plato's Timaeus* 3.18.20 ff.

62 Proclus, *Elements of Theology* 190–1; *Commentary on Plato's Timaeus* 127.26 ff.; *Commentary on Euclid's Elements* 4. In general see Charles-Saget (1982).

63 Proclus, *Commentary on Euclid's Elements* 19–20, 44.

64 Charles-Saget (1982).

65 Proclus, *Commentary on Euclid's Elements* 168, tr. G.R. Morrow, Princeton University Press 1970, reproduced with permission.

66 Proclus, *ibid.* 236–7, 268–70, 403. Cf. Serenus, *Section of a Cylinder* 117, who claims that a certain problem, which properly belongs to optics, cannot be solved without geometry.

67 Proclus, *Commentary on Euclid's Elements* 297-298, translation as above.

68 See e.g. Proclus, *ibid.* 252 ff., 394–5, and cf. Netz (1999b).

69 Proclus, *ibid.* e.g. 377.

70 Proclus, *Commentary on Plato's Republic* 16 227.23–235.3; cf. also his *Sketch of Astronomy* and *Physics*.

71 Hyginus, *On the Division of Military Camps* 12. He mentions compasses at 55. The dating of this work has been very controversial: it could be second, late second or third century AD.

72 Hyginus, *ibid.* 45–7. About mathematics and the military in late antiquity, Vegetius (late fourth/early fifth century AD) implies that numeracy, in the form of arithmetical skills, was required, or was considered desirable, for soldiers and potential recruits, see *On Military Things* 2.19, 3.15.

73 Agennius Urbicus, *Controversies about Fields* 22.7–8.

74 Agennius Urbicus, *ibid.* 25.15–27, my translation.

75 Agennius Urbicus, *ibid.* 50.9–15, my translation.

76 Cassiodorus, *Letters* 3.52. Cf. also 7.5.

77 Serenus, *On the Section of a Cylinder* 3–5 (fourth century AD).

78 Mansfeld (1998).

79 Paul of Alexandria, *Elements of Astrology* e.g. 33.12, 79.10; Hephaestio of Thebes, *Astrology* e.g. 32.10 (the 'divine' Ptolemy), 88.20, 126.1. Both are late fourth century AD.

80 See Segonds (1981), 15.

81 Synesius, *Letters* 160 (*To Paeonius, on the Gift*), tr. A. Fitzgerald, Oxford 1926, with modifications.

82 Marinus, *Commentary on Euclid's Data* 252.14–18.

83 Marinus, *ibid.* 244.1–8 (that the spiral is 'feasible'), 248.3–8.

84 *Palatine Anthology* 14.4. The solution, according to the Loeb editor, is 240 (120+30+20+12+8+50).

85 *Palatine Anthology* 14.134. The solution, as above, is: 'The mother in a day 6/17, the daughters respectively 8/17 and 3/17'.

86 *Palatine Anthology* 14.126. The solution, as above, is: 'He was a boy for 14 years, a youth for 7, at 33 he married, at 38 he had a son born to him who died at the age of 42. The father survived him for 4 years, dying at the age of 84'.

87 See Morgan (1998).

accuracy. Therefore, Christian authors and authorities were interested in accurate time-keeping, by means of astronomical observations and instruments.[2] Another area where mathematics and the Christian faith intersected was in understanding God and his creation. As we have seen in chapter 5, Philo of Alexandria had analyzed the meaning of various numbers in the Bible: the seven days of creation, the forty days of the flood, the twelve tribes of Israel. Inspired by Philo and/or by the Platonism which partly inspired Philo himself, Christian interpreters often carried out the same hermeneutic operation, comforted in this by Scriptural phrases such as 'Everything You [God] ordered with measure, number and weight' (*Book of Wisdom* 11.21). If the world since its creation followed a mathematical order, then numerology was an important instrument of exegesis. Moreover, mathematics provided a type of reasoning that was certain and undisputable, while at the same time it did not crucially rely on the senses – it was abstract. It could work as a model for the knowledge of spiritual entities, such as God or the soul. Finally, mathematics was an integral part of the non-Christian educational systems or philosophies to which nearly all educated Christians had been exposed. Thus, to the extent to which mathematics represented, say, Platonist philosophy, or more generally a liberal education, and to the extent to which a liberal education and Platonist philosophy represented the old pagan order, talking about mathematics was part of wider discourses about the relation between new faith and old values, between Hellenic institutions and the new Jewish-born, in principle cosmopolitan, religion and its own institutions. I will explore these issues through two examples: Clement of Alexandria and Augustine of Hippo.

Clement of Alexandria is a bit early for our purposes, because he died in AD 216, but he offers some interesting insights into the dialectic between traditional and new Christian values. One of his works, for instance, is devoted to arguing that the evangelic pronunciation against wealthy people (it would be easier for a camel to go through a needle's eye than for a rich man to enter heaven) is actually not to be taken literally, and that the wealthy need not worry too much about their salvation, as long as they redistribute some of their possessions.[3] An example of how the new beliefs could be accommodated to extant and definitely unchanged social and economic hierarchies. Mathematics comes onto Clement's horizon for several reasons. First, as part of the pagan educational curriculum and thus as a Greek form of knowledge. In this guise, the potential dangers of some branches of mathematics are made amply clear: in the course of interpreting the prophet Enoch, Clement even concludes that humans were taught astronomy by the evil angels.[4] On a less dramatic note, astronomy and geometry (and dialectic) are seen as futile – they do not teach the real truth; or they (in this case geometry and music, and grammar and rhetoric) are represented as the handmaids of philosophy, which itself, while limited, at least is a quest for truth.[5] But Clement cuts the 'Greekness' of traditional knowledge down to size. He underlines that Greek philosophy is heavily indebted to non-Greek one, and that all *technai*, philosophy, geometry, astrology and the division of the year into months were invented by non-Greeks.[6] On the other hand, he appreciates the value of some philosophical authors, especially Plato and Pythagoras, and also of some mathematics. Moses, he tells us, studied arithmetic and geometry, and mathematics (following a common Platonist tenet) is useful to train the soul towards non-material, higher, realities. Moreover, mathematics represents order, in particular the harmony and regularity of God's creation: in line with what is stated in the Bible, He is the measure, weight and number of the universe, the one who has counted the depth of the oceans and the number of hairs on everyone's head. Clement states explicitly that God possesses mathematical knowledge; he draws a parallel between faith and science, which both start from undemonstrated first principles, and engages in quite a lot of numerology. The size of the ark of the covenant in the tabernacle in Jerusalem is discussed from four different mathematical points of view: arithmetical, geometrical, astronomical and musical.[7] Here is part of the arithmetical explanation:

> 'And the days of the men will be' it says '120 days' [...] the hundred-and-twenty is a triangular number and has been formed from the equality of the 64, of which the composition part by part gives birth to squares, 1 3 5 7 9 11 13 15, on the other hand from the unequality of the 56, seven of the even numbers starting from two, which give birth to the rectangles, 2 4 6 8 10 12 14. According to yet another interpretation the number hundred 20 has been formed from four <numbers>, one, the triangular <number> fifteen, another, the square <number> 25, the third the pentagonal <number> 35, the fourth the hexagonal <number> 45.[8]

For the definition of triangular, pentagonal and hexagonal numbers, as well as for an explanation of the 'birth' of squares and rectangles, the reader has to be referred to Nicomachus or Iamblichus – it is material that does not appear prominently in Euclid's *Elements*, where we only find definitions of square and rectangular numbers.

Yet more is found in Augustine. Some mathematical notions seem to have puzzled him for their philosophical consequences: for instance, he wondered whether number, measure and weight pre-existed creation and, if yes, in what sense, and addressed the question of whether the soul could be talked about as if it was a number or a geometrical object.[9] Augustine's interest in mathematics dated from his young days: as a child, he had had rhythmically to chant mathematical tables; later, he had received a good liberal education, which included music and arithmetic, and got interested in astrology when attracted to Manichaeism. Even in the troubled phase of his conversion, when he was ready to jettison his pagan education entirely, he could not rid his memory of mathematical notions, because they were abstract, neither Greek nor Latin, nor related to any material object in particular.[10] Augustine ended up a Christian, but, instead of getting rid of mathematics, he put it to Christian use. True, he often inveighed against curiosity, and insisted that knowledge without knowledge of God is worth nothing. That included mathematical knowledge:

> A man who knows that he owns a tree and gives thanks to you [God] for the use of it, even though he does not know exactly how many cubits high it is or what is the width of its spread, is better than the man who measures it and counts all its branches but does not own it, nor knows and loves its Creator. In an analogous way the believer has the whole world of wealth and possesses all things as if he had nothing by virtue of his attachment to you whom all things serve; yet he may know nothing about the circuits of the Great Bear. It is stupid to doubt that he is better than the person who measures the heaven and counts the stars and weighs the elements, but neglects you who have disposed everything by measure and number and weight.[11]

But Augustine also wrote a detailed treatise on music, with the last sections devoted to a distinction between different types of numbers and their relative order from the point of view of spiritual excellence.[12] Through numbers, for instance the numbers contained in music and rhythm, human beings could get closer to God, because if numbers were beautiful, ordered and balanced, they would induce admiration and love in all those who saw them. That love would then extend to the creator of things in which numbers were contained, and also creator of numbers themselves. Augustine also carefully re-calculated the number of generations before Christ, in order to refute criticisms according to which their number was different in different gospels,[13] and used numbers to elucidate theological concepts.[14] For instance, immutability: the simple propositions 'Two and four is six' and 'Four is the sum of two and two: this sum is not two: therefore, two is not four' were used as example of unchanging reasoning.[15] Or again, in the *Soliloquies*, which are an early work, Augustine engaged in dialogue with his own reason, about (among other things) the relation between the knowledge of God and other, more familiar, forms of knowledge such as sense knowledge or mathematical knowledge. The importance of this latter lay in the fact that it provided evidence for the existence of unassailable certainties about abstract, non-material objects. Augustine often compared the type of certainty offered by mathematical truths with the certainty that should be inspired by faith, and the existence of mathematics gave him ammunition to reject sceptic attacks on the possibility itself of knowing things like the soul. For instance, in *On the Quantity of the Soul*, he contended that understanding the soul can start from a process similar to the understanding of mathematical objects. It was however clear to Augustine that knowledge of God and mathematical knowledge were not the same thing (a notion entertained and then dismissed in the *Soliloquies*), because their objects and the respective value of those latter were different. Nonetheless, 'the dissimilarity rests on a difference of objects and not of understanding'.[16] Mathematics, for all its limitations, could indeed provide a step-ladder to divinity.

A further facet of mathematics for Augustine, as for Philo and Clement, was numerology. In one dazzling passage of *On the trinity*, he finds meanings in, among other things, the six days of creation (six is the first perfect number), the years a certain woman who was miraculously cured suffered from her illness, the years a certain fig tree was left alone before being punished for being unfruitful, the length of the year according to both the moon and the sun calendars, and other episodes in the life of Christ. He informs us that we live in precisely the sixth age of man, which has been inaugurated by the birth of Christ, and determines the exact day of Christ's conception. He concludes defiantly:

> And now a word about the reasons for putting these numbers in the Sacred Scriptures. Someone else may discover other reasons, and either those which I have given are to be preferred to them, or both are equally probable, or theirs may be even more probable than mine, but let no one be so foolish or so absurd as to contend that they have been put in the Scriptures for no purpose at all, and that there are no mystical reasons why these numbers have been mentioned there. But those which I have given have been handed down by the Fathers with the approval of the Church, or I have gathered them from the testimony of the divine Scriptures, or from the nature of numbers and analogies. No sensible person will decide against reason, no Christian against the Scriptures, no peaceful man against the Church.[17]

Augustine's firm belief in the mystical rationale for the numbers in the Bible should be seen in connection with his notion, mentioned above, that numbers and order in the universe may draw people closer to God. Numerology had a strong ethical component, which was not limited to intensely

Christian contexts. For instance, the author of the panegyric for Constantius (AD 297–8) first deploys mathematics for rhetoric purposes – he declares it impossible to do justice to Constantius' and his family's achievements, so he will limit himself to counting them, except that even counting proves impossible because of the sheer quantity of great things done by the emperor. Then he moves on to numerological speculations, which provide him with a device to connect cosmological and human/political order. Thus, the tetrarchy (two emperors, each with an appointed successor) reflects the number of the seasons, of the elements, of the divisions of the Earth, even of the 'lamps' in the sky.[18] Numerous other instances can be found in neo-Platonic/neo-Pythagorean authors:

> The number five is highly expressive of justice, and justice comprehends all the other virtues. [...] if we suppose that the row of numbers is some weighing instrument, and the mean number 5 is the hole of the balance, then all the parts towards the seven, starting with the six, will sink down because of their quantity, and those towards the one, starting with the four, will rise up because of their fewness, and the ones which have the advantage will altogether be triple the total of the ones over which they have the advantage, but 5 itself, as the hole in the beam, partakes of neither, but it alone has equality and sameness. [...] thanks to the fact that five is a point of distinction and reciprocal separation, if the disadvantaged one which is closest to the balance on that side is subtracted from the one which is furthest from the balance on the excessive side and added to the one which is furthest from the balance on the other side – if, to effect equalization, 4 is subtracted from 9 and added to 1; and from 8, 3 is subtracted, which will be the addition to 2; and from 7, 2 is subtracted, which is added to 3; and from 6, 1 is subtracted, which is the addition to 4 to effect equalization; then all of them equally, both the ones which have been punished, as excessive, and the ones which have been set right, as wronged, will be assimilated to the mean of justice. For all of them will be 5 each; and 5 alone remains unsubtracted and unadded, so that it is neither more nor less, but it alone encompasses by nature what is fitting and appropriate.[19]

The presence of numerology across religious divides, with holy men as different as Clement and [Iamblichus] attributing fundamentally similar meanings to simple arithmetical properties, or to simple numbers, raises a few questions about the religious divides themselves. Numerology presupposed a belief in the link between mathematics and the divine, whether located in the Christian God or in the neo-Platonic One/Monad. The fact that, despite the evident parallels, the Christian God and the NeoPlatonic One were not the same thing opened a potential ground for conflict. If numbers were the key to a higher, deeper and more stable form of knowledge, who was most qualified to read them? There are instances of direct competition between pagan and Christian holy men, and it has been suggested that the numerologist *par excellence*, Pythagoras, as depicted by above all Iamblichus, was 'the pagan response and counterpart to Christ'.[20] Keeping in mind Clement's dismissal of the Greekness of philosophy and of several mathematical disciplines, let us read a micro-history of mathematics by the rampantly anti-Christian emperor Julian, apparently an admirer of Iamblichus:

> The theory of the heavenly bodies was perfected among the Hellenes, after the first observations had been made among the barbarians in Babylon. And the study of geometry took its rise in the measurement of the land in Egypt, and from this grew to its present importance. Arithmetic began with the Phoenician merchants, and among the Hellenes in course of time acquired the aspect of a regular science.[21]

The work containing this passage is devoted to showing that the Jewish people, the people Jesus belonged to, never invented anything or did anything remarkable. In sum, there are elements to suggest that mathematics may have been one of the many battling grounds for late ancient discussions over the divine signification of history and of the universe in general. On both fronts, numbers were transmitted through sacred ancestral texts – the Pythagorean *symbola* in the *Arithmetical Theology*, the Scriptures in Clement and Augustine – which awaited interpretation. Augustine's appeal to a trinity of authorities (reason, written tradition and the Church) in order to defend the importance of numerology may then be seen as addressed not exclusively to fellow Christians who did not subscribe to that particular way of doing mathematics, or simply had different opinions about the meaning of particular numbers. It may also have served as a strong reminder that decoding of numbers, both those found in the Bible and those inscribed in the book of creation, was a Christian thing – mathematics was God's own activity. The hapless person who counts the stars but neglects God, the Manichean astrologers of Augustine's early days and the pagan believers in mathematical theurgy were all examples of bad mathematics: bad not because of its contents, nor even because of its methods, but because of the meanings that one ascribed, or failed to ascribe, to it, because of the use one made of it.

The problem of ancient histories of ancient mathematics

As we discussed in the section on the problem of the birth of a mathematical community in chapter 4 p. 135, the past was already an inevitable presence in the second century BC; not just for mathematics, but also for philosophy, medicine, grammar. As knowledge accumulated in the form especially of books, scrolls at first, codexes later, collected in the libraries of rich Romans or later of bishops and abbots, the presence of the past and its inevitability grew even larger. It is thus not surprising that late antiquity saw a proliferation of 'deuteronomic' texts, not only in mathematics but in practically any other form of knowledge in which the written medium played an important role. I will first try better to characterize how late ancient mathematicians behaved towards previous sources – this has been tackled in chapter 7, so here I will only say some more with reference to the same authors: Pappus, Proclus, Eutocius. Second, I will look at explicit statements these authors made about the past, at the histories they wrote. Third, I will address the question of why the past was used so extensively.

We have already pointed out that authors writing about other authors put a lot of emphasis on clarity and intelligibility, often for the benefit of students who had different levels of ability, or of an addressee who is described as excellently gifted but not a mathematical expert. Thus, late ancient deuteronomic texts supply missing steps in the demonstrations, redefine or footnote the terminology, explore the limits of possibility or determinacy of a solution through variations in the diagram. The presence of a tradition also leads to the examination of sub-cases and generalizations: Pappus extended the theorem of Pythagoras to any triangle; Proclus, the first problem of the first book of Euclid's *Elements* ('On a given finite straight line to construct an equilateral triangle') to isosceles and scalene triangles.[22] The point of, for instance, sub-cases is to ensure that the problem is fully determined and that no further manipulation of its elements will produce configurations for which the solution is not valid. Sub-cases also provide exercise for the student because of their variety; and allow the reader to grasp the full implications of the proposition, its complete ramifications in the scheme of things mathematical. Acquiring power in problem-solving through an extensive command of the material is a recurrent theme in the mathematics of this period.

Late ancient mathematicians classified, defined and systematized quite a lot; they attached great importance to the right order of exposition; occasionally they formulated rules and prescriptions. A case in hand is Pappus' treatment of the duplication of the cube in book 3 of the *Mathematical Collection*: first he criticizes a solution that had been sent to him, then launches into a classification of problems and of the procedures appropriate to solve them,[23] then appends three previous solutions to the problem (Eratosthenes', Nicomedes' and Hero's), all of which are in line with his prescriptions, and finally provides his own solution. Since Eutocius' anthology of solutions to the duplication of the cube includes those quoted by Pappus, a comparison can be made between the two reports, and it can be argued that Pappus modified the earlier sources to have them fit within his classificatory scheme and with his own solution.

But the ways in which late ancient mathematicians reworked their sources could be even more pervasive. Faced with several, somewhat diverging, manuscripts of the *Conics*, Eutocius thought that they might be the different editions that Apollonius mentioned in the preface to the first book, and set out to collate them into one text, which would correspond to Apollonius' original intentions.[24] In other words, Eutocius produced the 'original' text he was commenting on to an extent which can be assessed only partly. Some insight can be gained by analyzing his guiding criteria, for instance his notion of clarity,[25] a highly subjective notion which he emphasized over and over again, and corresponded to an ability to distinguish clear from obscure, and to understand this latter to the point of explaining it to others. At one point, Eutocius justifies his selection of a proof as Apollonius' proof over another from an alternative version of the *Conics* with the remark that he 'produces clear light for the readers'.[26]

Eutocius' interventions reflect a situation common to other late ancient authors, where they tried to collect as many manuscripts of the same texts as possible. They were aware of problems of scribal transmission both for the text and the diagrams, and assessed the antiquity and genuineness of a manuscript on the basis of, in the case of Archimedes, the use of Doric dialect (but note that this could and was forged in the production of Pythagorean writings); the use of old-fashioned terminology (principally the old terms for conic sections, but that could have been forged as well); the filling of significant gaps (again a good sign of forgery). Eutocius refused to include into his account a solution to the duplication of the cube attributed to Eudoxus, no less, because the manuscript he had claimed that curved lines would be used, and then they were not, and a discontinuous proportion was used as if it was continuous. Yet, at the same time, he unquestioningly accepted a solution to the same problem by Plato, of which there is no record elsewhere.

Eutocius' luck at times does seem remarkable: in the course of commenting on *SC* 2.4, he mentioned that Archimedes at some point promised to provide a certain result but never did, at least not in any manuscript that Eutocius had seen. The lacuna, he said, had been there for a long time: already Diocles and a Dionysodorus whose date is uncertain had allegedly tried to fill the gap in Archimedes' account. But now, Eutocius had come across a very old book, in very bad condition, full of errors, which happened to contain just the right material, complete with traces of Doric dialect and old-fashioned terminology. The contents of the old book are not reported as they stand, however: we are provided with the results of Eutocius' very careful examination, cleaning of the text's errors and rephrasing of it in more usual and clearer language.[27] Clearly, the commentator is the medium through whom the readers can appreciate the greatness of Archimedes and be guided through his difficulties: more than that, he retrieves the past on their behalf.[28]

So, the point is not just interpreting the texts and ordering their contents in a newly-arranged mathematical universe ruled by new classifications – deuteronomic practice is also about the correct interpretation of the author's intention, the possibility itself to make the past understandable and meaningful for the present. A case taken up both by Pappus and Eutocius is Apollonius' criticism of Euclid in the *Conics*. Pappus interpreted it in these terms:

> The locus on three and four lines that [Apollonius] says, in [his account of] the third [book], was not completed by Euclid, neither he nor anyone else would have been capable of; no, he could not have added the slightest thing to what was written by Euclid, at any rate using only the conics that had been proved up to Euclid's time, as he himself confesses when he says that it is impossible to complete it without what he was forced to establish first. But [...] Euclid, out of respect for Aristaeus as meritorious for the conics he had published already, did not anticipate him, [...] for he was the fairest of men, and kindly to everyone who was the slightest bit able to augment knowledge as one should, and he was not at all belligerent, and though exacting, not boastful, the way this man [Apollonius] was, – [...] [Apollonius] was able to add the missing part to the locus because he had Euclid's writings on the locus already before him in his mind, and had studied for a long time in Alexandria under the people who had been taught by Euclid.[29]

Instead, in Eutocius' view, Apollonius is simply referring to another book by Euclid, now lost, so that the real meaning of his sneer is in a sense for ever irretrievable.[30] The aspect of transmission and the second element I want to explore in this section, history-making, are strictly related again in Pappus' book 7:

> The so-called domain of analysis, Hermodorus my son, is, taken as a whole, a special resource that was prepared after the production

> of the common elements for those who want to acquire a power in geometry that is capable of solving problems set to them; and it is useful for this alone. It was written by three men: Euclid the elementarist, Apollonius of Perga, and Aristaeus the Elder.[31]

The domain of analysis as a body of work has been produced through successive accretions over time; the transmission of a text embodies the history of that text and of the materials it contains. Pappus' history, however, is of course not neutral, as already evidenced by his unlikely reconstruction of the relationship between Euclid and Aristaeus: the three men he singles out here, for instance, are not the only authors of texts cited in book 7, and there is no evidence that they planned to produce a body of work. Neo-Platonist histories of mathematics tend not to be neutral either. Iamblichus rewrote traditional accounts of discoveries in a Pythagorean light, as in this passage, reported second-hand through Simplicius, where Sextus the Pythagorean supplants Archimedes as the discoverer of the area of the circle:

> Aristotle did not know this [the formulation for the area of the circle] at all, it seems; but it was discovered by the Pythagoreans, says Iamblichus, as it is clear from the proofs of Sextus the Pythagorean, who at the beginning learnt the method of the proofs according to the tradition. And afterwards, he says, Archimedes by means of † [the names of other mathematicians follow: Nicomedes, Apollonius, Carpus], as Iamblichus reports. And it is most extraordinary that the most learned Porphyry omits this – in fact he seems to say that there is a proof, according to which there is a figure of square that approximates the circle, as for the other figures, but does not transmit in any way that it was discovered.[32]

Again, when discussing the discovery of different types of means, Iamblichus' main reference points are Pythagoras and the Pythagoreans (Archytas, Ippasus of Metapontus, Philolaus), and anyone else is either cut down to size or forcibly enlisted. Thus, Plato is mentioned, but Iamblichus often specifies that things Plato said had been said before him by Pythagoras or the Pythagoreans, and Eudoxus figures as 'the Pythagorean'.[33] When Pappus deals with the very same topic, *his* sources turn out to be Nicomachus 'the Pythagorean' and Plato, but also Ptolemy and Eratosthenes.[34] And, like most authors before or after him, Pappus reports that the squaring of the circle was first discovered by Archimedes.

It is clear already from these few examples that using the past in the sense of writing a history of mathematics was not only not neutral, but could conflict with someone else's non-neutral use of that very same past. In other words, the mathematical past could be contested ground, with alternative reconstructions being proposed. A further interesting example of this is Proclus' micro-history of mathematics, discussed in chapter 2, p. 54. The aspect I want to emphasize here is that, while its value as information about the fifth and fourth centuries BC may be not very much, it is of course precious evidence about Proclus' own time and approach. Whatever his source or sources, Proclus created a more or less seamless chain that leads towards greater and greater generalization and rigour. While geometry and arithmetic may have had their origin in necessity, and while Thales, importing mathematics to Greece, may have retained a measure of empiricism, Pythagoras started the process of 'scientification' of mathematics. Everybody else who is mentioned joins in a path whose ultimate end is never put in doubt: systematization into elements, increased rigour, increased generality. Mathematicians across centuries are presented as aiming at a common end, embodied by Euclid, who

> belonged to the persuasion of Plato and was at home in this philosophy; and this is why he thought the goal of the *Elements* as a whole to be the construction of the so-called Platonic figures.[35]

Proclus created a mathematical tradition, culminating in Euclid, and related it indissolubly to the philosophical tradition he saw himself as part of – Platonism. But again, Proclus' story is only one version of ancient history of ancient mathematics: Pappus had other heroes, with Archimedes and Ptolemy playing a role that in Proclus is basically non-existent. We could even go as far as saying that Augustine had his own mathematical tradition, with the Sacred Scriptures as key texts to a twin understanding of the mathematical texture of reality and divinity. Those different stories, or alternative traditions, (I repeat) were in potential or actual conflict: Iamblichus writing Archimedes and Euclid out of the history of mathematics on the one hand, Pappus writing them in and attacking philosophers for their incompetence in mathematics (in book 5 of the *Collection*) on the other, Proclus revealing the real Platonic motives of Euclid's *Elements*, Eutocius revealing the real practical motives of Archimedes' *Measurement of the Circle*.

With reference to the third point raised in this section, then, one possible motive for the use of the past was that it made a good polemical weapon. One's authority could be enhanced strategically against competing claims by creating continuities between oneself and the figures of the tradition.[36] Late antiquity was a period of radical change, with traditional points of reference – religious authority, number of emperors, state boundaries, capital of the Empire, organization of the administrative machine – shifting, and new structures emerging. The past at least could be presented as something fully mapped and properly ordered, a stabilizing presence, and establishing links with it was a way to promote oneself and obtain legitimation.

Deuteronomic or revivalist practices were so widespread and pervasive that they may appear to be the one defining feature of late ancient culture. Recourse to the past, however, was a strategy used already at earlier times, nor was it the only strategy available. For instance, as far as late ancient mathematicians were concerned, utility, benefit for the community, both variously defined, and ethical significance of their knowledge were also values with which many of them wished to be associated, and which were officially promoted. Along with the rhetoric of 'past is good', we also find a notion that 'new is better': we have seen what Synesius thought about the astronomy of his own time, and that Iamblichus criticized Euclid; Apollonius for his part was criticized by Pappus, Proclus and Eutocius; even the great Archimedes was reprimanded by Pappus because he had not followed his prescriptions about problem-solving (which, incidentally, he had no reason to be aware of). Occasionally, Pappus indicated that there had been progress in mathematics with respect to the past: for instance, discovery of better solutions, clarification of some issues, additions to the body of knowledge. On one occasion, he warned the reader that 'we do not have to trust the opinion of the men who <first> discovered <this>'.[37]

In sum, late ancient mathematicians used the past as it suited them and their present concerns: their accounts were geared to different audiences, and mixed old and new both in their deployment of the sources and in their appreciation of what constituted good and bad mathematical practice. They make a fitting conclusion for this book, in that they were among the first to hand down histories of mathematics. Their histories are a warning that no history, not even that of mathematics, can ever be neutral.

I have set forth my agenda in the introduction: here I just want to repeat that I have chosen to present a vast range of heterogeneous material, because of my belief that mathematical practices, in antiquity as at probably any other time, are extremely complex, multilayered and occasionally baffling things. Distinguishing between what is and is not 'real' mathematics is a choice one has to make, and I have chosen to be inclusive of everyday mathematics, of non-mathematical views of mathematics and of 'applied' mathematics. If that has meant cutting down on some topics, I hope I have still managed to convey some of the brilliance and rigour of the theorems, discoveries and insights of ancient mathematicians. More than anything, people in the nearly thousand years we have surveyed could get incredibly passionate about mathematical matters, be they the properties of incommensurable lines or the appropriateness of measuring the heavens – my wish is that some of that passion has rubbed over onto you, the reader.

Notes

1 The term in this sense is introduced in Netz (1998).
2 McCluskey (1990).
3 Clement of Alexandria, *The Rich Man's Salvation*.
4 Clement of Alexandria, *Eclogues on the Prophets* 53.4 (commenting on 1 *Enoch* 8.3).
5 Clement of Alexandria, *Stromata* 6.11.93 and 1.5.29, respectively.
6 Clement of Alexandria, *ibid.* 1.15.66–1.16.80, cf. also 1.21.101.
7 Moses: *Stromata* 1.23.153; God as measure of the universe: *Exhortation to the Greeks* 6.60; God having mathematical knowledge, interpretation of the ark of the covenant: *Stromata* 6.11; faith and science: *Stromata* 2.4.14.
8 Clement of Alexandria, *Stromata* 6.11.84–85, my translation. The quote is from *Genesis* 6.3.
9 Augustine, *The Genesis Interpreted Literally* 4.7, *On the Quantity of the Soul* 3 ff.
10 Augustine, *Confessions* 10.19.
11 Augustine, *Confessions* 4.7, tr. H. Chadwick, Oxford 1991.
12 Augustine, *On Music* esp. book 6.
13 Augustine, *On the Agreement of the Evangelists* 2.4.
14 Augustine, *On Free Will* 2.20–24; *On the Immortality of the Soul* 1.5.
15 Augustine, *On the Immortality of the Soul* 2.2, tr. G. Watson, Aris and Phillips 1990. Cf. also similar examples at e.g. *On Free Will* 2.8, 2.12; *On the Immortality of the Soul* 1, 6 (examples with geometry); *Against the Academics* 3.11; *Confessions* 6.6; *Letters* 14.4.
16 Augustine, *Soliloquies* 1.11, translation as above. Cf. also *ibid.* 2.33.
17 Augustine, *On the Trinity* 4.10, tr. S. McKenna, Washington 1963. Cf. also *On the Trinity* 4.7, *The City of God* 11.30, *The Genesis Interpreted Literally* 4.7.
18 *Panegyric 8* 1.3–5 and 4.1–2, respectively.
19 [Iamblichus], *Theologia Arithmetica* 35.6 ff., tr. R. Waterfield, Phanes Press 1988, with modifications. Cf. also Anatolius, *On the Decad*, a very similar work written by a Christian.
20 O'Meara (1989), 214.
21 Julian, *Against the Galileans* 178a–b.
22 Pappus, *Mathematical Collection* 4.176.9–178.13; Proclus, *Commentary on the First Book of Euclid's Elements* e.g. 218–19, 228 ff., 323–6. See Netz (1998), Cuomo (2000a).
23 The same classification is reported in book 4 and quoted in the previous chapter.
24 Eutocius, *Commentary on Apollonius' Conics* 1.176.17–22, 230.13–16, 246.15–17, 250.16–22.
25 On this see Decorps-Foulquier (1998).
26 Eutocius, *Commentary on Apollonius' Conics* 2.296.6–7, cf. also 4.354.5–7.
27 Eutocius, *Commentary on Archimedes' Sphere and Cylinder* 2.4 (88 ff. Mugler).
28 Eutocius, *ibid.* 2.4 (100 Mugler).
29 Pappus, *Collection* 7.674.20–682.23, tr. A. Jones, Springer 1986.
30 Eutocius, *Commentary on Apollonius' Conics* 1.186.1–10.
31 Pappus, *Collection* 7.634.3–10, translation as above with modifications.
32 Simplicius, *On Aristotle's Categories* VII 192.15–25 (the relative passage in Aristotle is 7b15), my translation.
33 E.g. Iamblichus, *On Nicomachus* 105.2–11.
34 Pappus, *Collection* 68.17 ff.
35 Proclus, *Commentary on Euclid* 68, tr. G.R. Morrow, Princeton University Press 1970, reproduced with permission. Cf. also *ibid.* 82.
36 Parallels of this in fields other than mathematics are a huge number, see for further references Maas (1992), Cuomo (2000a).
37 Pappus, *Collection* 4.254.23–24, my translation.

GLOSSARY

Application: the operation that consists in constructing a parallelogram or a rectangle under certain conditions and which has as side a given straight line.

Arithmetic mean, see **Mean, arithmetic.**

Binomial: consisting of two terms; an expression which contains the sum or difference of two terms (*OED*).

Chorobates: an instrument for finding the level of water, a ground-level (*LS*).

Delian problem, see **Duplication of the cube.**

Delos, problem of, see **Duplication of the cube.**

Diorismos (pl. *diorismoi*): statement of the limits of possibility of a problem; particular enunciation of a problem (*LS*).

Duplication of the cube: to find a cube double a given cube. Found to be equivalent to the problem of finding two mean proportionals between two given lines, i.e. given the two lines A and B, one is required to find X and Y such that A:X=X:Y=Y:B. Also called Delian problem or problem of Delos because of a story that the god Apollo asked for a new altar for his temple at Delos double the size of the extant cubic one.

Equestrian order: originally members of the Roman cavalry with a certain social eminence, from which officers and the staffs of governors and commanders were drawn. Under the Empire, they constituted the second aristocratic order, ranking below the senators, provided the officer corps of the Roman army and held a wide range of posts in the civil administration. In general (although precise criteria for membership of the order remain disputed) all Roman citizens of free birth who possessed the minimum census qualification of 400,000 sesterces automatically qualified as members of the order (*OCD s.v.*).

Extreme and mean ratio: a line segment is divided into extreme and mean ratio when the shorter segment resulting from the division is to the larger segment resulting from the division as the larger segment is to the whole original segment.

Geometric mean, see **Mean, geometric.**

Gnomon: pointer of a sundial; carpenter's square; number added to a figurate number to obtain the next number of the same figure (*LS*); in Euclid (*El.* 2. def. 2), any of the parallelograms about the diagonal of a parallelogrammic area with the two complements.

Harmonic mean, see **Mean, harmonic.**

Interval: the difference of pitch between two sounds (*OED*).

Incommensurable: having no common measure with another quantity (*OED*).

Isoperimetrism: set of mathematical problems that deals with two or more figures having equal perimeter (*OED*).

Locus (pl. *loci*): the figure composed of all the points which satisfy certain conditions, or are generated by a point, line, or surface moving in accordance with certain conditions (*OED*).

Mean, arithmetic: of three terms, the first exceeds the second by the same amount as the second exceeds the third. I.e. B is the arithmetic mean between A and C if A – B = B – C.

Mean, geometric or proportional: of three terms, the first is to the second as the second is to the third. I.e. B is the geometric mean between A and C if A : B = B : C.

Mean, harmonic: of three terms, by whatever part of itself the first term exceeds the second, the middle term exceeds the third by the same part of the third. I.e. B will be the harmonic mean between A and C if C : A = (B – A) : (C – B).

Medial: pertaining to or designating a mathematical mean, or a line or area which is a mean proportional (*OED*).

Maximum (pl. *maxima*): the greatest value which a variable may have; a point at which a continuously varying quantity ceases to increase and begins to decrease (*OED*).

Minimum (pl. *minima*): the least value which a variable may have; a point at which a continuously varying quantity ceases to decrease and begins to increase (*OED*).

Modus ponens: an argument employing the rule that the consequent *q* may be inferred from the conditional statement *if p then q* and the statement *p* (*OED*).

Modus tollens: an argument employing the rule that the negation of the antedecent *p* (i.e. *not-p*) may be inferred from the conditional statement *if p then q* and the consequence *not-q* (*OED*).

Neusis (pl. *neuseis*): lit. inclination; a line segment of a given length, whose terminal points have to lie on given straight lines or curves and whose extension is to pass through a given point.

Nome: administrative subdivision of the territory in Egypt.

Parapegma: astronomical and meteorological calendar, inscribed on stone, the days of the months being inserted on movable pegs at the side of the text (*LS*).

Parameter of a conic: also called *latus rectum*, it is the straight line which is in the parabola the constant height of a rectangle which has as base the abscisse of a point of the conic, and whose area is equal to the square of the ordinate of the same point. In the hyperbola it is the straight line which is in the parabola the constant height of a rectangle which has as base the abscisse of a point of the conic, and whose area, plus another given area, is equal to the square of the ordinate of the same point. In the ellipse, it is the straight line which is in the parabola the constant height of a rectangle which has as base the abscisse of a point of the conic, and whose area, minus another given area, is equal to the square of the ordinate of the same point.

Plane number: a number resulting from the multiplication of two other numbers.

Porism: 'A porism occupies a place between a theorem and a problem; it deals with something already existing, as a theorem does, but has to *find* it (e.g. the centre of a circle), and, as a certain operation is therefore necessary, it partakes to that extent of the nature of a problem, which requires us to construct or produce something not previously existing.' (Heath (1921), I 434).

Precession of the equinoxes: the earlier occurrence of the equinoxes in each successive sidereal year (due to precession of the Earth's axis) (*OED*).

Prime number: a number whose only integral factors are the number itself and the unity.

Problem of Delos, see **Duplication of the cube**.

Problem of the two mean proportionals, see **Duplication of the cube**.

Quadrature of the circle: the problem of expressing the ratio between circumference and diameter of a circle, so as to reduce its area to a known rectlinear area.

Rectangular number: a number which is the product of the multiplication of two different numbers.

Senatorial order: originally a body of wealthy men of aristocratic birth, most of them ex-magistrates, which supervised the magistrates' work and could invalidate laws. Under the Empire, an official property qualification of one million sesterces was introduced, as well as a strong hereditary element (*OCD s.v.*).

Setting-out: the stage in a geometrical proof where the general enunciation is repeated with reference to a particular diagram.

Similar numbers: two numbers are similar if a mean proportional exists between them.

Similar polygons: they are similar when they contain the same angles and have the same shape (*OED*).

Solid number: a number which is the product of the multiplication of three other numbers.

Square number: a number which is the product of the multiplication of a number by itself.

Squaring the circle: see **Quadrature of the circle.**

Techne (pl. *technai*): art, craft, knowledge, activity which can be taught and learnt, and which produces some kind of artefact or result. Examples of *technai* include medicine, rhetoric, hunting, architecture, poetry, cooking, mathematics, mechanics.

BIBLIOGRAPHY

Primary sources

Aeschines, *Against Ctesiphon,* Engl. tr. C. Darwin Adams, London/Cambridge, MA: Heinemann and Harvard University Press 1958.

—— *Against Timarchus*, Engl. tr. C. Darwin Adams, London/Cambridge, MA: Heinemann and Harvard University Press 1958.

Aeschylus, *Prometheus Bound*, Engl. tr. J. Scully and C.J. Herington, Oxford: Oxford University Press 1975.

Agennius Urbicus, *Controversies about Fields*, ed. K. Thulin, in *Corpus agrimensorum romanorum*, vol.1 fasc.1, Leipzig: Teubner 1913.

Alcinous, *The Handbook of Platonism*, Engl. tr. J. Dillon, Oxford: Clarendon Press 1993.

Anatolius, *On the Decad*, French tr. P. Tannery, in *Mémoires scientifiques* 3, Paris/Toulouse: Gauthier-Villars and Privat 1915 (first publ. 1900), 12–31.

Palatine Anthology, Engl. tr. W.R. Paton, London/Cambridge, MA: Heinemann and Harvard University Press 1918, 5 vols.

Apollonius, *Conics*, ed. J.L. Heiberg, Leipzig: Teubner 1891–3.

Aratus, *Phaenomena*, Engl. tr. G.R. Mair, London/Cambridge, MA: Heinemann and Harvard University Press 1955.

Archimedes, *Cattle Problem*, ed. C. Mugler, Paris: Les Belles Lettres 1970–2.

—— *Conoids and Spheroids*, ed. C. Mugler, Paris: Les Belles Lettres 1970–2.

—— *Equilibria of Planes I and II*, ed. C. Mugler, Paris: Les Belles Lettres 1970–2.

—— *Measurement of the Circle*, ed. C. Mugler, Paris: Les Belles Lettres 1970–2.

—— *Method*, ed. C. Mugler, Paris: Les Belles Lettres 1970–2.

—— *Quadrature of the Parabola*, ed. C. Mugler, Paris: Les Belles Lettres 1970–2.

—— *Sand-Reckoner*, ed. C. Mugler, Paris: Les Belles Lettres 1970–2.

—— *Sphere and Cylinder I and II*, ed. C. Mugler, Paris: Les Belles Lettres 1970–2.

—— *Spirals*, ed. C. Mugler, Paris: Les Belles Lettres 1970–2.

Aristarchus of Samos, *On the Sizes and Distances of the Sun and Moon*, Engl. tr. T.L. Heath in *id., Aristarchus of Samos. The Ancient Copernicus*, Oxford: Clarendon Press 1913.

Aelius Aristides, *Sacred Discourse*, Engl. tr. C.A. Behr, Leiden: E.J. Brill 1981.

Aristophanes, *Birds*, ed. N. Dunbar, Oxford: Clarendon Press 1995.

—— *Clouds*, Engl. tr. A.H. Sommerstein, Warminster: Aris and Phillips 1982.
—— *Wasps*, Engl. tr. B. Bickley Rogers, Cambridge, MA/London: Harvard University Press and Heinemann 1978.
Aristotle, *Constitution of Athens*, Engl. tr. H. Rackham, Cambridge, MA/London: Harvard University Press and Heinemann 1952.
—— *Metaphysics*, Engl. tr. H. Tredennick, London/Cambridge, MA: Heinemann and Harvard UP 1935, 2 vols.
—— *Nicomachean Ethics*, Engl. tr. W.D. Ross, rev. by J.O. Urmson, Princeton: Princeton University Press 1984.
—— *Politics*, Engl. tr. H. Rackham, Cambridge, MA/London: Harvard University Press and Heinemann 1944.
—— *Politics*, Engl. tr. B. Jowett, Princeton: Princeton University Press 1984.
—— *Posterior Analytics*, Engl. tr. J. Barnes, Princeton: Princeton University Press 1984.
—— *Prior Analytics*, Engl. tr. H. Tredennick, Cambridge, MA: Harvard University Press 1938.
[Aristotle], *On Indivisible Lines*, Engl. tr. W.S. Hett, London/Cambridge, MA: Heinemann and Harvard University Press 1955.
Aristoxenus, *Elements of Harmonics*, Engl. tr. A. Barker, Cambridge 1989.
Athenaeus, *Deipnosophistae*, Engl. tr. C. Burton Gulick, London/New York: Heinemann and Putnam 1928.
Athenaeus Mechanicus, *On Machines*, in R. Schneider, 'Griechische Poliorketiker III', in *Abhandlungen der königlichen Gesellschaft der Wissenschaften zu Göttingen*, Philol.-hist. Kl. N.F. 12, heft 5 (1912).
Augustine, *Against the Academics*, ed. J.-P. Migne, Paris 1865, vol.1.
—— *The City of God*, Engl. tr. G.E. Mc Cracken *et al.*, London/Cambridge, MA: Heinemann and Harvard University Press 1957–72.
—— *Confessions*, Engl. tr. H. Chadwick, Oxford: Oxford University Press 1991.
—— *On the Quantity of the Soul*, ed. J.-P. Migne, Paris 1865, vol.1.
—— *The Genesis Interpreted Literally*, ed. and French tr. P. Agaësse and A. Solignac, Desclée de Brouwer 1970–2, 2 vols.
—— *On the Immortality of the Soul*, Engl. tr. G. Watson, Warminster: Aris and Phillips 1990.
—— *On Free Will*, ed. J.-P. Migne, Paris 1865, vol.1.
—— *On Music*, ed. J.-P. Migne, Paris 1865, vol.1.
—— *On the Trinity*, Engl. tr. S. McKenna, Washington: Catholic University of America Press 1963.
—— *Letters*, ed. J.-P. Migne, 1861, vol.2.
—— *On the Agreement of the Evangelists*, ed. J.-P. Migne, Paris 1865, vol.3.
—— *Soliloquia*, Engl. tr. G. Watson, Warminster: Aris and Phillips 1990.
Autolycus of Pitane, *La Sphère en Movement. Levers et Couchers héliaques. Testimonia*, ed. and French tr. G. Aujac, Paris: Les Belles Lettres 1979.
Balbus, *Explanation and Account of all Measures*, in *Gromatici Veteres*, eds F. Blume, K. Lachmann, A. Rudorff, Berlin: Reimer 1848–52, vol.1.
Biton, *Construction of War Instruments*, Engl. tr. E.W. Marsden, in *Greek and Roman Artillery: Technical Treatises*, Oxford: Clarendon Press 1971.

Callimachus, *Aetia*, ed. and Engl. tr. C.A. Trypanis, London/Cambridge, MA: Heinemann and Harvard University Press 1968.

Cassiodorus, *Letters*, ed. Th. Mommsen, Berlin: Weidmann 1894 (Monumenta Germaniae Historica 12).

M. Porcius Cato, *On Agriculture*, Engl. tr. W.D. Hooper and H. Boyd Ash, London/Cambridge, MA: Heinemann and Harvard University Press 1934.

Cicero, *Against L. Calpurnius Piso*, Engl. tr. N.H. Watts, Cambridge, MA/London: Harvard University Press and Heinemann 1931.

—— *Against Verres*, Engl. tr. L.H. Greenwood, Cambridge, MA: Harvard University Press 1935.

—— *Letters to Atticus*, Engl. tr. E.O. Winstedt, London/Cambridge, MA: Heinemann and Harvard University Press 1912–18.

—— *Letters to his Brother Quintus*, Engl. tr. W. Glynn Williams, Cambridge, MA/London: Harvard University Press and Heinemann 1972.

—— *On the Commonwealth*, Engl. tr. C.W. Keyes, London/Cambridge, MA: Heinemann and Harvard University Press 1948.

—— *On Divination*, Engl. tr. by W.A. Falconer, London/Cambridge, MA: Heinemann and Harvard University Press 1923.

—— *On the Greatest Good and Bad*, Engl. tr. H. Rackham, Cambridge, MA/London: Harvard University Press and Heinemann 1971.

—— *On the Agrarian Law*, Engl. tr. J.H. Freese, Cambridge, MA/ London: Harvard University Press and Heinemann 1967.

—— *On the Orator*, Engl. tr. H. Rackham, London/Cambridge, MA: Heinemann and Harvard University Press 1967, 2 vols.

—— *Philippicae*, Engl. tr. W.C.A. Ker, London/Cambridge, MA: Heinemann and Harvard University Press 1963.

—— *Tusculan Disputations*, Engl. tr. J.E. King, London/Cambridge, MA: Heinemann and Harvard University Press 1950.

Clement of Alexandria, *Eclogues on the prophets*, ed. O. Stählin, Berlin: Akademie 1970.

—— *Exhortation to the Greeks*, Engl. tr. G.W. Butterworth, Cambridge, MA/London: Harvard University Press and Heinemann 1919.

—— *Stromata*, ed. O. Stählin, Berlin: Akademie 1985, 2 vols.

—— *Stromata 5*, ed. A. Le Boulluec, Paris: Éditions du Cerf 1981.

—— *The Rich Man's Salvation*, Engl. tr. G.W. Butterworth, Cambridge, MA/London: Harvard University Press and Heinemann 1919.

Cleomedes, *On the Circular Motion of Heavenly Bodies*, French tr. R. Goulet as *Théorie élémentaire*, Paris: Vrin 1980.

Codex of Justinian, ed. E. Herrmann, Leipzig: Baumgartner 1875[5].

Codex of Theodosius, eds Th. Mommsen and P.M. Meyer, Berlin: Weidmann 1954, 3 vols.

L. Junius Moderatus Columella, *On Agriculture*, Engl. tr. H. Boyd Ash, London/Cambridge, MA: Heinemann and Harvard University Press 1941.

Demosthenes, *Against Aphobus I*, Engl. tr. A.T. Murray, London/Cambridge, MA: Heinemann and Harvard University Press 1958.

—— *Against Leptines*, Engl. tr. J.H. Vince, London/Cambridge, MA: Heinemann and Harvard University Press 1954.

—— *For Phormio*, Engl. tr. A.T. Murray, London/Cambridge, MA: Heinemann and Harvard University Press 1958.

—— *On the Crown*, Engl. tr. C.A. Vince and J.H. Vince, Cambridge, MA/London: Harvard University Press and Heinemann 1971.

Digesta of Justinian, eds Th. Mommsen and P. Krueger; Engl. tr. ed. by A. Watson, Philadelphia: University of Pennsylvania Press 1985, 4 vols.

Dio Cassius, *History*, Engl. tr. E. Cary, Cambridge, MA/ London: Harvard University Press and Heinemann 1969, 9 vols.

Diocles, *On Burning Mirrors*, ed. and Engl. tr. G.J. Toomer, Berlin/Heidelberg/ New York: Springer 1976.

Diodorus, *Historical library*, Engl. tr. C.H. Oldfather, London/Cambridge, MA: Heinemann and Harvard University Press 1933.

Diogenes Laertius, *Lives of the Philosophers*, Engl. tr. R.D. Hicks, London/ Cambridge, MA: Heinemann and Harvard University Press 1959.

Diophantus, *Arithmetica*, ed. P. Tannery, Leipzig: Teubner 1893.

—— *On Polygonal Numbers*, ed. P. Tannery, Leipzig: Teubner 1893.

—— *Les arithmétiques IV-VII*, ed. and French tr. R. Rashed, Paris: Les Belles Lettres 1984, 2 vols.

Euclid, *Data*, ed. H. Menge, Leipzig: Teubner 1896.

—— *Elements*, ed. J.L. Heiberg, revised by E.S. Stamatis, with *Scholia*, Leipzig: Teubner 1969–73, 5 vols.

—— *Optics*, ed. J.L. Heiberg, Leipzig: Teubner 1895.

—— *Phenomena*, Engl. tr. J.L. Berggren and R.S.D. Thomas, New York/London: Garland 1996.

[Euclid], *Sectio Canonis*, Engl. tr. A. Barker, Cambridge 1989.

Eunapius, *Lives of the Philosphers and Sophists*, Engl. tr. W. Cave Wright, London/ Cambridge, MA: Heinemann and Harvard University Press 1968.

Eutocius, *Commentary on Archimedes' Equilibria of Planes*, ed. and French tr. C. Mugler, Paris: Les Belles Lettres 1972 (in Archimedes, *Works*, vol.4).

—— *Commentary on Archimedes' Measurement of the Circle*, ed. and French tr. C. Mugler, Paris: Les Belles Lettres 1972 (in Archimedes, *Works*, vol.4).

—— *Commentary on Archimedes' Sphere and Cylinder*, ed. and French tr. C. Mugler, Paris: Les Belles Lettres 1972 (in Archimedes, *Works*, vol.4).

—— *Commentary on Apollonius' Conics*, ed. J.L. Heiberg, Leipzig: Teubner 1893 (in Apollonius, *Works*, vol. II).

Siculus Flaccus, *Fields Regulation*, ed. and French tr. M. Clavel-Lévêque, D. Conso, F. Favory, J.-Y. Guillaumin and P. Robin, Napoli: Jovene 1993.

Frontinus, *On the Aqueducts of the City of Rome*, Engl. tr. C.E. Bennett, London/ New York: Heinemann and Putnam 1925.

—— *On Land Surveying*, in *Corpus Agrimensorum Romanorum*, ed. K. Thulin, Leipzig: Teubner 1913.

Fronto, *Letters to the Emperor*, Engl. tr. C.R. Haines, London/Cambridge, MA: Heinemann and Harvard University Press 1919–20.

Galen, *An Exhortation to Study the Arts*, Engl. tr. P.N. Singer, *Selected Works*, Oxford: Oxford University Press 1997.
—— *On his Own Books*, ed. K.G. Kühn, Leipzig 1830 (repr. Hildesheim: Olms 1965), vol.19, Engl. tr. P.N. Singer, *Selected Works*, Oxford: Oxford University Press 1997.
—— *On the Sentences of Hippocrates and Plato*, Engl. tr. P. de Lacy, Berlin: Akademie 1981–4, 3 vols. (*CMG* 5.4.1.2).
—— *On Prognosis*, Engl. tr. V. Nutton, Berlin: Akademie 1979 (*CMG* 5.8.1).
—— *Logical Institution*, Engl. tr. J. Spangler Kieffer, Baltimore: The Johns Hopkins University Press 1964.
—— *The Affections and Errors of the Soul*, Engl. tr. P.N. Singer, *Selected Works*, Oxford: Oxford University Press 1997.
—— *The Best Doctor is also a Philosopher*, Engl. tr. P.N. Singer, *Selected Works*, Oxford: Oxford University Press 1997.
Geminus, *Introduction to the Phenomena*, ed. and French tr. G. Aujac, Paris: Les Belles Lettres 1975.
Hephaestio Thebanus, *Astrology*, ed. D. Pingree, Leipzig: Teubner 1973.
Hero of Alexandria, *Automata*, ed. W. Schmidt, Leipzig: Teubner 1899, vol.1.
—— *Belopoiika*, Engl. tr. E.W. Marsden, in *Greek and Roman Artillery: Technical Treatises*, Oxford: Clarendon Press 1971.
—— *Dioptra*, ed. H. Schöne, Leipzig: Teubner 1903, vol.3.
—— *Mechanica*, eds L. Nix and W. Schmidt, Leipzig: Teubner 1900.
—— *Metrica*, ed. H. Schöne, Leipzig: Teubner 1903.
—— *Pneumatica*, ed. W. Schmidt, Leipzig: Teubner 1899, vol.1.
[Hero], *Definitions*, ed. J.L. Heiberg, Leipzig: Teubner 1912.
Herodotus, *Histories*, Engl. tr. A.D. Godley, Cambridge, MA/London: Heinemann and Harvard University Press 1926, 4 vols.
Hipparchus, *Commentary to the Phenomena of Eudoxus and Aratus*, ed. and German tr. K. Manitius, Leipzig: Teubner 1894.
Homer, *Iliad*, Engl. tr. A.T. Murray, London/Cambridge, MA: Heinemann and Harvard University Press 1924, 2 vols.
Hyginus Astronomicus, *On Astronomy*, ed. and French tr. A. Le Bœuffle, Paris: Les Belles Lettres 1983.
Hyginus, *Categories of Fields*, in *Gromatici Veteres*, eds F. Blume, K. Lachmann and A. Rudorff, Berlin: Reimer 1848–52.
Hyginus 3, *On the Division of Military Camps*, ed. A. Grillone, Leipzig: Teubner 1977.
Hyginus Gromaticus, *On Boundary Regulation*, in *Gromatici Veteres*, eds F. Blume, K. Lachmann and A. Rudorff, Berlin: Reimer 1848–52, vol.1.
Hypsicles, *Book 14 of the Elements*, in Euclid, *Elements*, ed. J.L. Heiberg, Leipzig: Teubner 1888, vol.5.
Iamblichus, *De Mysteriis Aegyptorum*, ed. and French tr. É. des Places, Paris: Les Belles Lettres 1966.
—— *Exhortation to Philosophy*, Engl. tr. Th. Moore Johnson, Grand Rapids: Phanes Press 1988.

—— *Life of Pythagoras* (*On the Pythagorean Life*), Engl. tr. G. Clark, Liverpool: Liverpool University Press 1989.

—— *On Common Mathematical Knowledge*, ed. N. Festa, Leipzig: Teubner 1891.

—— *On Nicomachus' Introduction to Arithmetic*, ed. E. Pistelli, Leipzig: Teubner 1894.

[Iamblichus], *Theologia Arithmetica*, Engl. tr. R. Waterfield, Grand Rapids: Phanes Press 1988.

Isaeus, *On the Estate of Hagnias*, Engl. trans. E.S. Forster, London/New York: Heinemann and Putnam 1927.

Isocrates, *Against the Sophists,* Engl. tr. G. Norlin, London/New York: Heinemann and Putnam 1929.

—— *Antidosis*, Engl. tr. G. Norlin, London/New York: Heinemann and Putnam 1929.

—— *Helen*, Engl. tr. L. van Hook, London/Cambridge, MA: Heinemann and Harvard University Press 1945.

—— *Panathenaicus*, Engl. tr. G. Norlin, London/New York: Heinemann and Putnam 1929.

Julian, *Against the Galileans*, Engl. tr. W. Cave Wright, London/New York: Heinemann and Macmillan 1913.

[Julian], *Letters*, Engl. tr. W. Cave Wright, London/New York: Heinemann and Macmillan 1913.

Livy, *From the Foundation of the City*, Engl. tr. F. Gardner Moore, London/Cambridge, MA: Heinemann and Harvard University Press 1951.

Lucian, *Icaromenippus*, Engl. tr. A.M. Harmon, London/New York: Heinemann and Macmillan 1915.

—— *Philosophies for Sale*, Engl. tr. A.M. Harmon, London/New York: Heinemann and Macmillan 1915.

Lysias, *Against Diogeiton*, Engl. tr. W.R.M. Lamb, London/New York: Heinemann and Putnam 1930.

—— *Against Nicomachus*, Engl. tr. W.R.M. Lamb, London/New York: Heinemann and Putnam 1930.

—— *Defence Against a Charge of Taking a Bribe*, Engl. tr. W.R.M. Lamb, London/New York: Heinemann and Putnam 1930.

—— *On the Property of Aristophanes: Against the Treasury*, Engl. tr. W.R.M. Lamb, London/New York: Heinemann and Putnam 1930.

L. Volusius Maecianus, *Distribution*, ed. F. Hultsch, in *Metrologicorum Scriptorum Reliquiae II*, Leipzig: Teubner 1866.

Manilius, *Astronomica*, Engl. tr. G.P. Goold, Cambridge, MA/London: Harvard University Press 1992.

Marinus, *Commentary on Euclid's Data*, in Euclid, *Opera*, vol.VI, ed. J.L. Heiberg and H. Menge, Leipzig: Teubner 1896.

Martial (Marcus Valerius Martialis), *Epigrammata*, Engl. tr. D.R. Shackleton Bailey, London/Cambridge, MA: Harvard University Press 1993, 3 vols.

Nicomachus, *Introduction to Arithmetic*, Engl. tr. M.L. D'Ooge, in R. Maynard Hutchins (ed.), *Great Books of the Western World,* Chicago: Benton 1952 (the translation first publ. 1926), vol.11.

Marcus Junius Nipsus, *Measurement of a River, Replacing a Boundary, Measurement of an Area*, in *Gromatici Veteres*, eds F. Blume, K. Lachmann and A. Rudorff, Berlin: Reimer 1848–52, vol.1.

Notitia dignitatum, ed. O. Seeck, 1876, reprinted Frankfurt a. M.: Minerva 1962.

Panegyric 8, in C.E.V. Nixon and B. Saylor Rodgers, *In Praise of Later Roman Emperors. The Panegyrici Latini*, Engl. tr. of the Latin texts R.A.B. Mynors, Berkeley/Los Angeles/Oxford: University of California Press 1994.

Pappus of Alexandria, *Mathematical Collection*, ed. and Latin tr. F. Hultsch, 3 vols, Berlin: Weidmann 1876–8.

—— *Book VII of the Collection*, ed. and Engl. tr. A. Jones, New York/Berlin/Heidelberg/Tokyo: Springer 1986, 2 vols.

Paulus Alexandrinus, *Elements of Astrology*, ed. E. Boer, Leipzig: Teubner 1958.

Philo of Alexandria, *On Mating with the Preliminary Studies*, Engl. tr. F.H. Colson and G.H. Whitaker, London/New York: Heinemann and Putnam 1932.

—— *On the Creation of the World According to Moses*, Engl. tr. F.H. Colson and G.H. Whitaker, London/New York: Heinemann and Putnam 1929.

—— *On the Migration of Abraham*, Engl. tr. F.H. Colson and G.H. Whitaker, London/New York: Heinemann and Putnam 1932.

—— *Who is the Heir of Divine Things and on the Division into Equals and Opposites*, Engl. tr. F.H. Colson and G.H. Whitaker, London/New York: Heinemann and Putnam 1932.

Philo of Byzantium, *Book 8 of the Mechanical Syntaxis (Poliorketika)*, ed. and French tr. by Y. Garlan, in Garlan (1974), 279–404.

—— *Mechanica IV (Belopoiika)*, Engl. tr. E.W. Marsden, in *Greek and Roman Artillery: Technical Treatises*, Oxford: Clarendon Press 1971.

Philostratus, *Lives of the Sophists*, with Engl. tr. W. Cave Wright, London/New York: Heinemann and Putnam 1922.

Plato, *Euthydemus*, Engl. tr. W.R.M. Lamb, Cambridge, MA/London: Harvard University Press and Heinemann 1924.

—— *Euthyphro*, Engl. tr. H.N. Fowler, London/New York: Heinemann and Putnam 1914.

—— *Greater Hippias*, Engl. trans. H.N. Fowler, London/New York: Heinemann and Putnam 1926.

—— *Laws*, Engl. tr. T.J. Saunders, Indianapolis/Cambridge: Hackett 1997.

—— *Lesser Hippias*, Engl. tr. N.D. Smith, Indianapolis/Cambridge: Hackett 1997.

—— *Meno*, Engl. tr. G.M.A. Grube, Indianapolis/Cambridge: Hackett 1997.

—— *Parmenides*, Engl. tr. H.N. Fowler, London/Cambridge, MA: Heinemann and Harvard University Press 1926.

—— *Philebus*, Engl. tr. H.N. Fowler, London/Cambridge, MA: Heinemann and Harvard University Press 1925.

—— *Philebus*, Engl. tr. D. Frede, Indianapolis/Cambridge: Hackett 1997.

—— *Protagoras*, Engl. tr. W.R.M. Lamb, Cambridge, MA/London: Harvard University Press and Heinemann 1924.

—— *Protagoras*, Engl. tr. S. Lombardo and K. Bell, Indianapolis/Cambridge: Hackett 1997.

—— *Republic*, Engl. tr. P. Shorey, London/Cambridge, MA: Heinemann and Harvard University Press 1946, 2 vols.

—— *Republic*, Engl. tr. Robin Waterfield, Oxford: Oxford University Press 1993.

—— *Statesman*, Engl. tr. C.J. Rowe, Indianapolis/Cambridge: Hackett 1997.

—— *Theaetetus*, Engl. tr. M.J. Levett and rev. M.F. Burnyeat, Indianapolis/Cambridge: Hackett 1997.

—— *Timaeus*, Engl. tr. R.G. Bury, London/Cambridge, MA: Heinemann and Harvard University Press 1966.

[Plato], *Epinomis*, Engl. tr. A.E. Taylor, in *The Collected Dialogues of Plato*, eds E. Hamilton and H. Cairns, Princeton: Princeton University Press 1963.

Plautus, *Poenulus*, Engl. tr. P. Nixon, London/New York: Heinemann and Putnam 1932.

Pliny Jr., *Letters*, Engl. tr. W. Melmoth and W.M.L. Hutchinson, London/Cambridge, MA: Heinemann and Harvard University Press 1947, 2 vols.

Pliny Sr, *Natural History*, Engl. tr. by D.E. Eichholz, London/Cambridge, MA: Heinemann and Harvard University Press 1962 (10 vols).

Plutarch, *Against the Stoics on Common Conceptions*, Engl. tr. H. Cherniss, Cambridge, MA/London: Harvard University Press and Heinemann 1976.

—— *Life of Cicero*, Engl. tr. R. Warner in *Fall of the Roman Republic. Six Lives by Plutarch*, London: Penguin 1972.

—— *Life of Marcellus*, Engl. tr. B. Perrin, London/Cambridge, MA: Heinemann and Harvard University Press 1968.

—— *On the E in Delphi*, Engl. tr. F.C. Babbitt, London/Cambridge, MA: Heinemann and Harvard University Press 1969.

—— *On the Genius of Socrates*, Engl. tr. P.H. De Lacy and B. Einarson, London/Cambridge, MA: Heinemann and Harvard University Press 1968.

—— *Table-Talk*, Engl. tr. E.L. Minar, London/Cambridge, MA: Heinemann and Harvard University Press 1961.

Polybius, *Histories*, Engl. tr. W.R. Paton, London/Cambridge, MA: Heinemann and Harvard University Press 1960.

Proclus, *Commentary on the First Book of Euclid's Elements*, Engl. tr. G.R. Morrow, Princeton: Princeton University Press 1970.

—— *Commentary on Plato's Republic*, French tr. A.J. Festugière, Paris: Vrin 1970.

—— *A Sketch of the Astronomical Positions*, ed. and German tr. K. Manitius, Leipzig: Teubner 1909.

—— *Physics*, ed. and German tr. A. Ritzenfeld, Leipzig: Teubner 1912.

—— *The Elements of Theology*, ed. and Engl. tr. E.R. Dodds, Oxford: Clarendon Press 1963.

—— *Commentary on Plato's Timaeus*, ed. E. Diehl, Leipzig: Teubner 1903–6, 3 vols.

Ptolemy, *Astrology* (*Tetrabiblos*), Engl. tr. F.E. Robbins, London/Cambridge, MA: Heinemann and Harvard University Press 1940.

—— *For the Use of the Tables*, in *Opera astronomica minora*, ed. J.L. Heiberg, Leipzig: Teubner 1907.

—— *Geography*, Engl. tr. E.L. Stevenson, New York: Dover 1991 (first publ. 1932).

—— *Harmonics*, ed. I. Düring, Göteborg 1930, repr. New York/London: Garland 1980.

—— *On the Criterion,* Engl. tr. by the Liverpool/Manchester Seminar on Ancient Philosophy, in P. Huby and G. Neals (eds), *The Criterion of Truth*, Liverpool: Liverpool University Press 1989, 179–230.

—— *Optica,* French tr. from Latin tr. of Arabic tr. A. Lejeune, Leiden: Brill 1989.

—— *Planetary Hypotheses*, in *Opera astronomica minora,* ed. J.L. Heiberg, Leipzig: Teubner 1907.

—— *Syntaxis*, Engl. tr. G.J. Toomer, London: Duckworth 1984.

Quintilian (Marcus Fabius Quintilianus), *Handbook of Oratory*, Engl. tr. H.E. Butler, London/Cambridge, MA: Heinemann and Harvard University Press 1963, 4 vols.

Seneca, *Letters to Lucilius*, Engl. tr. C.D.N. Costa, Warminster: Aris and Phillips 1988.

Select Papyri, Engl. tr. A.S. Hunt and C.C. Edgar, Cambridge, MA/London: Harvard University Press and Heinemann 1932–4, 5 vols.

Serenus of Antinopolis, *On the Section of a Cylinder, On the Section of a Cone*, ed. J.L. Heiberg, Leipzig: Teubner 1896.

Sextus Empiricus, *Against the Arithmeticians*, Engl. tr. R.G. Bury, London/ Cambridge, MA: Heinemann and Harvard University Press 1936.

—— *Against the Astrologers*, Engl. tr. R.G. Bury, London/Cambridge, MA: Heinemann and Harvard University Press 1936.

—— *Against the Geometers*, Engl. tr. R.G. Bury, London/Cambridge, MA: Heinemann and Harvard University Press 1936.

—— *Against the Logicians*, Engl. tr. R.G. Bury, London/Cambridge, MA: Heinemann and Harvard University Press 1935.

—— *Against the physicists*, Engl. tr. R.G. Bury, London/Cambridge, MA: Heinemann and Harvard University Press 1936.

Simplicius, *Commentary on Aristotle's Categories*, ed. K. Kalbfleisch, CAG 8, Berlin: Reimer 1907.

—— *Commentary on Aristotle's Physics*, ed. H. Diels, CAG 9–10, Berlin: Reimer 1882, 1895.

Sophocles, *Antigone*, Engl. tr. H. Lloyd-Jones, Cambridge, MA/London: Harvard University Press 1994.

Stobaeus Johannes, *Anthology*, eds C. Wachsmuth and O. Hense, Berlin: Weidmann 1884–1923, 6 vols.

Strabo, *Geography*, Engl. tr. H.L. Jones, London/New York: Heinemann and Putnam 1917.

Synesius of Cyrene, *Letters*, Engl. tr. A. Fitzgerald, London/Oxford: Milford and Oxford University Press 1926.

Theocritus, *Idylls*, Engl. tr. A.S.F. Gow, Cambridge: Cambridge University Press 1952.

Theodosius Tripolites, *On Stations. On Days and Nights*, ed. R. Fecht, in *Abhandlungen der Gesellschaft der Wissenschaften zu Göttingen*, phil.-hist. Kl., n.s. 19 no. 4, 1927.

—— *Sphaerica*, ed. J.L. Heiberg, in *Abhandlungen der Gesellschaft der Wissenschaften zu Göttingen*, phil.-hist. Kl., n.s. 19 no. 3, 1927.

Theognis, *Works* in *Greek Lyric Poetry*, Engl. tr. M.L. West, Oxford: Clarendon Press 1993.

Theon of Alexandria, *Commentary on Ptolemy's Syntaxis*, ed. A. Rome, Cittá del Vaticano: Biblioteca Apostolica Vaticana 1936 (Studi e Testi 72).

—— *Greater Commentary on Ptolemy's Handy Tables*, eds J. Mogenet and A. Tyhon, Citta' del Vaticano 1985–91.

—— *Lesser Commentary on Ptolemy's Handy Tables*, ed. A. Tyhon, Citta' del Vaticano 1978.

Theon of Smyrna, *Account of Mathematics Useful to Reading Plato*, ed. E. Hiller, Leipzig: Teubner 1878.

Thucydides, *The Peloponnesian War*, Engl. tr. C. Forster Smith, London/Cambridge, MA: Heinemann and Harvard University Press 1928.

Ulpian, *De excusatione* in *Iurisprudentiae anteiustinianae reliquias*, eds P.E. Huschke, E. Seckel and B. Kuebler, Leipzig: Teubner 1911, 2 vols.

M. Terentius Varro, *On Agriculture*, Engl. tr. W.D. Hooper and H. Boyd Ash, London/Cambridge. MA: Heinemann and Harvard University Press 1935.

Vegetius, *On military things*, Engl. tr. L.F. Stelten, New York/Bern/Frankfurt a.M./ Paris: Lang 1990.

Vettius Valens, *Anthology*, ed. D. Pingree, Leipzig: Teubner 1986.

Vitruvius, *On Architecture*, Engl. tr. F. Granger, London/Cambridge, MA: Heinemann and Harvard University Press 1962.

Xenophon, *Memorabilia*, Engl. tr. E.C. Marchant, London/New York: Heinemann and Putnam 1923.

Secondary sources

Alföldi G. (1975), *Römische Sozialgeschichte*, Engl. tr. by D. Braund and F. Pollock as *The Social History of Rome*, London/Sydney: Croom Helm 1985.

Andreau J. (1999), *Banking and Business in the Roman World*, Engl. tr. J. Lloyd, Cambridge: Cambridge University Press.

Angeli A. and Dorandi T. (1987), 'Il pensiero matematico di Demetrio Lacone', *Cronache ercolanesi* 17, 89–103.

Asheri D. (1966), 'Distribuzioni di terre nell'antica Grecia', *Memorie dell'Accademia di scienze di Torino, classe di scienze morali, storiche e filologiche*, serie 4, 10, 1–127.

—— (1975), 'Osservazioni sulle origini dell'urbanistica ippodamea', *Rivista storica italiana* 87, 5–16.

Austin M.M. (ed.) (1981), *The Hellenistic World from Alexander to the Roman Conquest. A Selection of Ancient Sources in Translation*, Cambridge: Cambridge University Press.

Badian E. (1972), *Publicans and Sinners. Private Enterprise in the Service of the Roman Republic*, Ithaca/London: Cornell University Press.

Beard M. and Crawford M. (1999), *Rome in the Late Republic. Problems and Interpretations*, London: Duckworth (2nd edn).

Becker O. (1933), 'Eudoxos-Studien I. Eine voreudoxische Proportionenlehre und ihre Spuren bei Aristoteles und Euklid', *Quellen und Studien zur Geschichte der Mathematik, Astronomie und Physik* Abt. B 2, 369–87.

—— (ed.) (1965), *Zur Geschichte der griechischen Mathematik*, Darmstadt: Wissenschaftliche Buchgesellschaft.

Bernal M. (1992), 'Animadversions on the Origins of Western Science', *Isis* 83, 596–607.

Bogaert R. (ed.) (1976), *Epigraphica. III: Texts on Bankers, Banking and Credit in the Greek World*, Leiden: Brill.

Bowen A.C. and Goldstein B.R. (1988), 'Meton of Athens and astronomy in the late fifth century BC', in E. Leichty, M. de J. Ellis and P. Gerardi (eds), *A Scientific Humanist: Studies in Memory of Abraham Sachs*, Philadelphia, 39–81.

Boyaval B. (1973), 'Tablettes mathématiques du Musée du Louvre', *Revue Archéologique*, 243–60.

Boyd T.D. and Jameson M.H. (1981), 'Urban and land division in ancient Greece', *Hesperia* 50, 327–42.

Brashear W. (1985), 'Holz- und Wachstafeln der Sammlung Kiseleff', *Enchoria* 13, 13–23.

Bruins E.M., Liesker W.H.M. and Sijpesteijn P.J. (1988), 'A Ptolemaic Papyrus from the Michigan Collection', *Zeitschrift für Papyrologie und Epigraphik* 74, 23–8.

Buchner E. (1982), *Die Sonnenuhr des Augustus*, Mainz a.R.: Philipp von Zabern.

Burford A. (1969), *The Greek Temple Builders at Epidauros. A Social and Economic Study of Building in the Asklepian Sanctuary, During the Fourth and Early Third Centuries B.C.*, Liverpool: Liverpool University Press.

Burkert W. (1972), *Weisheit und Wissenschaft: Studien zu Pythagoras, Philolaos und Platon*, Nürnberg: H. Carl 1962, revised Engl. tr. E.L. Minar Jr., *Lore and Science in Ancient Pythagoreanism*, Cambridge, MA: Harvard University Press.

Burns A. (1971), 'The Tunnel of Eupalinus and the Tunnel Problem of Hero of Alexandria', *Isis* 62, 172–85.

Cambiano G. (1991), *Platone e le tecniche*, Roma/Bari: Laterza (2nd edn).

Carrié J.-M. (1994), 'Diocletien et la fiscalité', *Antiquité tardive* 2, 33–64.

Carter J. Coleman (1990), 'Dividing the chora: Metaponto in the sixth century BC', *American Journal of Archaeology* 94, 320.

Cartledge P. (1996), 'Comparatively equal', in J. Ober and C. Hedrick (eds), *Demokratia. A Conversation on Democracies, Ancient and Modern,* Princeton: Princeton University Press, 175–85.

Castagnoli F. (1956), *Ippodamo di Mileto e l'urbanistica a pianta ortogonale*, Roma: De Luca.

Caveing M. (1990), 'Introduction générale', in B. Vitrac (ed.), *Les éléments de Euclide*, Paris: PUF, 13–148.

Charles-Saget A. (1982), *L'architecture du divin. Mathématique et philosophie chez Plotin et Proclus*, Paris: Les Belles Lettres.

Chisholm K. and Ferguson J. (eds) (1981), *Rome. The Augustan Age. A Sourcebook*, Oxford: Oxford University Press.

Chouquer G. and Favory F. (1992), *Les arpenteurs romains. Théorie et pratique*, Paris: Errance.

Clagett M. (1956), *Greek Science in Antiquity*, New York: Abelard-Schuman.
Clarke K. (1999), *Between Geography and History. Hellenistic Constructions of the Roman World*, Oxford: Clarendon Press.
Cohen E.E. (1992), *Athenian Economy and Society. A Banking Perspective*, Princeton: Princeton University Press.
Coulton J.J. (1977), *Greek Architects at Work. Problems of Structure and Design*, London: P. Elek.
Crawford D.S. (1953), 'A mathematical tablet', *Aegyptus* 33, 222–40.
Cuomo S. (2000a), *Pappus of Alexandria and the Mathematics of Late Antiquity*, Cambridge: Cambridge University Press.
—— (2000b), 'Divide and rule: Frontinus and Roman land-surveying', *Studies in History and Philosophy of Science* 31, 189–202.
Cuvigny H. (1985), *L'arpentage par espéces dans l'Égypte ptolémaïque d'après les papyrus grecs*, Bruxelles: Fondation Égyptologique Reine Élisabeth.
Decorps-Foulquier M. (1998), 'Eutocius d'Ascalon éditeur du traité des *Coniques* d'Apollonios de Pergé et l'exigence de 'clarté': un exemple des pratiques exégétiques et critiques des héritiers de la science alexandrine', in G. Argoud and J.-Y. Guillaumin (eds), *Sciences exactes et sciences appliquées à Alexandrie*, Saint-Étienne: Publications de l'Université de Saint-Étienne, 87–101.
DeLaine J. (1996), '"De aquis suis"?: The "Commentarius" of Frontinus', in C. Nicolet (ed.), *Les littératures techniques dans l'antiquité romaine. Statut, public et destination, tradition*, Genève: Vandœuvres, 117–45.
Déléage A. (1933), 'Les cadastres antiques jusqu'à Dioclétien', *Études de Papyrologie* 2, 73–228.
Della Corte M. (1954), *Case ed Abitanti di Pompei*, Roma: Bretschneider.
de Solla Price D.J. (1969), 'Portable sundials in antiquity, including an account of a new example from Aphrodisias', *Centaurus* 14, 242–66.
—— (1974), *Gears from the Greeks. The Antikythera Mechanism – A Calendar Computer from* c. *80 B.C.*, Philadelphia: The American Philosophical Society (Transactions of the American Philosophical Society N.S. vol.64, part 7).
Dicks D.R. (1971), 'Eratosthenes', *Dictionary of Scientific Biography*, New York: Scribner, vol.4, 388–93 .
Dijksterhuis E.J. (1956), *Archimedes*, Engl. tr. C. Dikshoorn, Princeton: Princeton University Press (reprinted 1987).
Dilke O.A.W. (1971), *The Roman Land Surveyors. An Introduction to the Agrimensores*, Newton Abbot: David and Charles.
—— (1987), *Mathematics and Measurement*, British Museum Publications.
Dörrie H. (1987), *Die geschichtlichen Wurzeln des Platonismus*, Stuttgart: Frommann Holzboog.
Dorandi T. (1994), 'La tradizione papiracea degli "Elementi" di Euclide', in W. Bülow-Jacobsen (ed.), *Proceedings of the XX International Congress of Papyrologists*, Copenhagen: Museum Tusculanum Press, 306–11.
Dufková M. and Pecírka J. (1970), 'Excavations of farms and farmhouses in the chora of Chersonesos in the Crimea', *Eirene* 8, 123–74.
Edgar C.C. (ed.) (1925), *Zenon Papyri (Catalogue général des antiquités égyptiennes du musée du Caire)*, Cairo: Institut Français d'archeologie orientale 1925–31, 4 vols.

Edwards C. (1993), *The Politics of Immorality in Ancient Rome*, Cambridge: Cambridge University Press.

Evans J. (1999), 'The material culture of Greek astronomy', *Journal for the History of Astronomy* 30, 237–307.

Fallu E. (1973), 'Les rationes du proconsul Cicéron. Un exemple de style administratif et d'interprétation historique dans la correspondance de Cicéron', in H. Temporini (ed.), *Aufstieg und Niedergang der Römischen Welt*, Berlin/New York: De Gruyter, vol.1.3, 209–38.

Fellmann R. (1983), 'Römische Rechentafeln aus Bronze', *Antike Welt* 14, 36–40.

Ferrari G.A. (1984), 'Meccanica allargata', *Atti del convegno, La scienza ellenistica, Pavia 1982*, Napoli: Bibliopolis 1984, 227–96.

Ferrary J.-L. (1988), *Philhellénisme et impérialisme. Aspects idéologiques de la conquête romaine du monde hellénistique, de la seconde guerre de Macédoine à la guerre contre Mithridate*, Rome: École Française de Rome.

Fowler D.H. (1988), 'A catalogue of tables', *Zeitschrift für Papyrologie und Epigraphik* 75, 273–80.

—— (1995), 'Further arithmetical tables', *Zeitschrift für Papyrologie und Epigraphik* 105, 225–8.

—— (1999), *The Mathematics of Plato's Academy. A New Reconstruction*, Oxford: Clarendon Press (2nd edn).

Gabba E. (1984), 'Per un'interpretazione storica della centuriazione romana' in *Misurare la terra: centuriazione e coloni nel mondo romano*, Modena: Panini, 20–7.

Gaiser K. (1988), *Philodems Academica. Die Berichte über Platon und die alte Akademie in zwei herkulanensischen Papyri*, Stuttgart: Cannstatt.

Garlan Y. (1974), *Recherches de poliorcétique grecque*, Athens: École Française d'Athènes.

Gerstinger H. and Vogel K. (1932), 'Eine stereometrische Aufgabensammlung im Papyrus Graecus Vindobonensis 19996', *Mitteilungen aus der Papyrussammlung der Nationalbibliothek in Wien (Papyrus Erzherzog Rainer)*, N.S. 1 Wien, 11–76.

Gibbs S.L. (1976), *Greek and Roman Sundials*, New Haven/London: Yale University Press.

Gigon O. (1973), 'Posidoniana-Ciceroniana-Lactantiana', in *Romanitas et Christianitas. Studia I. H. Waszink*, Amsterdam/London: North Holland, 145–80.

Grenfell B.P., Hunt A.S. and Gilbart Smyly J. (eds) (1902), *The Tebtunis Papyri*, London: Frowde.

Grodzynski D. (1974), 'Par la bouche de l'empereur', in J.-P. Vernant *et al.*, *Divination et rationalité*, Paris: Seuil, 267–94.

Guéraud O. and Jouguet P. (1938), *Un livre d'écolier du iiie siècle avant J.-C.*, Cairo: Institut Français d'archéologie orientale.

Guillaumin J.-Y. (1992), 'La signification des termes *contemplatio* et *observatio* chez Balbus et l'influence héronienne sur le traité', in J.-Y. Guillaumin (ed.), *Mathématiques dans l'antiquité*, Saint-Étienne: Publications de l'Université de Saint-Étienne, 205–14.

—— (1997), 'L'éloge de la géometrie dans la préface du livre 3 des *Metrica* d'Héron d'Alexandrie', *Revue des Études anciennes* 99, 91–9.

Harvey F.D. (1965), 'Two kinds of equality', *Classica et Mediaevalia* 26, 101–46.

Heath T.L. (1921), *A History of Greek Mathematics*, Oxford: Clarendon Press, 2 vols.

—— (1926), *The Thirteen Books of Euclid's Elements*, New York: Dover 1956 (first publ. 1926).

—— (1949), *Mathematics in Aristotle*, Oxford: Clarendon Press.

Hinrichs F.T. (1974), *Die Geschichte der Gromatischen Institutionen*, Wiesbaden: Steiner.

—— (1992), 'Die "agri per extremitatem mensura comprehensi". Diskussion eines Frontinstextes und der Geschichte seines Verständnisses', in O. Behrends and L. Capogrossi Colognesi (eds), *Die römische Feldmeßkunst. Interdisziplinäre Beiträge zu ihrer Bedeutung für die Zivilisationsgeschichte Roms*, Göttingen: Vandenhoeck and Ruprecht, 348–74.

Hopkins K. (1978), *Conquerors and Slaves. Sociological Studies in Roman History*, Cambridge: Cambridge University Press.

Huffman C.A. (1993), *Philolaus of Croton*, Cambridge: Cambridge University Press.

Hurst A. (1998), 'Géographes et poètes: le cas d'Apollonios de Rhodes', in G. Argoud and J.-Y. Guillaumin (eds), *Sciences exactes et sciences appliquées à Alexandrie*, Saint-Étienne: Publications de l'Université de Saint-Étienne, 279–88.

Huxley G. (1963), 'Friends and Contemporaries of Apollonius of Perga', *Greek, Roman and Byzantine Studies* 4, 100–5.

Ioannidou G. (ed.) (1996), *Catalogue of Greek and Latin Literary Papyri in Berlin*, Mainz a. R.: von Zabern.

Irigoin J. (1998), 'Les éditions de poètes à Alexandrie', in G. Argoud and J.-Y. Guillaumin (eds), *Sciences exactes et sciences appliquées à Alexandrie*, Saint-Étienne: Publications de l'Université de Saint-Étienne, 405–13.

Jones A. (1997a), 'Studies in the astronomy of the Roman period, I. The standard lunar scheme', *Centaurus* 39, 1–36.

—— (1997b), 'Studies in the astronomy of the Roman period, II. Tables for solar longitude', *Centaurus* 39, 211–29.

—— (1998), 'Studies in the astronomy of the Roman period, III. Planetary epoch tables', *Centaurus* 40, 1–41.

Kahn C.H. (1991), 'Some remarks on the origins of Greek science and philosophy', in A.C. Bowen (ed.), *Science and Philosophy in Classical Greece*, New York/ London: Garland, 1–10.

Kienast H.J. (1995), *Die Wasserleitung des Eupalinos auf Samos*, Bonn: Habelt (Samos 19).

Knorr W.R. (1975), *The Evolution of the Euclidean Elements*, Dordrecht/Boston: Reidel.

—— (1986), *The Ancient Tradition of Geometric Problems*, Boston/Basel/Berlin: Birkhäuser.

—— (1993), 'Arithmêtikê stoicheîosis: On Diophantus and Hero of Alexandria', *Historia Mathematica* 20, 180–92.

—— (1996), 'The Wrong Text of Euclid: On Heiberg's Text and its Alternatives', *Centaurus* 38, 208–76.

Kurke L. (1999), *Bodies, Games, Coins, and Gold. The Politics of Meaning in Archaic Greece*, Princeton: Princeton University Press.

Lang M. (1956), 'Numerical notation on Greek vases', *Hesperia* 25, 1–24.

—— (1968), 'Abaci from the Athenian agora', *Hesperia* 37, 241–3.

Lawrence A.W. (1946), 'Archimedes and the design of Euryalus fort', *Journal of Hellenic Studies* 66, 99–107.

—— (1979), *Greek Aims in Fortification*, Oxford: Clarendon Press.

Leonardos B. (1925–6), 'Ἀμφιαρείου ἐπιγραφαί', *Αρχαιομογική ἐφημερις*, 9–45.

Levick B. (1985), *The Government of the Roman Empire. A Sourcebook*, London/Sidney: Croom Helm.

Lewis N. (1986), *Greeks in Ptolemaic Egypt. Case Studies in the Social History of the Hellenistic World*, Oxford: Clarendon Press.

Lewis N. and Reinhold M. (1990), *Roman Civilization. Selected Readings*, New York: Columbia University Press, 2 vols, (3rd edn).

Lloyd G.E.R. (1963), 'Who is attacked in *On ancient medicine*?', *Phronesis* 8, 108–26, reprinted in *Methods and Problems in Greek Science. Selected Papers*, Cambridge: Cambridge University Press 1991, 49–69.

—— (1972), 'The social background of early Greek philosophy and science', in D. Daiches and A. Thorlby (eds), *Literature and Western Civilization, I: The Classical World*, London: Aldus 1972, 381–95, reprinted in *Methods and Problems in Greek Science. Selected Papers*, Cambridge: Cambridge University Press 1991, 121–40.

—— (1979), *Magic, Reason, and Experience*, Cambridge: Cambridge University Press.

—— (1987a), 'The Alleged Fallacy of Hippocrates of Chios', *Apeiron* 20, 103–28.

—— (1987b), *The Revolutions of Wisdom. Studies in the Claims and Practice of Ancient Greek Science*, Berkeley/Los Angeles/London: University of California Press.

—— (1990), *Demystifying Mentalities*, Cambridge: Cambridge University Press.

Lyons H. (1927), 'Ancient surveying instruments', *The Geographical Journal* 69, 132–43.

Maas M. (1992), *John Lydus and the Roman Past. Antiquarianism and Politics in the Age of Justinian*, London/New York: Routledge.

MacMullen R. (1966), *Enemies of the Roman Order. Treason, Unrest and Alienation in the Empire*, London/New York: Routledge 1992 (first publ. Harvard: Harvard University Press).

—— (1971), 'Social History in Astrology', *Ancient Society* 2, reprinted in *Changes in the Roman Empire: Essays in the Ordinary*, Princeton/London: Princeton University Press 1991, 218–24.

Mansfeld J. (1998), *Prolegomena Mathematica. From Apollonius of Perga to the Late Neoplatonists With an Appendix on Pappus and the History of Platonism,* Leiden/Boston/Köln: Brill.

Mau J. and Müller W. (1962), 'Mathematische Ostraka aus der Berliner Sammlung', *Archiv für Papyrusforschung* 17, 1–10.

McCluskey S.C. (1990), 'Gregory of Tours, monastic timekeeping, and early Christian attitudes to astronomy', *Isis* 81, 9–22.

McNicoll A.W. and Milner N.P. (1997), *Hellenistic Fortifications from the Aegean to the Euphrates*, Oxford: Clarendon Press.

Meiggs R. and Lewis D. (eds) (1989), *A Selection of Greek Historical Inscriptions to the End of the Fifth Century B.C.*, Oxford: Clarendon Press.

Mendell H. (1984), 'Two geometrical examples from Aristotle's *Metaphysics*', *Classical Quarterly* 34, 359–72.

Meritt B.D. (1932), *Athenian Financial Documents of the Fifth Century*, Ann Arbor: University of Michigan Press.

Meritt B.D. and Brown West A. (1934), *The Athenian Assessment of 425 B.C.*, Ann Arbor: University of Michigan Press.

Meritt B.D., Wade-Gery H.T. and McGregor M.F. (1939), *The Athenian Tribute Lists*, Cambridge, MA: Harvard University Press, vol.1.

Moatti C. (1993), *Archives et partage de la terre dans le monde romain (*IIe *siècle avant - 1er siècle après J.-C.),* Roma: École Française.

Montanari F. (1993), 'L'erudizione, la filologia e la grammatica', in G. Cambiano, L. Canfora and D. Lanza (eds), *Lo spazio letterario della Grecia antica. Vol.1: La produzione e la circolazione del testo. Tomo 2: l'Ellenismo*, Roma: Salerno, 235–81.

Morgan T. (1998), *Literate Education in the Hellenistic and Roman Worlds*, Cambridge: Cambridge University Press.

Morris I. (1996), 'The strong principle of equality and the archaic origins of Greek democracy', in J. Ober and C. Hedrick (eds), *Demokratia: A Conversation on Democracies, Ancient and Modern*, Princeton, 19–48.

Mueller I. (1981), *Philosophy of Mathematics and Deductive Structure in Euclid's Elements*, Cambridge, MA/London: MIT Press.

—— (1982), 'Geometry and Scepticism', in J. Barnes, J. Brunschwig, M. Burnyeat and M. Schofield (eds), *Science and Speculation. Studies in Hellenistic Theory and Practice*, Cambridge/Paris: Cambridge University Press and Maison des Sciences de l'Homme, 69–95.

—— (1987a), 'Mathematics and Philosophy in Proclus' Commentary on Book I of Euclid's Elements', in J. Pépin and H.D. Saffrey (eds), *Proclus. Lecteur et Interprète des Anciens*, Paris: CNRS, 305–18.

—— (1987b), 'Iamblichus and Proclus' Euclid Commentary', *Hermes* 115, 334–48.

—— (1992), 'Mathematical method and philosophical truth', in R. Kraut (ed.), *The Cambridge Companion to Plato*, Cambridge: Cambridge University Press, 170–99.

Murdoch J. (1971), 'Euclid: Transmission of the Elements', *Dictionary of Scientific Biography*, New York: Scribner, vol. 4, 437–59.

Napolitano Valditara L.M. (1988), *Le idee, i numeri, l'ordine: la dottrina della 'mathesis universalis' dall'Accademia antica al Neoplatonismo*, Napoli: Bibliopolis.

Netz R. (1998), 'Deuteronomic texts: late antiquity and the history of mathematics', *Revue d'Histoire des mathématiques* 4, 261–88.

—— (1999a), *The Shaping of Deduction in Greek Mathematics. A Study in Cognitive History*, Cambridge: Cambridge University Press.
—— (1999b), 'Proclus' division of the mathematical proposition into parts: how and why was it formulated?', *Classical Quarterly* 49, 282–303.
—— (forthcoming), 'Counter-culture and early Greek numeracy'.
Neugebauer O. and Van Hoesen H.B. (1959), *Greek Horoscopes*, Philadelphia: American Philosophical Society.
Nicolet C. (1970), 'Les finitores ex equestri loco de la loi Servilia', *Latomus* 29, 72–103.
—— (1988), *L'inventaire du monde: géographie et politique aux origines de l'empire romain*, Fayard, Engl. tr. by H. Leclerc, *Space, Geography, and Politics in the Early Roman Empire*, Ann Arbor: University of Michigan Press 1991.
Nussbaum M. (1986), *The Fragility of Goodness. Luck and Ethics in Greek Tragedy and Philosophy*, Cambridge: Cambridge University Press.
Ober J. (1992), 'Towards a typology of Greek artillery towers: the first and the second generations (*c.* 375–275 BC)', in S. van de Maele and J.M. Fossey (eds), *Fortificationes antiquae*, Amsterdam: Gieben, 147–69.
O'Meara D. (1989), *Pythagoras Revived: Mathematics and Philosophy in Late Antiquity*, Oxford: Clarendon Press.
Osborne R. (1994), 'Introduction. Ritual, finance, politics: an account of Athenian democracy', in R. Osborne and S. Hornblower (eds), *Ritual, Finance, Politics: Athenian Democratic Accounts Presented to David Lewis*, Oxford: Clarendon Press, 1–21.
—— (1996), *Greece in the Making 1200–479* BC, London/New York: Routledge.
Owens E.J. (1991), *The City in the Greek and Roman World*, London: Routledge.
Pamment Salvatore J. (1996), *Roman Republican Castrametation. A Reappraisal of Historical and Archaeological Sources*, Oxford: BAR.
Panerai M.C. (1984), 'Gli strumenti: un agrimensore a Pompei', in *Misurare la terra: centuriazione e coloni nel mondo romano*, Modena: Panini, 115–19.
Parker R.A. (1972), *Demotic Mathematical Papyri*, Providence, RI/London: Brown University Press and Lund Humphreys.
Pritchett W.K. (1965), 'Gaming tables and IG I^2 324', *Hesperia* 34, 131–47.
Purcell N. (1983), 'The *apparitores*: a study in social mobility', *Papers of the British School at Rome* 51, 125–73.
Rathbone D. (1991), *Economic Rationalism and Rural Society in Third-Century A.D. Egypt. The Heroninos Archive and the Appianus Estate*, Cambridge: Cambridge University Press.
Rhodes P.J. (1972), *The Athenian Boule*, Oxford: Clarendon Press.
—— (1981), *A Commentary on the Aristotelian Athenaion Politeia*, Oxford: Clarendon Press.
Rhodes P.J. and Lewis D.M. (1997), *The Decrees of the Greek States*, Oxford: Clarendon Press.
Rice E.E. (1983), *The Grand Procession of Ptolemy Philadelphus*, Oxford: Oxford University Press.
Rihll T.E. and Tucker J.V. (1995), 'Greek engineering. The case of Eupalinus' tunnel', in A. Powell (ed.), *The Greek World*, London: Routledge, 403–31.

Ritter J. (1989a), 'Babylon – 1800', in M. Serres (ed.), *A History of Scientific Thought*, Oxford/Cambridge, MA: Blackwell 1995, 17–43, Engl. tr. of *Eléments d'histoire des sciences*, Paris: Bordas.

—— (1989b), 'Measure for measure: mathematics in Egypt and Mesopotamia', in M. Serres (ed.), *A History of Scientific Thought*, Oxford/Cambridge, MA: Blackwell 1995, 44–72, Engl. tr. of *Eléments d'histoire des sciences*, Paris: Bordas.

Roochnik D. (1996), *Of Art and Wisdom. Plato's Understanding of Techne*, University Park, PA: Pennsylvania State University Press.

Sachs E. (1917), *Die fünf Platonische Körper*, Berlin: Weidmann.

Salmon E.T. (1985), 'La fondazione delle colonie latine', in *Misurare la terra: centuriazione e coloni nel mondo romano. Cittá, agricoltura, commercio: materiali da Roma e dal suburbio*, Modena: Panini, 13–19.

Salviat F. and Vatin Cl. (1974), 'Le cadastre de Larissa', *Bulletin de correspondance hellénique* 98, 247–62.

Salzman M.R. (1990), *On Roman Time. The Codex-Calendar of 354 and the Rhythms of Urban Life in Late Antiquity*, Berkeley/Los Angeles/Oxford: University of California Press.

Schütz M. (1990), 'Zur Sonnenuhr des Augustus auf dem Marsfeld', *Gymnasium* 97, 432–57.

Sedley D. (1976), 'Epicurus and the Mathematicians of Cyzicus', *Cronache ercolanesi* 6, 23–54.

Segonds A. (1981), 'Introduction', in Jean Philopon, *Traité de l'Astrolabe*, Paris: Astrolabica.

Shaw G. (1995), *Theurgy and the Soul. The Neoplatonism of Iamblichus*, University Park, PA: Pennsylvania State University Press.

Sherk R.K. (1988), *The Roman Empire: Augustus to Hadrian*, Cambridge: Cambridge University Press.

Smallwood E.M. (1966), *Documents Illustrating the Principates of Nerva, Trajan and Hadrian*, Cambridge: Cambridge University Press.

Smith D.E. (1951), *History of Mathematics*, New York: Dover 1951–3 (first edn 1923–5), 2 vols.

Stancic Z. and Slapsak B. (1999), 'The Greek field system at Pharos: a metric analysis', *Revue des études anciennes* 101, 115–24.

Straub J. (1970), 'Severus Alexander und die Mathematici', in J. Straub (ed.), *Bonner Historia-Augusta-Colloquium* 1968/1969, Bonn: Habelt, 247–72.

Svenson-Evers H. (1996), *Die griechischen Architekten archaischer und klassischer Zeit*, Frankfurt a.M.: Peter Lang.

Swain S. (1990b), 'Plutarch's lives of Cicero, Cato and Brutus', *Hermes* 118, 192–203.

—— (1997), 'Plutarch, Plato, Athens, and Rome', in J. Barnes and M. Griffin (eds), *Philosophia Togata II. Plato and Aristotle at Rome*, Oxford: Clarendon Press, 165–87.

Szabó Á. (1969), *Anfänge griechischen Mathematik*, Engl. tr. A.M. Ungar as *The Beginnings of Greek Mathematics*, Dordrecht: Reidel 1978.

Tannery P. (1884), 'Domninos de Larissa', in *Mémoires scientifiques* II, Toulouse/Paris: Privat and Gauthier-Villars 1912, 105–17.

—— (1885), 'Notes critiques sur Domninos', *Revue de Philologie* 9, 129–37.

Thesleff H. (1961), *An Introduction to the Pythagorean Writings of the Hellenistic Period*, Åbo: Åbo Akademi.

—— (ed.) (1965), *The Pythagorean Texts of the Hellenistic Period*, Åbo: Åbo Akademi.

Thomas I. (ed.) (1967), *Greek Mathematical Works*, 2 vols, London/Cambridge, MA: Heinemann and Harvard University Press.

Thomas R. (1992), *Literacy and Orality in Ancient Greece*, Cambridge: Cambridge University Press.

Thompson D.J. (writing as Crawford) (1971), *Kerkeosiris. An Egyptian Village in the Ptolemaic Period*, Cambridge: Cambridge University Press.

Tod M.N. (1911–12), 'The Greek numeral notation', *The Annual of the British School at Athens* 18, 98–132.

Tolbert Roberts J. (1982), *Accountability in Athenian Government*, Madison/London: University of Wisconsin Press.

Toomer G.J. (1970), 'Apollonius of Perga', *Dictionary of Scientific Biography*, New York: Scribner, vol.1, 179–93.

—— (ed.) (1990), *Apollonius. Conics – Books V to VII*, New York/Berlin/etc.: Springer, 2 vols.

Uguzzoni A. and Ghinatti F. (1968), *Le tavole greche di Eraclea*, Roma: Bretschneider.

Untersteiner M. (1961), *Sofisti. Testimonianze e Frammenti*, eds and Italian tr. M. Untersteiner and A. Battegazzore, Firenze: La Nuova Italia 1961–2, 4 vols.

van der Waerden B.L. (1954), *Science Awakening*, Engl. tr. by A. Dresden, Groningen: P. Noordhoff.

Vernant J.-P. (1965), *Mythe et pensée chez les Grecs. Etudes de psychologie historique*, Paris: Maspero, Engl. tr. as *Myth and Thought among the Greeks*, London: Routledge 1983.

Vitrac B. (1996), 'Mythes (et réalités?) dans l'histoire des mathématiques grecques anciennes', in C. Goldstein, J. Gray and J. Ritter (eds), *L'Europe mathématique. Histoires, mythes, identités*, Paris: Éditions de la Maison des Sciences de l'Homme, 33–51.

von Reden S. (1995), *Exchange in Ancient Greece*, London: Duckworth.

—— (1997), 'Money, law and exchange: coinage in the Greek polis', *Journal of Hellenic Studies* 117, 154–76.

von Staden H. (1998), 'Andréas de Caryste et Philon de Byzance: médecine et mécanique à Alexandrie', in G. Argoud and J.-Y. Guillaumin (eds), *Sciences exactes et sciences appliquées à Alexandrie*, Saint-Étienne: Publications de l'Université de Saint-Étienne, 147–72.

Wallace-Hadrill A. (1998), 'To be Roman, go Greek. Thoughts on Hellenization at Rome', in M. Austin, J. Harries and C. Smith (eds), *Modus Operandi. Essays in Honour of Geoffrey Rickman*, London: Institute of Classical Studies, 79–91.

Wasowicz A. (1972), 'Traces de lotissements anciens en Crimée', *Mélanges de l'école Française de Rome* 84, 199–229.

Waterhouse W.C. (1972), 'The Discovery of the Regular Solids', *Archive for History of Exact Sciences* 9, 212–21.

Wehrli F. (ed.) (1959), *Die Schule des Aristoteles*, Basel: Schwabe 1944–59.

Westermann W.L. and Sayre Hasenoehrl L. (eds) (1934), *Zenon Papyri. Business Papers of the Third Century B.C. Dealing with Palestine and Egypt*, New York: Columbia University Press 1934–40, 2 vols.

Winter F.E. (1963), 'The chronology of the Euryalos fortress at Syracuse', *American Journal of Archaeology* 67, 363–87.

—— (1971), *Greek Fortifications*, London: Routledge and Kegan Paul.

Zhmud L. (1997), *Wissenschaft, Philosophie und Religion im frühen Pythagoreismus*, Berlin: Akademie.

INDEX